Problems and Solutions in
**Differential Geometry,
Lie Series, Differential Forms,
Relativity and Applications**

Problems and Solutions in
Differential Geometry,
Lie Series, Differential Forms,
Relativity and Applications

Willi-Hans Steeb

University of Johannesburg, South Africa

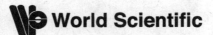

NEW JERSEY · LONDON · SINGAPORE · BEIJING · SHANGHAI · HONG KONG · TAIPEI · CHENNAI · TOKYO

Published by

World Scientific Publishing Co. Pte. Ltd.
5 Toh Tuck Link, Singapore 596224
USA office: 27 Warren Street, Suite 401-402, Hackensack, NJ 07601
UK office: 57 Shelton Street, Covent Garden, London WC2H 9HE

Library of Congress Cataloging-in-Publication Data
Names: Steeb, W.-H., author.
Title: Problems and solutions in differential geometry, Lie series,
    differential forms, relativity, and applications / by Willi-Hans Steeb
    (University of Johannesburg, South Africa).
Description: New Jersey : World Scientific, 2017. |
    Includes bibliographical references and index.
Identifiers: LCCN 2017047297| ISBN 9789813230828 (hardcover : alk. paper) |
    ISBN 9789813232969 (pbk. : alk. paper)
Subjects: LCSH: Geometry, Differential. | Differential forms.
Classification: LCC QA641 .S65 2017 | DDC 516.3/6--dc23
LC record available at https://lccn.loc.gov/2017047297

British Library Cataloguing-in-Publication Data
A catalogue record for this book is available from the British Library.

Printed in Singapore

Preface

The purpose of this book is to supply a collection of problems in differential geometry. Each chapter contains an introduction with the essential definitions and explanations to tackle the problems. If necessary, other concepts are explained directly with the problem. Thus the material in this book is self-contained. The topics range in difficulty from elementary to advanced. Students can learn important principles and strategies required for problem solving. Lecturers will also find this text useful, either as a supplement or textbook, since concepts and techniques are developed in the problems. In Chapter 1, curves, surfaces and manifolds are studied starting with curves in the plane. Curvature, torsion, arc length, tangent line, tangent surface, normal line, normal surface and coordinate systems are introduced. Chapter 2 is devoted to vector fields and Lie series to solve systems of ordinary differential equations as well as to Lie algebras provided by the vector fields. First integrals are also covered. Chapter 3 introduces metric tensor fields for n-dimensional Riemann manifolds and pseudo-Riemann manifolds. Euclidean and Minkowski spaces are considered. Geodesic flows are also included. Chapter 4 deals with differential forms and matrix-valued differential forms and various applications. Stokes theorem is also added. Finally Chapter 5 provides exercises for the Lie derivative of tensor fields and differential forms with various applications such as invariance, conformal invariance, first integrals, Killing vector fields and symmetries of differential equations.

For some solutions of the problems, software such as SymbolicC++, Maxima, Gnuplot and TikZ are utilized.

The International School for Scientific Computing (ISSC) provides certificate courses for this subject. Contact the author if you want to do this course or other courses of the ISSC.

e-mail addresses of the author:

`steebwilli@gmail.com`
`steeb_wh@yahoo.com`

Home page of the author: `http://issc.uj.ac.za`

Contents

Notation

$:=$	is defined as
$\in$	belongs to (a set)
$\notin$	does not belong to (a set)
$\cap$	intersection of sets
$\cup$	union of sets
$\emptyset$	empty set
$T \subset S$	subset T of set S
$S \cap T$	the intersection of the sets S and T
$S \cup T$	the union of the sets S and T
$\mathbb{N}$	set of natural numbers
$\mathbb{Z}$	set of integers
$\mathbb{Q}$	set of rational numbers
$\mathbb{R}$	set of real numbers
$\mathbb{R}^+$	set of nonnegative real numbers
$\mathbb{C}$	set of complex numbers
i	$\sqrt{-1}$
$\Re(z)$	real part of the complex number z
$\Im(z)$	imaginary part of the complex number z
$\|z\|$	modulus of complex number z $(\|x + iy\| = (x^2 + y^2)^{1/2})$, $x, y \in \mathbb{R}$
$\mathbb{R}^n$	space of column vectors with n real components
$\mathbb{E}^n$	n-dimensional Euclidean space
$\mathbb{C}^n$	n-dimensional complex linear space space of column vectors with n complex components
$\mathbb{S}^n$	n-sphere $\{(x_1, \ldots, x_{n+1} : x_1^2 + \cdots + x_{n+1}^2 = 1\}$
$\mathbb{T}^n$	torus
$\mathbb{CP}^n$	complex projective space
M	differentiable manifold

$f(S)$	image of set S under mapping f
$f \circ g$	composition of two mappings $(f \circ g)(x) = f(g(x))$.
$\mathbf{x}$	column vector in $\mathbb{C}^n$
$\mathbf{x}^T$	transpose of $\mathbf{x}$ (row vector)
$\mathbf{0}$	zero (column) vector
$\|\cdot\|$	norm
$\mathbf{x} \cdot \mathbf{y} \equiv \mathbf{x}^*\mathbf{y}$	scalar product (inner product) in $\mathbb{C}^n$
$\mathbf{x} \times \mathbf{y}$	vector product in $\mathbb{R}^3$
S_n	symmetric group
τ	dimensionless parameter
u_1, u_2	dimensionless parameters
$\det(A)$	determinant of a square matrix A
$\mathrm{tr}(A)$	trace of a square matrix A
$\mathrm{rank}(A)$	rank of matrix A
A^T	transpose of matrix A
$\overline{A}$	conjugate of matrix A
A^*	conjugate transpose of matrix A
A^{-1}	inverse of square matrix A (if it exists)
I_n	$n \times n$ unit matrix
I	unit operator
0_n	$n \times n$ zero matrix
$[A, B] := AB - BA$	commutator for square matrices A and B
$[A, B]_+ := AB + BA$	anticommutator for square matrices A and B
$\otimes$	tensor product
g	metric tensor field
V	vector field
$[V, W]$	commutator of vector fields V and W
$\wedge$	exterior product, Grassmann product, wedge product
d	exterior derivative
∇	gradient
$\rfloor$	contraction of vector fields and differential forms, interior product
$L_V(.)$	Lie derivative with vector field V
$\star$	Hodge duality operator
δ_{jk}	Kronecker delta with $\delta_{jk} = 1$ for $j = k$ and $\delta_{jk} = 0$ for $j \neq k$
Ω	volume differential form in $\mathbb{R}^n$
$\mathbb{H}^2_+$	Poincare upper half plane
λ	eigenvalue
ϵ	real parameter

t	time variable
τ	proper time
c	speed of light in vacuum
G	gravitational constant
L	Lagrange function
H	Hamilton function
$\hat{H}$	Hamilton operator

Chapter 1

Curves, Surfaces and Manifolds

1.1 Notations and Definitions

Curves in the Euclidean plane $\mathbb{E}^2$ can be given in parametric form $x_1(\tau)$, $x_2(\tau)$ (τ is a dimensionless parameter). We assume that $x_1(\tau)$, $x_2(\tau)$ are smooth functions. Other representations would be

$$x_2 = f(x_1)$$

or (implicit curve)

$$f(x_1, x_2) = 0$$

where it is assumed that $f(x_1)$ and $f(x_1, x_2)$ are smooth functions. If f is a polynomial in the two variables x_1, x_2 the curve is called an algebraic curve. Let $\mathbb{E}^n$ be the n-dimensional Euclidean space and $n \geq 2$. Let I be a non-empty interval of real numbers and $\tau \in I$ with τ a dimensionless parameter. A smooth vector-valued function

$$\mathbf{x} : I \to \mathbb{R}^n$$

is called a smooth parametric curve $\mathbf{x}(\tau)$. τ is called the parameter of the curve $\mathbf{x}(\tau)$. If I is a closed interval $[a, b]$, then $\mathbf{x}(a)$ the starting point and

$\mathbf{x}(b)$ is the endpoint of the curve $\mathbf{x}(\tau)$. If $\mathbf{x} : (a, b) \to \mathbb{R}^n$ is injective, we call the curve simple. If $\mathbf{x}(\tau)$ is a parametric curve which can be locally described as a power series, we call the curve analytic or of class C^ω. A smooth curve

$$\mathbf{x} : [a, b] \to \mathbb{R}^n$$

is called *regular* if $d\mathbf{x}/d\tau \neq \mathbf{0}$ for all $\tau \in I$.

A surface in $\mathbb{E}^3$ can be given in parametric form

$$\mathbf{x}(u_1, u_2) = \begin{pmatrix} x_1(u_1, u_2) \\ x_2(u_1, u_2) \end{pmatrix}$$

or as $x_3 = f(x_1, x_2)$ or $f(x_1, x_2, x_3) = 0$ (implicit form). It is assumed that the functions are smooth.

Let $n \geq 1$ and $m \geq 1$. Any smooth n-dimensional surface in the Euclidean space $\mathbb{E}^{n+m}$ is an n-dimensional manifold.

Let $\mathbf{f} : \mathbb{R}^n \mapsto \mathbb{R}^m$ be a smooth function. If

$$M := \{ \mathbf{x} \in \mathbb{R}^n \; : \; \mathbf{f}(\mathbf{x}) = \mathbf{0}, \; \text{rank}(\partial \mathbf{f}(\mathbf{x}/\partial\mathbf{x}) = m \}$$

is nonempty, then M is an $n - m$ dimensional manifold.

The *general linear group* $GL(n, \mathbb{R})$ is the smooth manifold for which every point is an $n \times n$ matrix over $\mathbb{R}$ with nonzero determinant.

Let $\mathbf{x}_0 \in \mathbb{R}^n$. The *tangent space* $T_{\mathbf{x}_0} \mathbb{R}^n$ is the n-dimensional vector space whose elements are given by the pairs $(\mathbf{x}_0, \mathbf{x}) \in \{\mathbf{x}_0\} \times \mathbb{R}^n$. The vector space structure is given by the bijection $T_{\mathbf{x}_0} \mathbb{R}^n \to \mathbb{R}^n$, $(\mathbf{x}_0, \mathbf{x}) \mapsto \mathbf{x}$. This means

$$(\mathbf{x}_0, \mathbf{x}) + (\mathbf{x}_0, \mathbf{y}) = (\mathbf{x}_0, \mathbf{x} + \mathbf{y}), \quad c(\mathbf{x}_0, \mathbf{x}) = (\mathbf{x}_0, c\mathbf{x})$$

with $c \in \mathbb{R}$.

Quadric surfaces (or quadratic surfaces) are the graphs of any equations of the form

$$a_{11}x_1^2 + a_{22}x_2^2 + a_{33}x_3^2 + a_1x_1 + a_2x_2 + a_3x_3 + a_{12}x_1x_2 + a_{13}x_1x_3 + a_{23}x_2x_3 + b = 0.$$

1.2 Solved Problems

Problem 1. (i) Consider the rational curve in the plane

$$x_2^2 = x_1^2 + x_1^3.$$

Find the parameter representation $x_1(\tau)$, $x_2(\tau)$.
(ii) Consider the rational curve in the plane (unit circle) $x_1^2 + x_2^2 = 1$.
Find the parameter representation $x_1(\tau)$, $x_2(\tau)$.
(iii) Show that the *Lemniscate of Gerono* $x_1^4 = x_1^2 - x_2^2$ can be parametrized by

$$\mathbf{x}(\tau) = \begin{pmatrix} x_1(\tau) \\ x_2(\tau) \end{pmatrix} = \begin{pmatrix} \sin(\tau) \\ \sin(\tau)\cos(\tau) \end{pmatrix}$$

where $0 \leq \tau \leq \pi$. Is the Lemniscate of Gerono regular?

Solution 1. (i) We obtain $x_1(\tau) = \tau^2 - 1$, $x_2(\tau) = \tau(\tau^2 - 1)$. Note that $x_1(1) = x_2(1) = 0$ and $dx_1/d\tau = 2\tau$, $dx_2/d\tau = 3\tau^2 - 1$.
(ii) Let $\tau \in \mathbb{R}$. We find

$$x_1(\tau) = \frac{2\tau}{1 + \tau^2}, \qquad x_2(\tau) = \frac{1 - \tau^2}{1 + \tau^2}.$$

Note that $x_1(0) = 0$, $x_2(0) = 1$ and $x_1(-1) = -1$, $x_2(-1) = 0$.
(iii) We have

$$x_1^4(\tau) = \sin^4(\tau), \quad x_2^2(\tau) = \sin^2(\tau)\cos^2(\tau).$$

Applying $\sin^2(\tau) + \cos^2(\tau) = 1$ we obtain

$$x_1^2(\tau) - x_2^2(\tau) = \sin^2(\tau) - \sin^2(\tau)\cos^2(\tau) = \sin^2(\tau)(1 - \cos^2(\tau))$$
$$= \sin^4(\tau).$$

We have

$$\frac{dx_1}{d\tau} = \cos(\tau), \qquad \frac{dx_2}{d\tau} = \cos^2(\tau) - \sin^2(\tau) = \cos(2\tau).$$

There is no τ such that $\cos(\tau) = 0$ and $\cos(2\tau) = 0$.

Problem 2. The *Maclaurin trisectrix* in the plane is given by the parameter representation ($\tau \in \mathbb{R}$)

$$x_1(\tau) = \frac{\tau^2 - 3}{\tau^2 + 1}, \qquad x_2(\tau) = \frac{\tau(\tau^2 - 3)}{\tau^2 + 1}.$$

The curve is self-intersecting. For $\tau = \sqrt{3}$ we have

$$(x_1(\tau = \sqrt{3}) = 0, x_2(\tau = \sqrt{3}) = 0)$$

and for $\tau = -\sqrt{3}$ we have

$$(x_1(\tau = -\sqrt{3}) = 0, x_2(\tau = \sqrt{3}) = 0).$$

Find the Cartesian representation.

Solution 2.   With

$$x_1^2 = \frac{(\tau^2 - 3)^2}{(\tau^2 + 1)^2}, \qquad x_2^2 = \frac{\tau^2(\tau^2 - 3)^2}{(\tau^2 + 1)^2}$$

we obtain $x_2^2(1 - x_1) = x_1^2(x_1 + 3)$.

Problem 3.   Let $a > 0$ and $b > 0$. Consider the curve in the plane (*trochoid*)

$$x_1(\tau) = a\tau - b\sin(\tau), \quad x_2(\tau) = a - b\cos(\tau).$$

Is the curve *regular*? Note that for $a = b$ one has the *cycloid*.

Solution 3.   We have

$$\frac{dx_1(\tau)}{d\tau} = a - b\cos(\tau), \quad \frac{dx_2(\tau)}{d\tau} = b\sin(\tau).$$

Hence $dx_1(0)/d\tau = a - b$, $dx_2(0)/d\tau = 0$. Thus the curve is not regular for $a = b$.

Problem 4.   Let $\tau \in \mathbb{R}$. Consider the curve in the plane

$$x_1(\tau) = 3\cos(\tau) - \cos(3\tau), \quad x_2(\tau) = 3\sin(\tau) - \sin(3\tau).$$

Study the properties of this curve.

Solution 4.   We have the properties

$$x_1(\tau + 2\pi) = x_1(\tau), \quad x_2(\tau + 2\pi) = x_2(\tau)$$

$$x_1(-\tau) = x_1(\tau), \quad x_2(-\tau) = -x_2(\tau)$$

$$x_1(\pi/2 + \tau) = -x_1(\pi/2 - \tau), \qquad x_2(\pi/2 + \tau) = x_2(\pi/2 - \tau).$$

Since $|\sin(\alpha)| \leq 1$, $|\cos(\alpha)| \leq 1$ for $\alpha \in \mathbb{R}$ we have $|x_1(\tau)| \leq 4$, $|x_2(\tau)| \leq 4$. We find $dx_1/d\tau = -3\sin(\tau) + 3\sin(3\tau)$, $dx_2/d\tau = 3\cos(\tau) - 3\cos(3\tau)$ and thus $dx_1(\tau = 0)/d\tau = 0$, $dx_2(\tau = 0)/d\tau = 0$.

Problem 5.   A curve $f(x, y) = 0$ admits a *rational parametrization* if there exist rational functions α and β such $f(\alpha(\tau), \beta(\tau)) = 0$ i.e.

$$f(x, y) = 0 \Leftrightarrow \begin{cases} x = \alpha(\tau) \\ y = \beta(\tau) \end{cases}.$$

Such a curve $f(x, y) = 0$ is called *unicursal*. The curve $x^n + y^n - 1 = 0$ is not unicursal for $n \geq 3$, otherwise *Fermat-Wiles* would be wrong.
(i) Find the rational parametrization for the unit circle $x^2 + y^2 - 1 = 0$.
(ii) Find the rational parametrization for the *nodal cubic* $y^3 + x^3 - xy = 0$.

Solution 5.   (i) The circle admits the rational parametrization

$$\alpha(\tau) = \frac{1 - \tau^2}{1 + \tau^2}, \qquad \beta(\tau) = \frac{2\tau}{1 + \tau^2}.$$

With $\tau = \tan(\theta/2)$ we obtain $\alpha(\theta) = \cos(\theta)$ and $\beta(\theta) = \sin(\theta)$.
(ii) The nodal cubic admits the rational parametrization

$$\alpha(\tau) = \frac{\tau}{1 + \tau^3}, \qquad \beta(\tau) = \frac{\tau^2}{1 + \tau^3}.$$

Problem 6.   Consider the equation

$$(y - \sin(\tau)) \cos(2\tau) = (x - \cos(\tau)) \sin(2\tau).$$

Take the derivative with respect to τ of this equation and then solve the system of equations with respect to x and y to find a curve in the plane.

Solution 6.   Differentiation with respect to τ yields

$$-y \sin(2\tau) + \frac{1}{2} \cos(\tau) \cos(2\tau) = x \cos(2\tau) - \frac{1}{2} \sin(\tau) \sin(2\tau).$$

The system of equations can now be written in matrix form

$$\begin{pmatrix} \sin(2\tau) & -\cos(2\tau) \\ \cos(2\tau) & \sin(2\tau) \end{pmatrix} \begin{pmatrix} x \\ y \end{pmatrix} = \begin{pmatrix} \sin(\tau) \\ \frac{1}{2}\cos(\tau) \end{pmatrix}.$$

The matrix on the left-hand side is invertible (determinant $+1$) and hence

$$\begin{pmatrix} x \\ y \end{pmatrix} = \begin{pmatrix} \sin(2\tau) & \cos(2\tau) \\ -\cos(2\tau) & \sin(2\tau) \end{pmatrix} \begin{pmatrix} \sin(\tau) \\ \frac{1}{2}\cos(\tau) \end{pmatrix}.$$

It follows that

$$x(\tau) = \cos(\tau) - \frac{1}{2}\cos(\tau)\cos(2\tau), \quad y(\tau) = \sin(\tau) - \frac{1}{2}\cos(\tau)\sin(2\tau).$$

The curve is known as the *nephroid*.

Problem 7.   The straight line *Hough transform* maps a line in the plane $\mathbb{R}^2$ into a point in the Hough transform space. The polar definition of the Hough transform is based on the representation of the lines by the parameters (ρ, θ) via the equation

$$\rho = x_j \cos(\theta) + y_j \sin(\theta).$$

All points (x_j, y_j) of a given line in the plane correspond to a point (ρ, θ) in the Hough transform space. Any point (x_j, y_j) is mapped to a sinusoidal curve in the Hough transform space. Consider the two points $(x_0, y_0) = (1, 0)$, $(x_1, y_1) = (0, 1)$ on a line. Find ρ, θ. Give a geometric interpretation for ρ and θ.

Solution 7.   For the two points (x_0, y_0) and (x_1, y_1) we find the two equations

$$\rho = \cos(\theta), \qquad \rho = \sin(\theta).$$

Hence $1 = \tan(\theta)$ and therefore $\theta = \pi/4$. It follows that

$$\rho = \cos(\pi/4) = \sin(\pi/4) = \frac{1}{\sqrt{2}}.$$

Consequently $(\rho, \theta) = (1/\sqrt{2}, \pi/4)$. Thus ρ is the shortest distance between the straight line and the origin $(0, 0)$. The angle θ is the angle between the distance vector and the positive x-direction.

Problem 8.   For a plane curve given by $f(x_1, x_2) = 0$ ($f : \mathbb{R}^2 \to \mathbb{R}$ and we assume that f is analytic) the equation of the *tangent line* at the point (p_1, p_2) ($f(p_1, p_2) = 0$) is given by

$$(x_1 - p_1)\frac{\partial f}{\partial x_1}(x_1 = p_1, x_2 = p_2) + (x_2 - p_2)\frac{\partial f}{\partial x_2}(x_1 = p_1, x_2 = p_2) = 0.$$

The *normal line* at the point (p_1, p_2) is given by

$$(x_1 - p_1)\frac{\partial f}{\partial x_2}(x_1 = p_1, x_2 = p_2) - (x_2 - p_2)\frac{\partial f}{\partial x_1}(x_1 = p_1, x_2 = p_2) = 0.$$

It is assumed that the curve is not self-crossing at the point (p_1, p_2). Consider the curve in the plane

$$x_1^4 - 6x_1^2 x_2 + 25x_2^2 - 16x_1^2 = 0.$$

Note the invariance of the curve under $x_1 \leftrightarrow -x_1$. If $x_2 = 4$, then the solutions of the quartic equation

$$x_1^4 - 40x_1^2 + 400 = 0$$

are given by $x_{2,1} = 2\sqrt{5}$, $x_{2,2} = -2\sqrt{5}$. Consider the point $(p_1, p_2) = (2\sqrt{5}, 4)$. Find the tangent line at this point. Find the normal line at this point. Where is the curve self-crossing?

Solution 8. We have

$$\frac{\partial f}{\partial x_1} = 4x_1^3 - 12x_1 x_2 - 32x_1 \quad \Rightarrow \quad \frac{\partial f}{\partial x_1}(p_1, p_2) = 0$$

$$\frac{\partial f}{\partial x_2} = -6x_1^2 + 50x_2 \quad \Rightarrow \quad \frac{\partial f}{\partial x_2}(p_1, p_2) = 80.$$

Hence the tangent line $80(x_2 - 4) = 0$ or $x_2 = 4$. The normal line is $80(x_1 - 2\sqrt{5}) = 0$ or $x_1 = 2\sqrt{5}$. The curve is self-crossing at $(0, 0)$.

Problem 9. Given the curve in the plane $F(x, y) \equiv y - x^2 = 0$. The curve intersects with the line $x + y = 2$ at

$$x_0 = \frac{1}{2}(\sqrt{5} - 1), \qquad y_0 = \frac{1}{2}(5 - \sqrt{5}).$$

There is also another intersecting point. Find the equation for the tangent line at (x_0, y_0). Find the equation for the normal line at (x_0, y_0). The equation for the *tangent line* is given by

$$(x - x_0)\frac{\partial F(x = x_0, y = y_0)}{\partial x} + (y - y_0)\frac{\partial F(x = x_0, y = y_0)}{\partial y} = 0.$$

The equation for the *normal line* is

$$(x - x_0)\frac{\partial F(x = x_0, y = y_0)}{\partial y} - (y - y_0)\frac{\partial F(x = x_0, y = y_0)}{\partial x} = 0.$$

Solution 9. With $\partial F/\partial x = -2x$, $\partial F/\partial y = 1$ and inserting x_0, y_0 provides the tangent line

$$x(1 - \sqrt{5}) + y + \frac{1}{2}(1 - \sqrt{5}) = 0.$$

With $\partial F/\partial x = -2x$, $\partial F/\partial y = 1$ and inserting x_0, y_0 provides the normal line

$$x + y(\sqrt{5} - 1) + \frac{11 - 7\sqrt{5}}{2} = 0.$$

Problem 10. Consider the curve (circle) in the plane

$$F(x, y) = x^2 + y^2 - 2 = 0$$

with $x_0 = -1$, $y_0 = 1$ satisfying this equation.
(i) Find the equation for the tangent line at this point.
(ii) Find the equation for the normal line at this point.

Solution 10. (i) We have $\partial F/\partial x = 2x$, $\partial F/\partial y = 2y$. The equation of the tangent line at the point $x_0 = -1$, $y_0 = 1$ is given by

$$x - y + 2 = 0 \;\Rightarrow\; (1 \quad -1) \begin{pmatrix} x \\ y \end{pmatrix} = -2.$$

(ii) Utilizing $\partial F/\partial x = 2x$, $\partial F/\partial y = 2y$ we find the normal line

$$x + y = 0 \;\Rightarrow\; (1 \quad 1) \begin{pmatrix} x \\ y \end{pmatrix} = 0$$

at the point $x_0 = -1$, $y_0 = 1$.

Problem 11. Let $a > 0$. Consider the curve (*folium of Descartes*) in the two-dimensional Euclidean plane

$$f(x_1, x_2) \equiv x_1^3 + x_2^3 - 3ax_1x_2 = 0.$$

Note that curve is invariant under $x_1 \mapsto x_2$, $x_2 \mapsto x_1$ and $f(0,0) = 0$. For $x_1 = x_2$ we have $2x_1^3 - 3ax_1^2 = 0$ with the solutions $x_1 = 0$ and $x_1 = 3a/2$.
(i) Show that this curve has a *singular point* at $(0,0)$.
(ii) Show that

$$x_1(\tau) = \frac{3a\tau}{1 + \tau^3}, \qquad x_2(\tau) = \frac{3a\tau^2}{1 + \tau^3}$$

is a parameter representation of the curve.

(iii) Find the derivative dx_2/dx_1 at $x_1 = x_2 = 3a/2$.

(iv) Find the area of the loop enclosed by the curve applying *Green's theorem*, i.e. we integrate

$$\frac{1}{2}(x_1(\tau)dx_2(\tau) - x_2(\tau)dx_1(\tau)).$$

Solution 11. (i) We have

$$\frac{\partial f}{\partial x_1} = 3x_1^2 - 3ax_2, \qquad \frac{\partial f}{\partial x_2} = 3x_2^2 - 3ax_1.$$

Thus $f(x_1 = 0, x_2 = 0) = 0$,

$$\frac{\partial f(x_1 = 0, x_2 = 0)}{\partial x_1} = 0, \qquad \frac{\partial f(x_1 = 0, x_2 = 0)}{\partial x_2} = 0.$$

Hence $(0, 0)$ is a singular point.

(ii) We apply the Maxima program

```
/* folium.mac */
x1: 3*a*tau/(1+tau*tau*tau);
x2: 3*a*tau*tau/(1+tau*tau*tau);
t1: x1*x1*x1; t2: x2*x2*x2; t3: 3*a*x1*x2;
r: t1+t2-t3;
r: ratsimp(r);
```

where `ratsimp` does the rational simplification.

(iii) We have

$$df = 3x_1^2 dx_1 + 3x_2^2 dx_2 - 3ax_2 dx_1 - 3ax_1 dx_2$$
$$= (3x_2^2 - 3ax_1)dx_2 + (3x_1^2 - 3ax_2)dx_1 = 0.$$

Hence

$$\frac{dx_2}{dx_1} = -\frac{x_1^2 - ax_2}{x_2^2 - ax_1} \Rightarrow \left.\frac{dx_2}{dx_1}\right|_{x_1 = x_2 = 3a/2} = -1.$$

(iv) With

$$dx_1(\tau) = \frac{3a(1 - 2\tau)}{(1 + \tau^3)^2}d\tau, \qquad dx_2(\tau) = \frac{3a\tau(2 - \tau^3)}{(1 + \tau^3)^2}d\tau$$

we obtain

$$x_1(\tau)dx_2(\tau) - x_2(\tau)dx_1(\tau) = \frac{9a\tau^2}{(1 + \tau^3)^2}.$$

Now
$$\int_0^\infty \frac{s^2}{1+2s^3+s^6}ds = -\frac{1}{3}\frac{1}{1+s^3}\Big|_0^\infty = \frac{1}{3}.$$
Thus the area of the loop enclosed by the curve is
$$\left|\frac{1}{2}\int_{\tau=0}^{\tau=\infty}(x_1(\tau)dx_2(\tau) - x_2(\tau)dx_1(\tau))\right| = \frac{3}{2}a^2.$$

Problem 12. Let α be a real parameter. The *envelope* of the family of curves $f(x_1, x_2, \alpha) = 0$ is the solution of the system of equations
$$f(x_1, x_2, \alpha) = 0, \qquad \frac{\partial f(x_1, x_2, \alpha)}{\partial \alpha} = 0.$$
Apply it to $f(x_1, x_2, \alpha) \equiv x_2^2 - 2\alpha x_1 = 0$ with $\alpha > 0$. The parametric form of this curve is $x_1(\tau) = 2\alpha\tau^2$, $x_2(\tau) = 2\alpha\tau$ (*parabola*).

Solution 12. We have $\partial f/\partial\alpha = -2x_1 = 0$. Thus $x_1 = 0$ and $x_2 = 0$.

Problem 13. The *arc length* of the equilateral *hyperbola*
$$h(\tau) = \sqrt{\tau^2 - 1}, \qquad \tau \geq 1$$
starting at $\tau = 1$ is given by
$$L_h(x) = \int_{\tau=1}^{\tau=x}\sqrt{\frac{2\tau^2 - 1}{\tau^2 - 1}}d\tau$$
as a function of the end point $\tau = x$. The tangent line to the hyperbola at $\tau = x$ is
$$T_h(\tau) = \sqrt{x^2 - 1} + \frac{x}{\sqrt{x^2 - 1}}(\tau - x)$$
whose intersection with the τ-axis is $\tau = 1/x$ ($\tau \in (0,1)$). The line
$$N_h(\tau) = -\frac{\sqrt{x^2 - 1}}{x}\tau$$
is perpendicular to T_h passing through the origin.
(i) Find the point P_h where the lines T_h and N_h intersect.
(ii) Calculate the distance from $(x, h(x))$ to the common point P_h.

Solution 13. (i) The lines T_h and N_h intersect at the point
$$P_h = \left(\frac{x}{2x^2 - 1}, -\frac{\sqrt{x^2 - 1}}{2x^2 - 1}\right).$$

(ii) The distance from $(x, h(x))$ to the common point P_h is

$$g_h(x) = 2x\sqrt{\frac{x^2 - 1}{2x^2 - 1}}.$$

Problem 14. The *implicit function theorem* is given by: Let U be an open subset of the vector space $\mathbb{R}^n \times \mathbb{R}^m$ and $\mathbf{F} : U \mapsto \mathbb{R}^m$ is a map of class C^ℓ with $1 \leq \ell \leq \infty$. Suppose that at the point $(\mathbf{x}_0, \mathbf{y}_0)$ we have $\mathbf{F}(\mathbf{x}_0, \mathbf{y}_0) = \mathbf{0}$ and that the $m \times m$ matrix $\partial F_k(\mathbf{x}_0, \mathbf{y}_0)/\partial y_j$ $(j, k = 1, \ldots, m)$ is non-singular (i.e. the determinant of this $m \times m$ matrix is nonzero). Then there exist open neighbourhoods V of $\mathbf{x}_0$ in $\mathbb{R}^n$ and W of $\mathbf{y}_0$ in $\mathbb{R}^m$ and a C^ℓ-map $g : V \mapsto W$ such that $V \times W \subset U$ and for each $(\mathbf{x}, \mathbf{y}) \in V \times W$ we have

$$\mathbf{F}(\mathbf{x}, \mathbf{y}) = \mathbf{0} \iff \mathbf{y} = \mathbf{g}(\mathbf{x}).$$

(i) Apply it to $F(x, y) \equiv xe^y - ye^x + x = 0$, i.e. $m = n = 1$ and $f(0, 0) = 0$.
(ii) Apply it to the surface of genus 2

$$F(x_1, x_2, y) = 2x_1^2(x_2^2 - 3x_1^2)(1 - y^2) + (x_1^2 + x_2^2)^2 - (9y^2 - 1)(1 - y^2) = 0$$

with $F(0, 0, 1) = 0$, i.e. $n = 2$ and $m = 1$.

Solution 14. (i) Since $\partial F/\partial y = xe^y - e^x$ we have

$$\frac{\partial F(0, 0)}{\partial y} = -1 \neq 0$$

we can solve with respect to y in a neighbourhood of $(0, 0)$. However the explicit form $y = f(x)$ cannot be given. The derivative of f at $x_0 = 0$ is given by

$$\frac{df(0)}{dx} = -\frac{\partial F(0, 0)/\partial x}{\partial F(0, 0)/\partial y} = 2.$$

(ii) We have

$$\frac{\partial F(0, 0, 1)}{\partial y} = -52 \neq 0.$$

For fixed x_1, x_2 we have a quartic equation in y. The four solutions are

$$y_1 = \frac{1}{3}\sqrt{\sqrt{T_1} + T_2}, \quad y_2 = -\frac{1}{3}\sqrt{\sqrt{T_1} + T_2}$$

$$y_3 = \frac{1}{3}\sqrt{-\sqrt{T_1} + T_2}, \quad y_4 = -\frac{1}{3}\sqrt{-\sqrt{T_1} + T_2}$$

with

$$T_1 = x_1^4 x_2^4 - 9x_2^4 - 6x_1^6 x_2^2 - 26x_1^2 x_2^2 + 9x_1^8 + 15x_1^4 + 16, \quad T_2 = x_1^2 x_2^2 - 3x_1^4 + 5.$$

Problem 15. Let $a > 0$. Consider the curve in the plane (*cycloid*)

$$x_1(\tau) = a(\tau - \sin(\tau)), \quad x_2(\tau) = a(1 - \cos(\tau)).$$

Find the *arc length*

$$L = \int_0^\pi \sqrt{\left(\frac{dx_1}{d\tau}\right)^2 + \left(\frac{dx_2}{d\tau}\right)^2} \, d\tau$$

from 0 to π. Note that for $\tau = 0$ we have $x_1(0) = x_2(0) = 0$ and for $\tau = \pi$ we have $x_1(\pi) = a\pi$, $x_2(\pi) = 2a$.

Solution 15. We have

$$\frac{dx_1}{d\tau} = a(1 - \cos(\tau)), \quad \frac{dx_2}{d\tau} = a\sin(\tau).$$

Hence

$$L = \int_0^\pi \sqrt{a^2 + a^2 \cos^2(\tau) - 2a^2 \cos(\tau) + a^2 \sin^2(\tau)}$$

$$= \int_0^\pi \sqrt{2a^2(1 - \cos(\tau))} d\tau$$

$$= \sqrt{2}a \int_0^\pi \sqrt{1 - \cos(\tau)} d\tau.$$

With $1 - \cos(\tau) \equiv 2\sin^2(\tau/2)$ and $\tau \in [0, \pi]$ we have

$$L = 2a \int_0^\pi \sin(\tau/2) d\tau = 4a.$$

Problem 16. Consider the ordinary differential equation

$$\left(\frac{dy}{dx}\right)^3 + x\frac{dy}{dx} - y = 0$$

with the solution $y(x) = Cx + C^3$. The *singular solution* is given by $4x^3 + 27y^2 = 0$ as can be seen as follows. Differentiation of $4x^3 + 27y^2 = 0$

yields $ydy/dx + (2/9)x^2 = 0$. Inserting this equation into the differential equation provides

$$-\frac{8x^6}{9^2} - 2x^3y^2 - 9y^4 = 0$$

which is satisfied with $y^2 = -4x^3/27$, i.e. we have the curve

$$F(x, y) = 4x^3 + 27y^2 = 0.$$

Find the equation of the tangent at $x_0 = -1$, $y_0 = 2/(3\sqrt{3})$. The equation for the tangent line is given by

$$(x - x_0)\frac{\partial F(x = x_0, y = y_0)}{\partial x} + (y - y_0)\frac{\partial F(x = x_0, y = y_0)}{\partial y} = 0.$$

Solution 16. With $\partial F/\partial x = 12x^2$, $\partial F/\partial y = 54y$. Inserting the values $x_0 = -1$, $y_0 = 2/(3\sqrt{3})$ provides the equation of the tangent line

$$x + \sqrt{3}y + \frac{1}{3} = 0.$$

Problem 17. Given the two curves in the plane

$$x^{2/3} + y^{2/3} = 1, \quad x^2 + y^2 = 4.$$

The curves do not intersect. Find a parameter representation for the curves. Find the shortest distance between the curves. Note the invariance of the curves under $x \mapsto y$, $y \mapsto x$. Are there other symmetries?

Solution 17. The first curve is a *hypocycloid* with four *cusps*. For $x = 0$ we have $y = \pm 1$ and for $y = 0$ we have $x = \pm 1$. The parameter representation is

$$x(\phi) = \cos^3(\phi), \quad y(\phi) = \sin^3(\phi).$$

The second curve is a circle with radius $r = 2$. The parameter representation is $x(\phi) = 2\cos(\phi)$, $y(\phi) = 2\sin(\phi)$. Owing to the symmetry of the problem the shortest distance is 1 along the x-axis (positive and negative) and along the y-axis (positive and negative).

Problem 18. The *curvature* $\kappa(\tau)$ of a curve in the plane given in parameter representation $x_1(\tau)$, $x_2(\tau)$ is given by

$$\kappa(\tau) = \frac{1}{(\|d\mathbf{x}/d\tau\|)^3} \det\begin{pmatrix} dx_1/d\tau & d^2x_1/d\tau^2 \\ dx_2/d\tau & d^2x_2/d\tau^2 \end{pmatrix}.$$

(i) Let $b \geq a > 0$. Consider the curve in the plane

$$\mathbf{x}(\tau) = \begin{pmatrix} x_1(\tau) \\ x_2(\tau) \end{pmatrix} = \begin{pmatrix} a\cos(\tau) \\ b\sin(\tau) \end{pmatrix}.$$

Find the curvature.

(ii) Consider the curve in the plane

$$\mathbf{x}(\tau) = \begin{pmatrix} x_1(\tau) \\ x_2(\tau) \end{pmatrix} = \begin{pmatrix} \tau \\ \cosh(\tau) \end{pmatrix}, \qquad \tau \in \mathbb{R}.$$

Show that the *curvature* is given by $\kappa(\tau) = \frac{1}{\cosh^2(\tau)}$.

Solution 18.   (i) Since

$$\frac{d\mathbf{x}}{d\tau} = \begin{pmatrix} -a\sin(\tau) \\ b\cos(\tau) \end{pmatrix}, \quad \frac{d^2\mathbf{x}}{d\tau^2} = \begin{pmatrix} -a\cos(\tau) \\ -b\sin(\tau) \end{pmatrix}$$

it follows that

$$\det \begin{pmatrix} -a\sin(\tau) & -a\cos(\tau) \\ b\cos(\tau) & -b\sin(\tau) \end{pmatrix} = ab(\sin^2(\tau) + \cos^2(\tau)) = ab$$

and

$$\|d\mathbf{x}/d\tau\|^3 = (a^2\cos^2(\tau) + b^2\sin^2(\tau))^{3/2}.$$

Therefore

$$\kappa(\tau) = \frac{ab}{(a^2\sin^2(\tau) + b^2\cos^2(\tau))^{3/2}}.$$

If $a = b$ the curvature is independent of τ and we obtain $\kappa(\tau) = 1/a$.

(ii) We have

$$\frac{dx_1}{d\tau} = 1, \qquad \frac{dx_2}{d\tau} = \sinh(\tau)$$

$$\frac{d^2x_1}{d\tau^2} = 0, \qquad \frac{d^2x_2}{d\tau^2} = \cosh(\tau).$$

Then

$$\det \begin{pmatrix} dx_1/d\tau & d^2x_1/d\tau^2 \\ dx_2/d\tau & d^2x_2/d\tau^2 \end{pmatrix} = \cosh(\tau)$$

and $\|d\mathbf{x}/d\tau\|^3 = \cosh^3(\tau)$. The result follows.

Problem 19.   (i) The *curvature* κ of a curve $(x_1(\tau), x_2(\tau))$ in the plane is given by

$$\kappa(\tau) = \frac{(dx_1/d\tau)(d^2x_2/d\tau^2) - (dx_2/d\tau)(d^2x_1/d\tau^2)}{((dx_1/d\tau)^2 + (dx_2/d\tau)^2)^{3/2}}.$$

Find the curvature for $(x_1(\tau) = \tau^2, x_2(\tau) = \tau^3)$.

(ii) For a curve in the plane given by $f(x_1, x_2) = 0$ the curvature is given by

$$\kappa(x_1, x_2) = \frac{-f_{x_2}^2 f_{x_1 x_1} + 2 f_{x_1} f_{x_2} f_{x_1 x_2} - f_{x_1}^2 f_{x_2 x_2}}{(f_{x_1}^2 + f_{x_2}^2)^{3/2}}$$

with $f_{x_1} \equiv \partial f / \partial x_1$ etc. Let $a > 0$. Find the curvature of the curve

$$f(x_1, x_2) = x_2^2 - x_1^2 - a^2 = 0.$$

Solution 19.   (i) With $dx_1/d\tau = 2\tau$, $d^2x_1/d\tau^2 = 2$, $dx_2/d\tau = 3\tau^2$, $d^2x_2/d\tau^2 = 6\tau$ we obtain

$$\kappa(\tau) = \frac{6}{\tau(4 + 9\tau^2)^{3/2}}.$$

(ii) With $f_{x_1} = -2x_1$, $f_{x_1 x_1} = -2$, $f_{x_2} = 2x_2$, $f_{x_2 x_2} = 2$, $f_{x_1 x_2} = 0$ we obtain

$$\kappa(x_1, x_2) = \frac{a^2}{(x_1^2 + x_2^2)^{3/2}}.$$

Problem 20.   Consider the analytic function $f : \mathbb{R}^2 \to \mathbb{R}$ given by

$$f(x_1, x_2) = \cos(x_1^2 + x_2^2).$$

Find the *level set* $\{ (x_1, x_2) : f(x_1, x_2) = 1 \}$.

Solution 20.   From the equation $\cos(x_1^2 + x_2^2) = 1$ we obtain

$$x_1^2 + x_2^2 = 2k\pi$$

for all $k \in \mathbb{N}_0$. For $k = 0$ we obtain the point $(0,0)$ and for $k \geq 1$ we obtain concentric circles centered at the origin $(0,0)$ with radius $\sqrt{2k\pi}$.

Problem 21.   (i) Consider the *circle* in the plane

$$\mathbb{S}^1 := \{(x_1, x_2) \in \mathbb{R}^2 : x_1^2 + x_2^2 = 1\}$$

and the *square* in the plane

$$I^2 = \{ (x_1, x_2) \in \mathbb{R} : (|x_1| = 1, |x_2| \leq 1), (|x_1| \leq 1, |x_2| = 1) \}.$$

Find a *homeomorphism*.

(ii) Show that an *open disc*

$$D^2 := \{\, (x_1, x_2) \in \mathbb{R}^2 \,:\, x_1^2 + x_2^2 < 1 \,\}$$

is *homeomorphic* to $\mathbb{R}^2$.

Solution 21.   (i) Let $r^2 = x_1^2 + x_2^2$. A homeomorphism is given by

$$f(x_1, x_2) = \left(\frac{x_1}{r}, \frac{x_2}{r} \right).$$

Since $r > 0$ the function f is invertible.
(ii) A homeomorphism $f : D^2 \to \mathbb{R}^2$ is

$$f(x_1, x_2) = \left(\frac{x_1}{\sqrt{1 - x_1^2 - x_2^2}}, \frac{x_2}{\sqrt{1 - x_1^2 - x_2^2}} \right).$$

Then the inverse function $f^{-1} : \mathbb{R}^2 \to D^2$ is given by

$$f^{-1}(x_1, x_2) = \left(\frac{x_1}{\sqrt{1 + x_1^2 + x_2^2}}, \frac{x_2}{\sqrt{1 + x_1^2 + x_2^2}} \right)$$

with $f \circ f^{-1} = \mathrm{id}_{\mathbb{R}^2}$ and $f^{-1} \circ f = \mathrm{id}_{D^2}$.

Problem 22.   Consider the two normalized vectors in $\mathbb{R}^2$

$$\mathbf{v}_1 = \begin{pmatrix} \cos(\theta_1) \\ \sin(\theta_1) \end{pmatrix}, \quad \mathbf{v}_2 = \begin{pmatrix} \cos(\theta_2) \\ \sin(\theta_2) \end{pmatrix}.$$

Find θ_1, θ_2 such that $\mathbf{v}_1^T \mathbf{v}_2 = 0$, i.e. the scalar product of $\mathbf{v}_1$ and $\mathbf{v}_2$ is equal to 0. Discuss.

Solution 22.   From $\mathbf{v}_1^T \mathbf{v}_2 = 0$ we obtain

$$\cos(\theta_1) \cos(\theta_2) + \sin(\theta_1) \sin(\theta_2) = 0.$$

Hence

$$\cos(\theta_1 - \theta_2) + \cos(\theta_1 + \theta_2) + \cos(\theta_1 - \theta_2) - \cos(\theta_1 + \theta_2) = 0.$$

It follows that $\cos(\theta_1 - \theta_2) = 0$. We select the solution $\theta_1 - \theta_2 = \pi/2$. Setting $\theta_1 = \theta$ we obtain the orthonormal basis in $\mathbb{R}^2$

$$\mathbf{v}_1 = \begin{pmatrix} \cos(\theta) \\ \sin(\theta) \end{pmatrix}, \quad \mathbf{v}_2 = \begin{pmatrix} \cos(\theta - \pi/2) \\ \sin(\theta - \pi/2) \end{pmatrix} = \begin{pmatrix} \sin(\theta) \\ -\cos(\theta) \end{pmatrix}.$$

Problem 23. Consider the compact differentiable manifold

$$\mathbb{S}^2 := \{ (x_1, x_2, x_3) \ : \ x_1^2 + x_2^2 + x_3^2 = 1 \}.$$

An element $\eta \in \mathbb{S}^2$ can be written as

$$\eta = (\cos(\phi)\sin(\theta), \sin(\phi)\sin(\theta), \cos(\theta))$$

where $\phi \in [0, 2\pi)$ and $\theta \in [0, \pi]$. The *stereographic projection* is a map

$$\Pi \ : \ \mathbb{S}^2 \setminus \{ (0, 0, -1) \} \to \mathbb{R}^2$$

given by

$$x_1(\theta, \phi) = \frac{2\sin(\theta)\cos(\phi)}{1 + \cos(\theta)}, \qquad x_2(\theta, \phi) = \frac{2\sin(\theta)\sin(\phi)}{1 + \cos(\theta)}.$$

(i) Let $\theta = 0$ and ϕ arbitrary. Find x_1, x_2. Give a geometric interpretation.
(ii) Find the inverse of the map, i.e. find $\Pi^{-1} \ : \ \mathbb{R}^2 \to \mathbb{S}^2 \setminus \{ (0, 0, -1) \}$.

Solution 23. (i) Since $\sin(0) = 0$ we find $x_1 = x_2 = 0$, i.e. the point $(0, 0, 1)$ is mapped to the origin $(0, 0)$.
(ii) Using division we find

$$\phi(x_1, x_2) = \arctan\left(\frac{x_2}{x_1}\right).$$

Since

$$x_1^2 + x_2^2 = \frac{4\sin^2(\theta)}{(1 + \cos(\theta))^2}, \qquad \tan\left(\frac{\theta}{2}\right) \equiv \frac{\sin(\theta)}{1 + \cos(\theta)}$$

we obtain

$$\theta(x_1, x_2) = 2\arctan\left(\frac{\sqrt{x_1^2 + x_2^2}}{2}\right).$$

Problem 24. Let $x, y \in \mathbb{R}$. Consider the map

$$\xi(x, y) = \frac{x}{1 + x^2 + y^2}, \qquad \eta(x, y) = \frac{y}{1 + x^2 + y^2}, \qquad \zeta(x, y) = \frac{x^2 + y^2}{1 + x^2 + y^2}.$$

Calculate

$$\xi^2 + \eta^2 + \left(\zeta - \frac{1}{2}\right)^2.$$

Discuss. Find $\xi(0,0)$, $\eta(0,0)$, $\zeta(0,0)$ and $\xi(1,1)$, $\eta(1,1)$, $\zeta(1,1)$.

Solution 24.   We obtain

$$\xi^2 + \eta^2 + \left(\zeta - \frac{1}{2}\right)^2 = \frac{1}{4}$$

and $\xi(0,0) = \eta(0,0) = \zeta(0,0) = 0$, $\xi(1,1) = 1/3$, $\eta(1,1) = 1/3$, $\zeta(1,1) = 2/3$.

Problem 25.   Let $S = \{\,(x,y) \in \mathbb{R}^2 \;:\; x^2 \le y \le 1\,\}$.
(i) Find the cone of feasible directions at $(1,1) \in S$.
(ii) Find a supporting hyperplane for $(1,1)$ at S.
(iii) Is $(0,1)$ a maximum/minimum point to $(1-x)^2 y^2$ on S? Prove or disprove.

Solution 25.   (i) Let (d_x, d_y) denote a feasible direction, and let $\lambda > 0$, $\lambda \in \mathbb{R}$. If $(1,1) + \lambda(d_x, d_y) \in S$ then

$$(1,1) + (\lambda d_x, \lambda d_y) \in S \quad \Rightarrow \quad (1 + \lambda d_x)^2 \le 1 + \lambda d_y \le 1.$$

From $1 + \lambda d_y \le 1$ and $\lambda > 0$ we find $d_y \le 0$. From $(1 + \lambda d_x)^2 \le 1 + \lambda d_y$ we find $d_x(2 + \lambda d_x) \le d_y$. If $d_x = 0$, then since $d_y \le 0$ it follows that $d_y = 0$. However, a feasible direction obeys $(d_x, d_y) \ne (0,0)$. Thus $d_x \ne 0$. It follows that

$$0 < \lambda \le \frac{d_y - 2d_x}{d_x^2}$$

where we require that $d_y > 2d_x$. Choosing (for $d_y \le 0$ and $2d_x < d_y$)

$$\delta = \frac{d_y - 2d_x}{d_x^2}$$

we find that $(1,1) + (\lambda d_x, \lambda d_y) \in S$ for all $0 \le \lambda < \delta$. Consequently $\{\,(a,b) \in \mathbb{R}^2 \;:\; b \le 0, 2a < b\,\}$ are the feasible directions at $(1,1)$.
(ii) The figure displays the supporting hyperplane

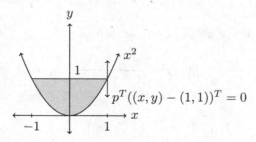

Here any supporting hyperplane could have been chosen. One such hyperplane is given by $p = (1,0)^T$. Then from $x^2 \leq 1$ it follows that $-2 \leq x - 1 \leq 0$ and

$$p^T((x,y) - (1,1))^T = p^T(x-1, y-1)^T = x - 1 \leq 0 \qquad \forall\, (x,y) \in S.$$

(iii) No (in both cases). Since at $x = 0, y = 1$ we have $(1-x)^2 y^2 = 1$, but at $x = 1, y = 1$ we have $(1-x)^2 y^2 = 0 < 1$. Hence the point does not provide a minimum. At $x = -1, y = 1$ we have $(1-x)^2 y^2 = 4 > 1$. Hence the point does not provide a maximum.

Problem 26.   Let $n \geq 2$. Consider the compact Lie group $SO(n)$. Let $M(\tau)$ be a smooth curve in $SO(n)$ with $M(0) = I_n$. Then $M(\tau)$ satisfies $\det(M(\tau)) = 1$ and $M(\tau)M^T(\tau) = I_n$. Obviously $M^T(\tau) = M^{-1}(\tau)$.
(i) Show that

$$\mathrm{tr}\left(M^{-1}(\tau)\frac{dM(\tau)}{d\tau}\right) = 0, \quad \frac{dM(\tau)}{d\tau}M^T(\tau) + M(\tau)\frac{dM^T(\tau)}{d\tau} = 0_n.$$

(ii) Give an example for such a matrix for $n = 2$.

Solution 26.   (i) Utilizing that

$$0 = \frac{d}{d\tau}1 = \frac{d}{d\tau}\det(M(\tau)) \equiv \det(M(\tau))\mathrm{tr}\left(M^{-1}(\tau)\frac{d}{d\tau}M(\tau)\right)$$

and

$$0_n = \frac{dI_n}{d\tau} = \frac{d}{d\tau}(M(\tau)M^T(\tau))$$

we find the equations.
(ii) An example is the *rotation matrix*

$$M(\tau) = \begin{pmatrix} \cos(\tau) & -\sin(\tau) \\ \sin(\tau) & \cos(\tau) \end{pmatrix} \Rightarrow M^{-1}(\tau) = \begin{pmatrix} \cos(\tau) & \sin(\tau) \\ -\sin(\tau) & \cos(\tau) \end{pmatrix}$$

with

$$\frac{dM(\tau)}{d\tau} = \begin{pmatrix} -\sin(\tau) & -\cos(\tau) \\ \cos(\tau) & -\sin(\tau) \end{pmatrix}.$$

Problem 27.   Consider the two manifolds (circles)

$$M_1 = \{\, (x_1, x_2) : x_1^2 + x_2^2 = 1 \,\},$$

$$M_2 = \{\, (x_1, x_2) : (x_1 - 2)^2 + (x_2 - 3/2)^2 = 1 \,\}$$

embedded in $\mathbb{R}^2$. Find the shortest distance between the non-intersecting circles

$$x_1^2 + x_2^2 = 1, \qquad (x_1 - 2)^2 + (x_2 - 3/2)^2 = 1 \tag{1}$$

applying the *Lagrange multiplier method*. Then find the *normal line* for the shortest distance vector and the two tangent lines at the two circles where the normal line crosses the circles.

Solution 27.   We rename the first circle to $x_{1,1}^2 + x_{1,2}^2 = 1$ and the second circle to $(x_{2,1} - 2)^2 + (x_{2,2} - 3/2)^2 = 1$. The square of the *Euclidean distance* for points on the two circles is given by

$$d^2 = (x_{1,1} - x_{2,1})^2 + (x_{1,2} - x_{2,2})^2$$

and thus the *Lagrange function* is

$$L = d^2 + \lambda_1(x_{1,1}^2 + x_{1,2}^2 - 1) + \lambda_2((x_{2,1} - 2)^2 + (x_{2,2} - 3/2)^2 - 1).$$

From $\partial L/\partial x_{1,1} = 0$, $\partial L/\partial x_{1,2} = 0$, $\partial L/\partial x_{2,1} = 0$, $\partial L/\partial x_{2,2} = 0$ we obtain the four equations

$$x_{2,1} = x_{1,1}(1 + \lambda_1), \quad x_{2,2} = x_{1,2}(1 + \lambda_1),$$

$$x_{1,1} - x_{2,1} = \lambda_2(x_{2,1} - 2), \quad x_{1,2} - x_{2,2} = \lambda_2(x_{2,2} - 3/2).$$

Together with eq.(1) we have six equations with six unknowns $x_{1,1}$, $x_{1,2}$, $x_{2,1}$, $x_{2,2}$, λ_1, λ_2. The solutions for the shortest distance is

$$x_{1,1} = \frac{4}{5}, \quad x_{1,2} = \frac{3}{5}, \quad x_{2,1} = \frac{6}{5}, \quad x_{2,2} = \frac{9}{10}, \quad \lambda_1 = \frac{1}{2}, \quad \lambda_2 = \frac{1}{2}.$$

From the two points $(x_{1,1}, x_{1,2})$, $(x_{2,1}, x_{2,2})$ it follows that the normal line is

$$x_1 = \frac{4}{3}x_2.$$

The tangent line at $x_{1,1} = 4/5$, $x_{1,2} = 3/5$ is given by

$$\frac{8}{5}x_1 + \frac{6}{5}x_2 = 2.$$

The tangent line at $x_{2,1} = 6/5$, $x_{2,2} = 9/10$ is given by

$$\frac{8}{5}x_1 + \frac{6}{5}x_2 = 3.$$

Problem 28.   A *Frenet frame* is a moving reference frame of n orthonormal vectors $e_j(\tau)$ $(j = 1, \ldots, n)$ in the Euclidean space $\mathbb{E}^n$. They describe a curve locally at each point $\mathbf{x}(\tau)$. Using the Frenet frame we describe local properties (e.g. curvature, torsion) in terms of a local reference system. Given a smooth curve $\mathbf{x}(\tau)$ in $\mathbb{E}^n$ which is regular of order n the Frenet frame for the curve is the set of orthonormal vectors (Frenet vectors)

$$\mathbf{e}_1(\tau), \ldots, \mathbf{e}_n(\tau).$$

They are constructed from the derivatives of $\mathbf{x}(\tau)$ using the Gram-Schmidt orthogonalization algorithm with

$$\mathbf{e}_1(\tau) = \frac{d\mathbf{x}(\tau)/d\tau}{\|d\mathbf{x}(\tau)/d\tau\|}, \quad \mathbf{e}_j(\tau) = \frac{\bar{\mathbf{e}}_j(\tau)}{\|\bar{\mathbf{e}}_j(\tau)\|}, \quad j = 2, \ldots, n$$

where

$$\bar{\mathbf{e}}_j(\tau) = \mathbf{x}^{(j)}(\tau) - \sum_{i=1}^{j-1} \langle \mathbf{x}^{(j)}(\tau), \mathbf{e}_i(\tau) \rangle \, \mathbf{e}_i(\tau)$$

where $\mathbf{x}^{(j)}$ denotes the j derivative with respect to τ and $\langle \, , \, \rangle$ denotes the scalar product in the Euclidean space $\mathbb{E}^n$. The Frenet frame is invariant under reparametrization. Find the Frenet frame for the curve ($\tau \in \mathbb{R}$)

$$\mathbf{x}(\tau) = \begin{pmatrix} \cos(\tau) \\ \tau \\ \sin(\tau) \end{pmatrix}$$

in $\mathbb{E}^3$.

Solution 28. We have

$$\frac{d\mathbf{x}(\tau)}{d\tau} = \begin{pmatrix} -\sin(\tau) \\ 1 \\ \cos(\tau) \end{pmatrix}, \quad \frac{d^2\mathbf{x}(\tau)}{d\tau^2} = \begin{pmatrix} -\cos(\tau) \\ 0 \\ -\sin(\tau) \end{pmatrix}, \quad \frac{d^3\mathbf{x}(\tau)}{d\tau^3} = \begin{pmatrix} \sin(\tau) \\ 0 \\ -\cos(\tau) \end{pmatrix}.$$

Thus

$$\mathbf{e}_1(\tau) = \frac{1}{\sqrt{2}} \begin{pmatrix} -\sin(\tau) \\ 1 \\ \cos(\tau) \end{pmatrix}, \quad \bar{\mathbf{e}}_2(\tau) = \mathbf{e}_2(\tau) = \begin{pmatrix} -\cos(\tau) \\ 0 \\ -\sin(\tau) \end{pmatrix},$$

$$\bar{\mathbf{e}}_3(\tau) = \frac{1}{2} \begin{pmatrix} \sin(\tau) \\ 1 \\ -\cos(\tau) \end{pmatrix} \Rightarrow \mathbf{e}_3(\tau) = \frac{1}{\sqrt{2}} \begin{pmatrix} \sin(\tau) \\ 1 \\ -\cos(\tau) \end{pmatrix}$$

and $\mathbf{e}_1(\tau), \mathbf{e}_2(\tau), \mathbf{e}_3(\tau)$ form an orthonormal basis in the Euclidean space $\mathbb{E}^3$.

Problem 29.   Consider the two manifolds

$$x_1^2 + x_2^2 = 1, \quad y_1^2 + y_2^2 = 1.$$

Show that $|x_1 y_1 + x_2 y_2| \leq 1$. Hint. Set

$$x_1(s) = \cos(s), \quad x_2(s) = \sin(s), \quad y_1(\tau) = \cos(\tau), \quad y_2(\tau) = \sin(\tau).$$

Solution 29.   We have

$$
\begin{aligned}
x_1 y_1 + x_2 y_2 &= \cos(s)\cos(\tau) + \sin(s)\sin(\tau) \\
&= \frac{1}{2}\left(\cos(s-\tau) + \cos(s+\tau) + \cos(s-\tau) - \cos(s+\tau)\right) \\
&= \cos(s-\tau).
\end{aligned}
$$

Since $|\cos(\alpha)| \leq 1$ for all $\alpha \in \mathbb{R}$ the assumption follows.

Problem 30.   Consider the two-dimensional unit sphere

$$\mathbb{S}^2 := \{\, \mathbf{x} \in \mathbb{R}^3 \ : \ x_1^2 + x_2^2 + x_3^2 = 1 \,\}.$$

Show that $\mathbb{S}^2$ is an orientable two-dimensional differentiable manifold. Use the following orientation-preserving *atlas*

$$U_1 = \{\, \mathbf{x} \in \mathbb{S}^2 \ : \ x_3 > 0 \,\}, \qquad U_2 = \{\, \mathbf{x} \in \mathbb{S}^2 \ : \ x_3 < 0 \,\},$$

$$U_3 = \{\, \mathbf{x} \in \mathbb{S}^2 \ : \ x_2 > 0 \,\}, \qquad U_4 = \{\, \mathbf{x} \in \mathbb{S}^2 \ : \ x_2 < 0 \,\},$$

$$U_5 = \{\, \mathbf{x} \in \mathbb{S}^2 \ : \ x_1 > 0 \,\}, \qquad U_6 = \{\, \mathbf{x} \in \mathbb{S}^2 \ : \ x_1 < 0 \,\}.$$

Solution 30.   The corresponding maps for the atlas are

$$\mathbf{h}_1(x_1, x_2, x_3) = (x_1, x_2), \qquad \mathbf{h}_1^{-1}(x_1, x_2) = (x_1, x_2, \sqrt{1 - x_1^2 - x_2^2})$$

$$\mathbf{h}_2(x_1, x_2, x_3) = (x_2, x_1), \qquad \mathbf{h}_2^{-1}(x_2, x_1) = (x_1, x_2, -\sqrt{1 - x_1^2 - x_2^2})$$

$$\mathbf{h}_3(x_1, x_2, x_3) = (x_3, x_1), \qquad \mathbf{h}_3^{-1}(x_3, x_1) = (x_1, \sqrt{1 - x_1^2 - x_3^2}, x_3)$$

$$\mathbf{h}_4(x_1, x_2, x_3) = (x_1, x_3), \qquad \mathbf{h}_4^{-1}(x_1, x_3) = (x_1, -\sqrt{1 - x_1^2 - x_3^2}, x_3)$$

$$\mathbf{h}_5(x_1, x_2, x_3) = (x_2, x_3), \qquad \mathbf{h}_5^{-1}(x_2, x_3) = (\sqrt{1 - x_2^2 - x_3^2}, x_2, x_3)$$

$$\mathbf{h}_6(x_1, x_2, x_3) = (x_3, x_2), \qquad \mathbf{h}_6^{-1}(x_3, x_2) = (-\sqrt{1 - x_2^2 - x_3^2}, x_2, x_3).$$

Next we calculate $\mathbf{h}_i \circ \mathbf{h}_j(\mathbf{x})$, $D(\mathbf{h}_i \circ \mathbf{h}_j^{-1}(\mathbf{x}))$. We find that

$$\det(D(\mathbf{h}_i \circ \mathbf{h}_j^{-1}(\mathbf{x}))) > 0$$

for all $\mathbf{x} \in \mathbf{h}_j(U_i \cap U_j)$, where $U_i \cap U_j \neq \emptyset$. For example we have

$$\mathbf{h}_1 \circ \mathbf{h}_3^{-1}(x_3, x_1) = (x_1, \sqrt{1 - x_1^2 - x_3^2}).$$

It follows that

$$D(\mathbf{h}_1 \circ \mathbf{h}_3^{-1}(x_3, x_1)) = \begin{pmatrix} 0 & 1 \\ \frac{-x_3}{\sqrt{1-x_1^2-x_3^2}} & \frac{-x_1}{\sqrt{1-x_1^2-x_3^2}} \end{pmatrix}$$

and

$$\det(D(\mathbf{h}_1 \circ \mathbf{h}_3^{-1}(x_3, x_1))) = \frac{x_3}{\sqrt{1 - x_1^2 - x_3^2}} > 0$$

for all

$$(x_3, x_1) \in \mathbf{h}_3(U_1 \cap U_3) = \{ (x_3, x_1) \in \mathbb{R}^2 : x_1^2 + x_3^2 < 1, \, x_3 > 0 \}.$$

Analogously we show it for the other maps. We only need to do this for $j > i$ since $U_1 \cap U_2 = U_3 \cap U_4 = U_5 \cap U_6 = \emptyset$. The atlas has been chosen to preserve outward normals.

Problem 31.   We know that $\mathbb{C}^{n+1}$ is an $n + 1$-dimensional complex manifold. The *complex projective space* $\mathbb{CP}^n$ is defined to be the set of lines through the origin in $\mathbb{C}^{n+1}$, i.e.

$$\mathbb{CP}^n := (\mathbb{C}^{n+1} \setminus \{\mathbf{0}\})/\sim$$

for the equivalence relation

$$(u_0, u_1, \ldots, u_n) \sim (v_0, v_1, \ldots, v_n) \Leftrightarrow \exists \gamma \in \mathbb{C}^* : \gamma u_j = v_j \,\, \forall \, 0 \leq j \leq n$$

where $\mathbb{C}^* := \mathbb{C} \setminus \{0\}$.
(i) Show that $\mathbb{CP}^1$ is a one-dimensional complex manifold.
(ii) Let f be the map that takes nonzero vectors in $\mathbb{C}^2$ to vectors in $\mathbb{R}^3$ by

$$f(z_1, z_2) = \left(\frac{z_1\overline{z}_2 + \overline{z}_1 z_2}{z_1\overline{z}_1 + \overline{z}_2 z_2}, \frac{z_1\overline{z}_2 - \overline{z}_1 z_2}{i(z_1\overline{z}_1 + \overline{z}_2 z_2)}, \frac{z_1\overline{z}_1 - \overline{z}_2 z_2}{z_1\overline{z}_1 + \overline{z}_2 z_2} \right)^T.$$

The map f defines a bijection between $\mathbb{CP}^1$ and the unit sphere in $\mathbb{R}^3$. Consider the normalized vectors in $\mathbb{C}^2$

$$\begin{pmatrix} 1 \\ 0 \end{pmatrix}, \quad \begin{pmatrix} 0 \\ 1 \end{pmatrix}, \quad \frac{1}{\sqrt{2}} \begin{pmatrix} 1 \\ 1 \end{pmatrix}, \quad \frac{1}{\sqrt{2}} \begin{pmatrix} 1 \\ -1 \end{pmatrix}, \quad \frac{1}{\sqrt{2}} \begin{pmatrix} i \\ -i \end{pmatrix}.$$

Apply f to these vectors in $\mathbb{C}^2$.

Solution 31.   (i) Denote by $[z_1, z_2]$ the equivalence class of the point

$$(z_1, z_2) \in \mathbb{C}^2 \setminus (0,0)$$

then $\mathbb{CP}^1$ can be covered by two open subsets

$$U_1 := \{ [z_1, z_2] : z_1 \neq 0 \} \quad \text{and} \quad U_2 := \{ [z_1, z_2] : z_2 \neq 0 \}.$$

The chart map ϕ_1 is given by the homeomorphism

$$U_1 \overset{\phi_1}{\to} \mathbb{C}, \qquad [z_1, z_2] \to \frac{z_2}{z_1}.$$

The map ϕ_1 is independent of the chosen representant in the equivalence class and is defined as $z_1 \neq 0$ on U_1. Analogously

$$U_2 \overset{\phi_2}{\to} \mathbb{C}, \qquad [z_1, z_2] \to \frac{z_1}{z_2}$$

is a homeomorphism. Consequently, $\mathbb{CP}^1$ is a manifold of complex dimension one. To prove that it is equipped with a holomorphic structure we consider the maps $\phi_2 \circ \phi_1^{-1}$ and $\phi_1 \circ \phi_2^{-1}$, where $\circ$ denotes function composition and ϕ_1^{-1}, ϕ_2^{-1} are the inverse maps. Since

$$U_1 \cap U_2 = \{ [z_1, z_2] : z_1 \neq 0, \, z_2 \neq 0 \}$$

it follows that

$$\phi_k(U_1 \cap U_2) = \mathbb{C}^* \equiv \mathbb{C} \setminus \{ 0 \}$$

for $k = 1, 2$. Let c be a local coordinate function for $\mathbb{C}^*$, then the maps are of the form

$$\mathbb{C}^* \to \mathbb{C} \quad \text{defined by} \quad c \mapsto \frac{1}{c}$$

which is a holomorphic function on $\mathbb{C}^*$ (a rational function without poles in $\mathbb{C}^*$). That is, $\mathbb{CP}^1$ is a complex one-dimensional manifold.

(ii) Since all vectors in $\mathbb{C}^2$ are normalized we have $z_1 \bar{z}_1 + z_2 \bar{z}_2 = 1$. We find

$$\begin{pmatrix} 0 \\ 0 \\ 1 \end{pmatrix}, \quad \begin{pmatrix} 0 \\ 0 \\ -1 \end{pmatrix}, \quad \begin{pmatrix} 1 \\ 0 \\ 0 \end{pmatrix}, \quad \begin{pmatrix} -1 \\ 0 \\ 0 \end{pmatrix}, \quad \begin{pmatrix} -1 \\ 0 \\ 0 \end{pmatrix}.$$

Problem 32.   Let

$$S^n := \{ (x_1, x_2, \ldots, x_{n+1}) : x_1^2 + x_2^2 + \cdots + x_{n+1}^2 = 1 \}.$$

(i) Show that $\mathbb{S}^3$ can be considered as a subset of $\mathbb{C}^2$ ($\mathbb{C}^2 \cong \mathbb{R}^4$)

$$\mathbb{S}^3 = \{\, (z_1, z_2) \in \mathbb{C}^2 \; : \; |z_1|^2 + |z_2|^2 = 1 \,\}.$$

(ii) The *Hopf map* $\pi : \mathbb{S}^3 \to \mathbb{S}^2$ is defined by

$$\pi(z_1, z_2) := (\overline{z}_1 z_2 + \overline{z}_2 z_1, -i\overline{z}_1 z_2 + i\overline{z}_2 z_1, |z_1|^2 - |z_2|^2).$$

Find the parametrization of $\mathbb{S}^3$, i.e. find $z_1(\theta, \phi)$, $z_2(\theta, \phi)$ and thus show that π maps $\mathbb{S}^3$ onto $\mathbb{S}^2$.

(iii) Show that $\pi(z_1, z_2) = \pi(z_1', z_2')$ if and only if $z_j' = e^{i\alpha} z_j$ $(j = 1, 2)$ and $\alpha \in \mathbb{R}$.

Solution 32. (i) Let $z_1 = x_1 + iy_1$ and $z_2 = x_2 + iy_2$, where $x_1, x_2, y_1, y_2 \in \mathbb{R}$. Then from $|z_1|^2 + |z_2|^2 = 1$ it follows that $x_1^2 + y_1^2 + x_2^2 + y_2^2 = 1$.

(ii) Since

$$|z_1|^2 + |z_2|^2 = 1$$

we have the parametrization

$$z_1(\theta, \phi) = \cos(\theta/2)e^{i\phi_1}, \qquad z_2(\theta, \phi) = \sin(\theta/2)e^{i\phi_2}$$

where $0 \le \theta \le \pi$ and $\phi_1, \phi_2 \in \mathbb{R}$. Thus

$$\pi(\cos(\theta/2)e^{i\phi_1}, \sin(\theta/2)e^{i\phi_2}) = (\sin(\theta)\cos(\phi_2 - \phi_1), \sin(\theta)\sin(\phi_2 - \phi_1), \cos(\theta)).$$

(iii) From $\pi(z_1, z_2) = \pi(z_1', z_2')$ we obtain the three equations

$$\overline{z}_1 z_2 + \overline{z}_2 z_1 = \overline{z}_1' z_2' + \overline{z}_2' z_1'$$
$$-\overline{z}_1 z_2 + \overline{z}_2 z_1 = -\overline{z}_1' z_2' + \overline{z}_2' z_1'$$
$$|z_1|^2 - |z_2|^2 = |z_1'|^2 - |z_2'|^2$$

with the solution $z_1' = e^{i\alpha} z_1$ and $z_2' = e^{i\alpha} z_2$, where $\alpha \in \mathbb{R}$.

Problem 33. Consider the solid torus $M = \mathbb{S}^1 \times D^2$, where D^2 is the unit disk in $\mathbb{R}^2$. On it we define coordinates (φ, x, y) such that $\varphi \in \mathbb{S}^1$ and $(x, y) \in D^2$, that is, $x^2 + y^2 \le 1$. Using these coordinates we define the map

$$f : M \to M, \qquad f(\varphi, x, y) = \left(2\varphi, \frac{1}{10}x + \frac{1}{2}\cos(\varphi), \frac{1}{10}y + \frac{1}{2}\sin(\varphi) \right).$$

(i) Show that this map is well-defined, that is, $f(M) \subset M$.

(ii) Show that f is *injective*.

Solution 33.   (i) Since $x, y \in [0, 1]$ and

$$|\sin(\varphi)| \le 1, \quad |\cos(\varphi)| \le 1, \quad \sin^2(\varphi) + \cos^2(\varphi) = 1$$

we have

$$\left(\frac{1}{10}x + \frac{1}{2}\cos(\varphi)\right)^2 + \left(\frac{1}{10}y + \frac{1}{2}\cos(\varphi)\right)^2 \le \frac{1}{100} + \frac{2}{10} + \frac{1}{4} < 1.$$

(ii) Suppose that $f(\varphi_1, x_1, y_1) = f(\varphi_2, x_2, y_2)$. Then

$$2\varphi_1 = 2\varphi_2 \pmod{2\pi}$$
$$\frac{1}{10}x_1 + \frac{1}{2}\cos(\varphi_1) = \frac{1}{10}x_2 + \frac{1}{2}\cos(\varphi_2)$$
$$\frac{1}{10}y_1 + \frac{1}{2}\sin(\varphi_1) = \frac{1}{10}y_2 + \frac{1}{2}\sin(\varphi_2).$$

If $\varphi_1 = \varphi_2$ we find that $x_1 = x_2$ and $y_1 = y_2$. If $\varphi_1 = \varphi_2 + \pi$, then

$$\frac{1}{10}x_1 + \frac{1}{2}\cos(\varphi_1) = \frac{1}{10}x_2 - \frac{1}{2}\cos(\varphi_1)$$
$$\frac{1}{10}y_1 + \frac{1}{2}\sin(\varphi_1) = \frac{1}{10}y_2 - \frac{1}{2}\sin(\varphi_1)$$

or

$$\frac{1}{10}(x_2 - x_1) = \cos(\varphi_1), \qquad \frac{1}{10}(y_2 - y_1) = \sin(\varphi_1)$$

which implies that $(x_2 - x_1)^2 + (y_2 - y_1)^2 = 100$. Since the left-hand side is bounded by 8, this is impossible. If we consider any cross section $C = \{\theta\} \times D^2$ of M, the image $f(M)$ will intersect C in two disjoint disks of radius $1/10$.

Problem 34.   Consider the *unit ball*

$$\mathbb{S}^2 := \{(x_1, x_2, x_3) \in \mathbb{R}^3 : x_1^2 + x_2^2 + x_3^2 = 1\}.$$

Let $\mathbf{x}(\tau) = (x_1(\tau), x_2(\tau), x_3(\tau))$ be a parametrized differentiable curve on $\mathbb{S}^2$. Show that the vector $(x_1(\tau), x_2(\tau), x_3(\tau))$ (τ fixed) is normal to the sphere at the point $(x_1(\tau), x_2(\tau), x_3(\tau))$.

Solution 34.   We have

$$\frac{d}{d\tau}(x_1^2(\tau) + x_2^2(\tau) + x_3^2(\tau)) = 2\left(x_1(\tau)\frac{dx_1}{d\tau} + x_2(\tau)\frac{dx_2}{d\tau} + x_3(\tau)\frac{dx_3}{d\tau}\right) = 0$$

or written as scalar product

$$(x_1(\tau) \quad x_2(\tau) \quad x_3(\tau)) \begin{pmatrix} dx_1/d\tau \\ dx_2/d\tau \\ dx_3/d\tau \end{pmatrix} = 0.$$

Problem 35. A generic *superquadric surface* can be defined as a closed surface in $\mathbb{R}^3$

$$\mathbf{x}(\eta,\omega) \equiv \begin{pmatrix} x_1(\eta,\omega) \\ x_2(\eta,\omega) \\ x_3(\eta,\omega) \end{pmatrix} = \begin{pmatrix} a_1 \cos^{\epsilon_1}(\eta) \cos^{\epsilon_2}(\omega) \\ a_2 \cos^{\epsilon_1}(\eta) \sin^{\epsilon_2}(\omega) \\ a_3 \sin^{\epsilon_1}(\eta) \end{pmatrix}$$

where $-\pi/2 \le \eta \le \pi/2$, $-\pi \le \omega < \pi$. There are five parameters ϵ_1, ϵ_2, a_1, a_2, a_3. Here ϵ_1 and ϵ_2 are the deformation parameters that control the shape with $\epsilon_1, \epsilon_2 \in (0, 2)$. The parameters a_1, a_2, a_3 define the size in x_1, x_2 and x_3 directions. Find the implicit representation.

Solution 35. We find

$$\left(\left(\frac{x_1}{a_1} \right)^{2/\epsilon_2} + \left(\frac{x_2}{a_2} \right)^{2/\epsilon_2} \right)^{\epsilon_2/\epsilon_1} + \left(\frac{x_3}{a_3} \right)^{2/\epsilon_1} = 1.$$

Thus we have an inside-outside function

$$f(x_1, x_2, x_3) = \left(\left(\frac{x_1}{a_1} \right)^{2/\epsilon_2} + \left(\frac{x_2}{a_2} \right)^{2/\epsilon_2} \right)^{\epsilon_2/\epsilon_1} + \left(\frac{x_3}{a_3} \right)^{2/\epsilon_1}.$$

Given any point (x_1, x_2, x_3) its position to the superquadric surface can be determined by

$$f(x_1, x_2, x_3) = 1 \Leftrightarrow \text{point on the surface}$$
$$f(x_1, x_2, x_3) < 1 \Leftrightarrow \text{point inside the surface}$$
$$f(x_1, x_2, x_3) > 1 \Leftrightarrow \text{point outside the surface.}$$

Problem 36. Let $U = \mathbb{R}^2 \setminus \{(0,0)\}$ and the map $\mathbf{f} : U \to U$

$$\mathbf{f}(\mathbf{x}) = \begin{pmatrix} f_1(\mathbf{x}) \\ f_2(\mathbf{x}) \end{pmatrix} = \begin{pmatrix} x_1^2 - x_2^2 \\ 2x_1 x_2 \end{pmatrix}.$$

(i) Find the functional matrix. Show that the map is not *injective*.

(ii) Find df_1, df_2 and $df_1 \wedge df_2$, where $\wedge$ is the exterior product.

Solution 36.   (i) We obtain for the *functional matrix*

$$\begin{pmatrix} \partial f_1/\partial x_1 & \partial f_1/\partial x_2 \\ \partial f_2/\partial x_1 & \partial f_2/\partial x_2 \end{pmatrix} = \begin{pmatrix} 2x_1 & -2x_2 \\ 2x_2 & 2x_1 \end{pmatrix}.$$

It follows that

$$\det \begin{pmatrix} 2x_1 & -2x_2 \\ 2x_2 & 2x_1 \end{pmatrix} = 4(x_1^2 + x_2^2) \neq 0.$$

We have $\mathbf{f}(1,0) = \mathbf{f}(-1,0) = (1,0)$ and thus $\mathbf{f}$ is not injective.
(ii) One obtains $df_1 = 2x_1 dx_1 - 2x_2 dx_2$, $df_2 = 2x_1 dx_2 + 2x_2 dx_1$ and
hence $df_1 \wedge df_2 = 4(x_1^2 + x_2^2)dx_1 \wedge dx_2$. The coefficient of $dx_1 \wedge dx_2$ is the
determinant of the functional matrix.

Problem 37.   Consider the function $\mathbf{f} : (0, \pi/2) \times (0, 2\pi) \mapsto \mathbb{R}^3$

$$\mathbf{f}(u_1, u_2) = \begin{pmatrix} f_1(u_1, u_2) \\ f_2(u_1, u_2) \\ f_3(u_1, u_2) \end{pmatrix} = \begin{pmatrix} \sin(u_1)\cos(u_2) \\ \sin(u_1)\sin(u_2) \\ \cos(u_1) \end{pmatrix}.$$

Find the vector

$$\frac{\partial \mathbf{f}(u_1, u_2)}{\partial u_1} \times \frac{\partial \mathbf{f}(u_1, u_2)}{\partial u_2}$$

where $\times$ is the vector product. Find the *normal vector*

$$\mathbf{n}(\mathbf{f}(u_1, u_2)) := \frac{\frac{\partial \mathbf{f}}{\partial u_1} \times \frac{\partial \mathbf{f}}{\partial u_2}}{\|\frac{\partial \mathbf{f}}{\partial u_1} \times \frac{\partial \mathbf{f}}{\partial u_2}\|}.$$

Solution 37.   We have

$$\begin{pmatrix} \partial f_1/\partial u_1 \\ \partial f_2/\partial u_1 \\ \partial f_3/\partial u_1 \end{pmatrix} = \begin{pmatrix} \cos(u_1)\cos(u_2) \\ \cos(u_1)\sin(u_2) \\ -\sin(u_1) \end{pmatrix}, \quad \begin{pmatrix} \partial f_1/\partial u_2 \\ \partial f_2/\partial u_2 \\ \partial f_3/\partial u_2 \end{pmatrix} = \begin{pmatrix} -\sin(u_1)\sin(u_2) \\ \sin(u_1)\cos(u_2) \\ 0 \end{pmatrix}.$$

Thus

$$\frac{\partial \mathbf{f}(u_1, u_2)}{\partial u_1} \times \frac{\partial \mathbf{f}(u_1, u_2)}{\partial u_2} = \begin{pmatrix} \sin^2(u_1)\cos(u_2) \\ \sin^2(u_1)\sin(u_2) \\ \sin(u_1)\cos(u_1) \end{pmatrix}.$$

The norm of the vector is given by $\sin(u_1)$. Hence the normal vector is given by

$$\mathbf{n}(\mathbf{f}(u_1, u_2)) = \begin{pmatrix} \sin(u_1)\cos(u_2) \\ \sin(u_1)\sin(u_2) \\ \cos(u_1) \end{pmatrix}.$$

Problem 38. Given the smooth surface in $\mathbb{R}^3$

$$\mathbf{x}(u_1, u_2) = \begin{pmatrix} x_1(u_1, u_2) \\ x_2(u_1, u_2) \\ x_3(u_1, u_2) \end{pmatrix} = \begin{pmatrix} u_1\cos(u_2) \\ u_1\sin(u_2) \\ u_2 \end{pmatrix}.$$

Find the *tangent plane* for $u_{10} = 1$, $u_{20} = 2\pi$. Note that

$$\mathbf{x}(1, 2\pi) = \begin{pmatrix} 1 \\ 0 \\ 2\pi \end{pmatrix}.$$

The tangent plane $T(s_1, s_2)$ is given by

$$T(s_1, s_2) = s_1 \frac{\partial \mathbf{x}}{\partial u_1}\Big|_{\substack{u_1=u_{10}\\u_2=u_{20}}} + s_2 \frac{\partial \mathbf{x}}{\partial u_2}\Big|_{\substack{u_1=u_{10}\\u_2=u_{20}}} + \mathbf{x}(u_1 = u_{10}, u_2 = u_{20}).$$

Solution 38. We have

$$\frac{\partial \mathbf{x}}{\partial u_1} = \begin{pmatrix} \cos(u_2) \\ \sin(u_2) \\ 0 \end{pmatrix}, \qquad \frac{\partial \mathbf{x}}{\partial u_2} = \begin{pmatrix} -u_1\sin(u_2) \\ u_1\cos(u_2) \\ 1 \end{pmatrix}$$

and

$$\frac{\partial \mathbf{x}}{\partial u_1}\Big|_{u_1=1, u_2=2\pi} = \begin{pmatrix} 1 \\ 0 \\ 0 \end{pmatrix}, \qquad \frac{\partial \mathbf{x}}{\partial u_2}\Big|_{u_1=1, u_2=2\pi} = \begin{pmatrix} 0 \\ 1 \\ 1 \end{pmatrix}.$$

Hence

$$T(s_1, s_2) = \begin{pmatrix} s_1 \\ 0 \\ 0 \end{pmatrix} + \begin{pmatrix} 0 \\ s_2 \\ s_2 \end{pmatrix} + \begin{pmatrix} 1 \\ 0 \\ 2\pi \end{pmatrix} = \begin{pmatrix} s_1 + 1 \\ s_2 \\ s_2 + 2\pi \end{pmatrix}.$$

Problem 39. Given a smooth surface $f(x_1, x_2, x_3) = 0$ in $\mathbb{R}^3$. Let $(x_{10}, x_{20}, x_{30}) \in \mathbb{R}^3$ with $f(x_{10}, x_{20}, x_{30}) = 0$. The *tangent plane* to $f(x_1, x_2, x_3) = 0$ at (x_{10}, x_{20}, x_{30}) is the plane with *normal vector*

$$\nabla f|_{x_1=x_{10}, x_2=x_{20}, x_3=x_{30}}$$

which passes through the point (x_{10}, x_{20}, x_{30}) and the equation for the tangent plane is

$$(x_1 - x_{10})\frac{\partial f(\mathbf{x_0})}{\partial x_1} + (x_2 - x_{20})\frac{\partial f(\mathbf{x_0})}{\partial x_2} + (x_3 - x_{30})\frac{\partial f(\mathbf{x_0})}{\partial x_3} = 0.$$

Find the tangent plane for the surface $f(x_1, x_2, x_3) = x_1^2 + 2x_2^2 + 3x_3^2 - 36 = 0$ at the point $(x_{10} = 1, x_{20} = 2, x_{30} = 3)$.

Solution 39.   We have

$$\frac{\partial f}{\partial x_1} = 2x_1, \quad \frac{\partial f}{\partial x_2} = 4x_2, \quad \frac{\partial f}{\partial x_3} = 6x_3.$$

It follows that the tangent surface is given by

$$2(x_1 - 1) + 8(x_2 - 2) + 18(x_3 - 3) = 0$$

or $2x_1 + 8x_2 + 18x_3 - 72 = 0$.

Problem 40.   Consider the smooth function $f : \mathbb{R}^2 \to \mathbb{R}$

$$f(x_1, x_2) = x_1^2 x_2 + x_2^3.$$

Let $\mathbf{p} = (1\ 2)^T$, i.e. $p_1 = 1$, $p_2 = 2$.
(i) Find $f(p_1, p_2)$.
(ii) Find the *gradient* $\nabla(f)$.
(iii) Find $df(x_1, x_2)$ and $df_\mathbf{p}(\mathbf{x} - \mathbf{p})$.
(iv) Find the *tangent plane* at the point $\mathbf{p}$.

Solution 40.   (i) We have $f(1, 2) = 10$.
(ii) With

$$\frac{\partial f}{\partial x_1} = 2x_1 x_2, \quad \frac{\partial f}{\partial x_2} = x_1^2 + 3x_2^2$$

we obtain

$$\nabla(f) = \begin{pmatrix} 2x_1 x_2 \\ x_1^2 + 3x_2^2 \end{pmatrix} \quad \Rightarrow \quad \nabla(f(\mathbf{p})) = \begin{pmatrix} 4 \\ 13 \end{pmatrix}.$$

(iii) We obtain the differential one-form

$$df(\mathbf{x}) = \frac{\partial f}{\partial x_1} dx_1 + \frac{\partial f}{\partial x_2} dx_2 = 2x_1 x_2 dx_1 + (x_1^2 + 3x_2^2) dx_2.$$

With $x_1 \mapsto p_1$, $x_2 \mapsto p_2$, $dx_1 \mapsto x_1 - p_1$, $dx_2 \mapsto x_2 - p_2$ we arrive at

$$df_{\mathbf{p}}(\mathbf{x} - \mathbf{p}) = 2p_1 p_2 (x_1 - p_1) + (p_1^2 + 3p_2^2)(x_2 - p_2).$$

(iv) Using the result from (iii) we find the tangent plane

$$x_3 = f(p_1, p_2) + df_{\mathbf{p}}(\mathbf{x} - \mathbf{p}) = f(p_1, p_2) + 2p_1 p_2 (x_1 - p_1) + (p_1^2 + 3p_2^2)(x_2 - p_2).$$

It follows that

$$x_3 = 10 + 4(x_1 - 1) + 13(x_2 - 2) = 4x_1 + 13x_2 - 20.$$

Problem 41.   Consider the analytic function $f : \mathbb{R}^3 \to \mathbb{R}$

$$f(x_1, x_2, x_3) = 3x_1^2 + 4x_2^2 + x_3$$

and the smooth surface in $\mathbb{R}^3$

$$S = \{\, (x_1, x_2, x_3) \,:\, f(x_1, x_2, x_3) = -2 \,\}.$$

(i) Show that $\mathbf{p} = (1, 1, -9) \in \mathbb{R}^3$ satisfies $f(x_1, x_2, x_3) = -2$.
(ii) Find the normal vector $\mathbf{n}$ at $\mathbf{p}$.
(iii) Let

$$\mathbf{v}^T = (\,v_1 \quad v_2 \quad v_3\,).$$

Calculate $\mathbf{v}^T (\nabla f)_{\mathbf{p}}$. Find the conditions on v_1, v_2, v_3 such that

$$\mathbf{v}^T (\nabla f)_{\mathbf{p}} = 0$$

and

$$T_{\mathbf{p}} := \{\, \mathbf{v} \,:\, \mathbf{v}^T (\nabla f)_{\mathbf{p}} = 0 \,\}.$$

Solution 41.   (i) We have $3 + 4 - 9 = -2$.
(ii) We obtain

$$\nabla f = \begin{pmatrix} 6x_1 \\ 8x_2 \\ 1 \end{pmatrix} \Rightarrow \nabla f|_{\mathbf{p}} = \begin{pmatrix} 6 \\ 8 \\ 1 \end{pmatrix} = \mathbf{n}.$$

(iii) We have

$$\mathbf{v}^T (\nabla f)_{\mathbf{p}} = (\,v_1 \quad v_2 \quad v_3\,) \begin{pmatrix} 6 \\ 8 \\ 1 \end{pmatrix} = 6v_1 + 8v_2 + v_3.$$

From $\mathbf{v}^T(\nabla f)_{\mathbf{p}} = 0$ we obtain $6v_1 + 8v_2 + v_3 = 0$ or $v_3 = -6v_1 - 8v_2$ which describes a plane in $\mathbb{R}^3$. Thus

$$T_{\mathbf{p}} = \left\{ \begin{pmatrix} v_1 \\ v_2 \\ -6v_1 - 8v_2 \end{pmatrix} : v_1, v_2 \in \mathbb{R} \right\}.$$

Problem 42. (i) Let $w \in \mathbb{C}$. Consider the *stereographic projection*

$$r(w) = \left(\frac{2\Re(w)}{|w|^2 + 1}, \frac{2\Im(w)}{|w|^2 + 1}, \frac{|w|^2 - 1}{|w|^2 + 1} \right).$$

Let $w = 1$. Find $r(w)$. Let $w = i$. Find $r(w)$. Let $w = e^{i\phi}$. Find $r(w)$. Let $w = 1/2$. Find $r(w)$.
(ii) Let $x_1, x_2, x_3 \in \mathbb{R}$ and $x_1^2 + x_2^2 + x_3^2 = 1$. Let $w \in \mathbb{C}$ with

$$w = \frac{x_1 + ix_2}{1 + x_3} \Rightarrow \overline{w} = \frac{x_1 - ix_2}{1 + x_3}.$$

Find x_1, x_2, x_3 as functions of w and $\overline{w}$.

Solution 42. (i) Since $\Re(1) = 1$ and $\Im(1) = 0$ we obtain $r(1) = (1, 0, 0)$. Since $\Re(i) = 0$ and $\Im(i) = 1$ we obtain $r(i) = (0, 1, 0)$. Since $e^{i\phi} \equiv \cos(\phi) + i\sin(\phi)$ and $\Re(e^{i\phi}) = \cos(\phi)$, $\Im(e^{i\phi}) = \sin(\phi)$ we find $r(e^{i\phi}) = (\cos(\phi), \sin(\phi), 0)$. Since $\Re(1/2) = 1/2$, $\Im(1/2) = 0$ we find $(4/5, 0, -3/5)$.
(ii) First we note that

$$w\overline{w} = \frac{x_1^2 + x_2^2}{1 + 2x_3 + x_3^2} = \frac{1 - x_3^2}{1 + 2x_3 + x_3^2}.$$

Thus we find

$$x_3 = \frac{1 - w\overline{w}}{1 + w\overline{w}}.$$

Looking at $w + \overline{w}$ and $w - \overline{w}$ we obtain

$$x_1 = \frac{2\Re(\omega)}{1 + w\overline{w}}, \quad x_2 = \frac{2\Im(\omega)}{1 + w\overline{w}}.$$

Problem 43. Let $\alpha \in \mathbb{R}$ be a parameter. Consider the plane

$$f(x_1, x_2, x_3; \alpha) \equiv 3\alpha^2 x_1 - 3\alpha x_2 + x_3 - \alpha^3 = 0 \tag{1}$$

in $\mathbb{R}^3$. Then from $\partial f/\partial\alpha = 0$ we obtain

$$\alpha^2 - 2\alpha x_1 + x_2 = 0. \tag{2}$$

(i) Solve the two equations for x_1, x_2, x_3.
(ii) Solve the two equations for α with x_1, x_2, x_3 fixed.

Solution 43.   (i) We find $x_1 = \gamma$, $x_2 = 2\gamma\alpha - \alpha^2$, $x_3 = 3\gamma\alpha^2 - 2\alpha^3$ with γ arbitrary. Thus we could set $\gamma = \alpha$ which provides $x_1 = \alpha$, $x_2 = \alpha^2$, $x_3 = \alpha^3$.
(ii) Multiplying the second equation by α and subtracting from the second equation provides

$$\alpha^2 x_1 - 2\alpha x_2 + x_3 = 0. \tag{3}$$

From (2) and (3) we find

$$\alpha = \frac{x_1 x_2 - x_3}{2(x_1^2 - x_2)}.$$

Inserting α now into eq.(1) provides the *envelope*

$$(x_1 x_2 - x_3)^2 = 4(x_1^2 - x_2)(x_2^2 - x_1 x_3).$$

Problem 44.   Consider the three-dimensional Euclidean space. The *monkey saddle* is given by

$$x_1(u_1, u_2) = u_1, \quad x_2(u_1, u_2) = u_2, \quad x_3(u_1, u_2) = u_1^3 - 3u_2^2 u_1$$

with $x_1(0,0) = x_2(0,0) = x_3(0,0) = 0$. Find the normal vector at $(u_1 = 0, u_2 = 0)$. The *normal vector* $\mathbf{n}(u_1, u_2)$ is given by

$$\mathbf{n}(u_1, u_2) := \frac{(\partial\mathbf{x}/\partial u_1) \times (\partial\mathbf{x}/\partial u_2)}{\|(\partial\mathbf{x}/\partial u_1) \times (\partial\mathbf{x}/\partial u_2)\|}.$$

Solution 44.   Since

$$\frac{\partial\mathbf{x}}{\partial u_1} = \begin{pmatrix} 1 \\ 0 \\ 3u_1^2 - 3u_2^2 \end{pmatrix}, \quad \frac{\partial\mathbf{x}}{\partial u_2} = \begin{pmatrix} 0 \\ 1 \\ -6u_1 u_2 \end{pmatrix}$$

we obtain

$$\frac{\partial\mathbf{x}}{\partial u_1} \times \frac{\partial\mathbf{x}}{\partial u_2} = \begin{pmatrix} -3u_1^2 + 3u_2^2 \\ 6u_1 u_2 \\ 1 \end{pmatrix}$$

and
$$\mathbf{n}(u_1, u_2) = \frac{1}{1 + 9(u_1^4 + u_2^4)} \begin{pmatrix} -3u_1^2 + 3u_2^2 \\ 6u_1 u_2 \\ 1 \end{pmatrix}.$$

It follows that $\mathbf{n}(0,0) = (\,0 \quad 0 \quad 1\,)^T$.

Problem 45. Let $n \geq 1$ and $f : \mathbb{R}^{n+1} \to \mathbb{R}$ be a smooth function and define the set
$$M_f := \{\, \mathbf{x} \in \mathbb{R}^{n+1} \,:\, f(\mathbf{x}) = 0 \,\}.$$
If $(df)(\mathbf{p}) \neq 0$ for all points $\mathbf{p} \in M_f$, then M_f is a smooth n-dimensional manifold. Sometimes also called the *hypersurface* given by the function f. Apply this definition to $f(\mathbf{x}) = x_1^2 + x_2^2 + \cdots + x_{n+1}^2 - 1$.

Solution 45. We have
$$df(\mathbf{x}) = 2x_1 dx_1 + 2x_2 dx_2 + \cdots + 2x_{n+1} dx_{n+1}.$$
Now $x_1^2 + x_2^2 + \cdots + x_{n+1}^2 = 1$. Thus $df(\mathbf{x})|_{\mathbf{p} \in M_f} \neq 0$.

Problem 46. (i) Consider the transformation in $\mathbb{R}^3$
$$x_0(a, \theta_1) = \cosh(a)$$
$$x_1(a, \theta_1) = \sinh(a) \sin(\theta_1)$$
$$x_2(a, \theta_1) = \sinh(a) \cos(\theta_1)$$
where $a \geq 0$ and $0 \leq \theta_1 < 2\pi$. Find $x_0^2 - x_1^2 - x_2^2$.
(ii) Consider the transformation in $\mathbb{R}^4$
$$x_0(a, \theta_1, \theta_2) = \cosh(a)$$
$$x_1(a, \theta_1, \theta_2) = \sinh(a) \sin(\theta_2) \sin(\theta_1)$$
$$x_2(a, \theta_1, \theta_2) = \sinh(a) \sin(\theta_2) \cos(\theta_1)$$
$$x_3(a, \theta_1, \theta_2) = \sinh(a) \cos(\theta_2)$$
where $a \geq 0$, $0 \leq \theta_1 < 2\pi$ and $0 \leq \theta_2 \leq \pi$. Find $x_0^2 - x_1^2 - x_2^2 - x_3^2$. Extend the transformation to $\mathbb{R}^n$.

Solution 46. (i) Since $\cos^2(\alpha) + \sin^2(\alpha) = 1$ and $\cosh^2(\alpha) - \sinh^2(\alpha) = 1$ we have $x_0^2 - x_1^2 - x_2^2 = 1$.
(ii) Since $\cos^2(\alpha) + \sin^2(\alpha) = 1$ and $\cosh^2(\alpha) - \sinh^2(\alpha) = 1$ we have
$$x_0^2 - x_1^2 - x_2^2 - x_3^2 = 1.$$

The extension of this transformation is

$$x_0(a, \theta_1, \ldots, \theta_{n-2}) = \cosh(a)$$
$$x_1(a, \theta_1, \ldots, \theta_{n-2}) = \sinh(a)\sin(\theta_{n-2})\cdots\sin(\theta_1)$$
$$x_2(a, \theta_1, \ldots, \theta_{n-2}) = \sinh(a)\sin(\theta_{n-2})\cdots\sin(\theta_2)\cos(\theta_1)$$

$$\vdots$$

$$x_{n-2}(a, \theta_1, \ldots, \theta_{n-2}) = \sinh(a)\sin(\theta_{n-2})\cos(\theta_{n-3})$$
$$x_{n-1}(a, \theta_1, \ldots, \theta_{n-2}) = \sinh(a)\cos(\theta_{n-2})$$

where $a \geq 0$, $0 \leq \theta_1 < 2\pi$ and $0 \leq \theta_j \leq \pi$ with $j = 2, 3, \ldots$. We have

$$x_0^2 - x_1^2 - \cdots - x_{n-1}^2 = 1.$$

Problem 47. (i) Consider the upper sheet of the *hyperboloid*

$$H^2 := \{\, \mathbf{v} \in \mathbb{R}^3 \, : \, \mathbf{v}^2 = v_0^2 - v_1^2 - v_2^2 = 1, \, v_0 > 0 \,\}.$$

Find a parametrization for $\mathbf{v}$.

(ii) Consider the *hyperboloid* embedded in $(2+1)$-dimensional *Minkowski space* given by

$$x_1^2 + x_2^2 - x_0^2 = -\frac{1}{4}, \quad x_0 \geq \frac{1}{2}.$$

Show that the hyperboloid can be parametrized with coordinates $\alpha \in [0, \infty)$, $\beta \in [0, 2\pi)$

$$x_1(\alpha, \beta) = \frac{1}{2}\sinh(\alpha)\cos(\beta), \quad x_2(\alpha, \beta) = \frac{1}{2}\sinh(\alpha)\sin(\beta),$$

$$x_0(\alpha, \beta) = \frac{1}{2}\cosh(\alpha).$$

Solution 47. (i) We have

$$\mathbf{v} = (\cosh(2|\zeta|), \sinh(2|\zeta|)\cos(\phi), \sinh(2|\zeta|)\sin(\phi))$$

where $\zeta = |\zeta|e^{i\phi}$ and $v_0 > 0$ since $\cosh(2|\zeta|) > 0$.

(ii) With

$$x_1^2 + x_2^2 = \frac{1}{4}\sinh^2(\alpha)\cos^2(\beta) + \frac{1}{4}\sinh^2(\alpha)\sin^2(\beta) = \frac{1}{4}\sinh^2(\alpha)$$

and $\cosh^2(\alpha) - \sinh^2(\alpha) = 1$ we obtain $x_1^2 + x_2^2 - x_0^2 = -1/2$.

Problem 48.   Consider the surface in $\mathbb{R}^3$, $x_1^2 + x_2^2 - x_3^2 = 1$. Show that a parametrization of this surface is given by

$$\mathbf{x}(u_1, u_2) = \begin{pmatrix} x_1(u_1, u_2) \\ x_2(u_1, u_2) \\ x_3(u_1, u_2) \end{pmatrix} = \begin{pmatrix} \cosh(u_1)\cos(u_2) \\ \cosh(u_1)\sin(u_2) \\ \sinh(u_1) \end{pmatrix}$$

where $-1 \le u_1 \le 1$ and $-\pi \le u_2 \le \pi$.

Solution 48.   Since $\sin^2(u_2) + \cos^2(u_2) = 1$, $\cosh^2(u_1) - \sinh^2(u_1) = 1$ we obtain that $x_1(u_1, u_2)$, $x_2(u_1, u_2)$, $x_3(u_1, u_2)$ is a parametrization.

Problem 49.   The *Hammer projection* is an equal-area cartographic projections that maps the entire surface of a sphere to the interior of an ellipse of semiaxis $\sqrt{8}$ and $\sqrt{2}$. The Hammer projection is given by the transformation between (θ, ϕ) and (x_1, x_2)

$$x_1(\theta, \phi) = \frac{\sqrt{8}\sin(\theta)\sin(\phi/2)}{\sqrt{1 + \sin(\theta)\cos(\phi/2)}}, \qquad x_2(\theta, \phi) = \frac{\sqrt{2}\cos(\theta)}{\sqrt{1 + \sin(\theta)\cos(\phi/2)}}$$

where $0 \le \theta \le \pi$ and $0 \le \phi < 2\pi$.
(i) Show that $x_1^2/8 + x_2^2/2 \le 1$.
(ii) Find $\theta(x_1, x_2)$ and $\phi(x_1, x_2)$.

Solution 49.   (i) We have

$$x_1^2 = \frac{8\sin^2(\theta)\sin^2(\phi/2)}{1 + \sin(\theta)\cos(\phi/2)}, \qquad x_2^2 = \frac{2\cos^2(\theta)}{1 + \sin(\theta)\cos(\phi/2)}.$$

Thus

$$\frac{x_1^2}{8} + \frac{x_2^2}{2} = \frac{\cos^2(\theta)(1 - \sin^2(\phi/2)) + \sin^2(\phi/2)}{1 + \sin(\theta)\cos(\phi/2)}.$$

(ii) We have

$$\theta(x_1, x_2) = \arccos\left(x_2\sqrt{1 - x_1^2/16 - x_2^2/4} \right)$$

$$\phi(x_1, x_2) = 2\arctan\left(\frac{x_1\sqrt{1 - x_1^2/16 - x_2^2/4}}{4(1 - x_1^2/16 - x_2^2/4) - 2} \right).$$

Problem 50.   Consider the two normalized vectors in $\mathbb{R}^3$

$$\mathbf{v}_1 = \begin{pmatrix} \sin(\theta_1)\cos(\phi_1) \\ \sin(\theta_1)\sin(\phi_1) \\ \cos(\theta_1) \end{pmatrix}, \qquad \mathbf{v}_2 = \begin{pmatrix} \sin(\theta_2)\cos(\phi_2) \\ \sin(\theta_2)\sin(\phi_2) \\ \cos(\theta_2) \end{pmatrix}.$$

Find the conditions on ϕ_1, ϕ_2, θ_1, θ_2 such that the scalar product of $\mathbf{v}_1$ and $\mathbf{v}_2$ vanishes, i.e. $\mathbf{v}_1^T \mathbf{v}_2 = 0$. Then calculate the vector product $\mathbf{v}_1 \times \mathbf{v}_2$.

Solution 50. Utilizing the identities

$$\sin(\alpha)\sin(\beta) \equiv \frac{1}{2}(\cos(\alpha - \beta) - \cos(\alpha + \beta))$$

$$\cos(\alpha)\cos(\beta) \equiv \frac{1}{2}(\cos(\alpha - \beta) + \cos(\alpha + \beta))$$

$$\sin(\alpha)\cos(\beta) \equiv \frac{1}{2}(\sin(\alpha - \beta) + \sin(\alpha + \beta))$$

we obtain the equation

$$\cos(\theta_1 - \theta_2)(1 + \cos(\phi_1 - \phi_2)) + \cos(\theta_1 + \theta_2)(1 - \cos(\phi_1 - \phi_2)) = 0.$$

Solutions to this equation are

$$(\theta_1 - \theta_2) = \pi/2, \quad (\phi_1 - \phi_2) = 0, \qquad (\theta_1 - \theta_2) = 3\pi/2, \quad (\phi_1 - \phi_2) = 0$$

$$(\theta_1 + \theta_2) = \pi/2, \quad (\phi_1 - \phi_2) = \pi, \qquad (\theta_1 + \theta_2) = 3\pi/2, \quad (\phi_1 - \phi_2) = \pi.$$

Selecting $\theta_1 + \theta_2 = \pi/2$, $\phi_1 - \phi_2 = \pi$ we obtain

$$\mathbf{v}_2 = \begin{pmatrix} -\cos(\theta_1)\sin(\phi_1) \\ -\cos(\theta_1)\sin(\phi_1) \\ \sin(\theta_1) \end{pmatrix}$$

and hence

$$\mathbf{v}_1 \times \mathbf{v}_2 = \begin{pmatrix} \sin(\phi_1) \\ -\cos(\phi_1) \\ 0 \end{pmatrix}.$$

Problem 51. Let $\alpha \in \mathbb{R}$. Consider the 2×2 matrix

$$F(\alpha) = \begin{pmatrix} f_{11}(\alpha) & f_{12}(\alpha) \\ f_{21}(\alpha) & f_{22}(\alpha) \end{pmatrix}$$

with $f_{jk} : \mathbb{R} \to \mathbb{R}$ be analytic functions. Let

$$X := \left. \frac{dF(\alpha)}{d\alpha} \right|_{\alpha=0} = \left. \begin{pmatrix} df_{11}(\alpha)/d\alpha & df_{12}(\alpha)/d\alpha \\ df_{21}(\alpha)/d\alpha & df_{22}(\alpha)/d\alpha \end{pmatrix} \right|_{\alpha=0}.$$

Find the conditions on the functions f_{jk} such that

$$\exp(\alpha X) = F(\alpha).$$

Apply the *Cayley-Hamilton theorem*. Set $f'_{jk}(0) := df_{jk}(\alpha)/d\alpha|_{\alpha=0}$ and

$$\text{tr} := f'_{11}(0) + f'_{22}(0), \qquad \det := f'_{11}(0)f'_{22}(0) - f'_{12}(0)f'_{21}(0).$$

Solution 51.   The eigenvalue equation for X is given by

$$\lambda^2 - \lambda \text{tr} + \det = 0$$

with the solution

$$\lambda_{\pm} = \frac{1}{2}\text{tr} \pm \sqrt{-\det + (\text{tr})^2/4}.$$

We assume that $\lambda_+ \neq \lambda_-$ and find the case $\lambda_+ = \lambda_-$ applying L'Hospital rule. The *Cayley-Hamilton theorem* tells us that

$$e^{\alpha X} = c_1 \alpha X + c_0 I_2 = \begin{pmatrix} c_1 \alpha f'_{11}(0) + c_0 & c_1 \alpha f_{12}(0) \\ c_1 \alpha f'_{21}(0) & c_1 \alpha f'_{22}(0) + c_0 \end{pmatrix}$$

and

$$e^{\alpha\lambda_+} = c_1 \alpha \lambda_+ + c_0, \quad e^{\alpha\lambda_-} = c_1 \alpha \lambda_- + c_0.$$

Hence

$$c_0 = \frac{e^{\alpha\lambda_-}\lambda_+ - e^{\alpha\lambda_+}\lambda_-}{\lambda_+ - \lambda_-}, \qquad c_1 = \frac{e^{\alpha\lambda_+} - e^{\alpha\lambda_-}}{\alpha(\lambda_+ - \lambda_-)}.$$

Inserting c_0 and c_1 into

$$f_{11}(\alpha) = c_1 \alpha f'_{11}(0) + c_0, \quad f_{12}(\alpha) = c_1 \alpha f'_{12}(0)$$
$$f_{21}(\alpha) = c_1 \alpha f'_{21}(0), \quad f_{22}(\alpha) = c_1 \alpha f'_{22}(0) + c_0$$

we obtain with $D := -\det + (\text{tr})^2/4$ that

$$f_{11}(\alpha) = \frac{1}{2}\frac{e^{(\alpha/2)\text{tr}}}{\sqrt{D}} \sinh(\alpha\sqrt{D})(f'_{11}(0) - f'_{22}(0)) + e^{(\alpha/2)\text{tr}}\cosh(\alpha\sqrt{D})$$

$$f_{12}(\alpha) = \frac{e^{(\alpha/2)\text{tr}}\sinh(\alpha\sqrt{D})}{\sqrt{D}}f'_{12}(0)$$

$$f_{21}(\alpha) = \frac{e^{(\alpha/2)\text{tr}}\sinh(\alpha\sqrt{D})}{\sqrt{D}}f'_{21}(0)$$

$$f_{22}(\alpha) = \frac{1}{2}\frac{e^{(\alpha/2)\text{tr}}}{\sqrt{D}}\sinh(\alpha\sqrt{D})(-f'_{11}(0) + f'_{22}(0)) + e^{(\alpha/2)\text{tr}}\cosh(\alpha\sqrt{D}).$$

If $D = 0$, i.e. $\lambda_+ = \lambda_-$ we arrive at

$$f_{11}(\alpha) = \frac{1}{2}e^{(\alpha/2)\mathrm{tr}\,\alpha}(f'_{11}(0) - f'_{22}(0)) + e^{(\alpha/2)\mathrm{tr}}$$
$$f_{12}(\alpha) = e^{(\alpha/2)\mathrm{tr}\,\alpha}f'_{12}(0)$$
$$f_{21}(\alpha) = e^{(\alpha/2)\mathrm{tr}\,\alpha}f'_{21}(0)$$
$$f_{22}(\alpha) = \frac{1}{2}e^{(\alpha/2)\mathrm{tr}\,\alpha}(-f'_{11}(0) + f'_{22}(0)) + e^{(\alpha/2)\mathrm{tr}}.$$

Problem 52. A four-dimensional *torus* $\mathbb{S}^3 \times \mathbb{S}^1$ can be defined as

$$(\sqrt{x_1^2 + x_2^2 + x_3^2 + x_4^2} - a)^2 + w^2 = 1$$

where $a > 1$ is the constant radius of $\mathbb{S}^3$. Show that the four-dimensional torus can be parametrized as

$$x_1(\psi, \rho, \phi_1, \phi_2) = (a + \cos(\psi))\rho\cos(\phi_1)$$
$$x_2(\psi, \rho, \phi_1, \phi_2) = (a + \cos(\psi))\rho\sin(\phi_1)$$
$$x_3(\psi, \rho, \phi_1, \phi_2) = (a + \cos(\psi))\sqrt{1 - \rho^2}\cos(\phi_2)$$
$$x_4(\psi, \rho, \phi_1, \phi_2) = (a + \cos(\psi))\sqrt{1 - \rho^2}\sin(\phi_2)$$
$$w(\psi, \rho, \phi_1, \phi_2) = \sin(\psi)$$

where $\phi_1 \in [0, 2\pi]$, $\phi_2 \in [0, 2\pi]$, $\psi \in [0, 2\pi]$, $\rho \in [0, 1]$.

Solution 52. We have

$$x_1^2 = (a + \cos(\psi))^2\rho^2\cos^2(\phi_1)$$
$$x_2^2 = (a + \cos(\psi))^2\rho^2\sin^2(\phi_1)$$
$$x_3^2 = (a + \cos(\psi))^2(1 - \rho^2)\cos^2(\phi_2)$$
$$x_4^2 = (a + \cos(\psi))^2(1 - \rho^2)\sin^2(\phi_2)$$

and $w^2 = \sin^2(\psi)$. Hence

$$x_1^2 + x_2^2 = (a + \cos(\psi))^2\rho^2, \quad x_3^2 + x_4^2 = (a + \cos(\psi))^2(1 - \rho^2).$$

It follows that $x_1^2 + x_2^2 + x_3^2 + x_4^2 = (a + \cos(\psi))^2$ and hence

$$\sqrt{x_1^2 + x_2^2 + x_3^2 + x_4^2} - a = \cos(\psi).$$

The metric tensor field is given by

$$g_T = d\psi \otimes d\psi$$
$$+ (a + \cos(\psi))^2 \left(\frac{d\rho \otimes d\rho}{1 - \rho^2} + \rho^2 d\phi_1 \otimes d\phi_1 + (1 - \rho^2) d\phi_2 \otimes d\phi_2 \right).$$

Problem 53.   The Lie group $SL(2, \mathbb{R})$ is the group of 2×2 real unimodular matrices. Consider the unitary matrix

$$V = \frac{1}{\sqrt{2}} \begin{pmatrix} 1 & -i \\ 1 & i \end{pmatrix} \quad \Rightarrow \quad V^* = \frac{1}{\sqrt{2}} \begin{pmatrix} 1 & 1 \\ i & -i \end{pmatrix}$$

with $\det(V) = i$. The Lie group $SL(2, \mathbb{R})$ is isomorphic to the Lie group $SU(1, 1)$ defined by

$$U = \begin{pmatrix} \alpha & \beta \\ \overline{\beta} & \overline{\alpha} \end{pmatrix}, \quad \alpha, \beta \in \mathbb{C}, \quad |\alpha|^2 - |\beta|^2 = 1$$

and the *isomorphism* is given by $SU(1, 1) = VSL(2, \mathbb{R})V^*$. The parameter space of $SU(1, 1)$ is the *hyperboloid* $\mathbb{H}_{2,2}$ of signature $(2, 2)$. Let $\gamma \in \mathbb{R}$ and

$$S = \begin{pmatrix} 1 & \gamma \\ 0 & 1 \end{pmatrix} \in SL(2, \mathbb{R}).$$

Find VSV^*.

Solution 53.   We obtain

$$VSV^* = \frac{1}{2} \begin{pmatrix} 2 + i\gamma & -i\gamma \\ i\gamma & 2 - i\gamma \end{pmatrix} \quad \Rightarrow \quad \det(VSV^*) = 1.$$

Problem 54.   Let $a > b > 0$. Then the equation of the *ellipse* in parametric form is given by $x(\phi) = a \sin(\phi)$, $y(\phi) = b \cos(\phi)$, where $0 \leq \phi < 2\pi$. The *metric tensor field* of the two-dimensional Euclidean space is given by

$$g = dx \otimes dx + dy \otimes dy. \tag{1}$$

Calculate the *arc length* of an *ellipse*.

Solution 54.   Since $dx = a \cos(\phi) d\phi$, $dy = -b \sin(\phi) d\phi$ we obtain

$$dx \otimes dx = a^2 \cos^2(\phi) d\phi \otimes d\phi, \qquad dy \otimes dy = b^2 \sin^2(\phi) d\phi \otimes d\phi.$$

Consequently, we obtain from (1)

$$g = (a^2 \cos^2(\phi) + b^2 \sin^2(\phi)) d\phi \otimes d\phi.$$

Therefore the *line element* is given by

$$\left(\frac{ds}{d\phi}\right)^2 = a^2 \cos^2(\phi) + b^2 \sin^2(\phi).$$

Using the identity $\sin^2(\phi) + \cos^2(\phi) \equiv 1$ we arrive at

$$\int_0^{2\pi} ds = a \int_0^{2\pi} \sqrt{1 - \frac{a^2 - b^2}{a^2} \sin^2(\phi)} d\phi.$$

This is a *complete elliptic integral of the second kind*, where

$$k^2 = \frac{a^2 - b^2}{a^2} =: e^2.$$

Here e is the *eccentricity* of the ellipse. We find

$$\int_0^{2\pi} ds = 4aE(k, \pi/2).$$

The complete elliptic integral of the second kind can be given as

$$E(k, \pi/2) = \int_0^{\pi/2} \sqrt{1 - k^2 \sin^2(\phi)} d\phi.$$

Thus

$$E(k, \pi/2) = \frac{\pi}{2} \left(1 - \left(\frac{1}{2}\right)^2 k^2 - \left(\frac{1 \cdot 3}{2 \cdot 4}\right)^2 \frac{k^4}{3} - \left(\frac{1 \cdot 3 \cdot 5}{2 \cdot 4 \cdot 6}\right)^2 \frac{k^6}{5} - \cdots \right)$$

where $|k^2| < 1$. We have used the *Taylor series expansion* around $x = 0$ of the function

$$f(x) = \sqrt{1 - x^2}$$

with $x^2 < 1$, i.e.

$$\sqrt{1 - x^2} = 1 - \frac{1}{2}x^2 - \frac{1}{2 \cdot 4}x^4 - \frac{1 \cdot 3}{2 \cdot 4 \cdot 6}x^6 - \cdots, \qquad -1 < x \leq 1.$$

Problem 55.   Consider $\mathbf{y} \in \mathbb{R}^2$ and $\mathbf{x} \in \mathbb{R}^3$. Let $\otimes$ be the Kronecker product. From the conditions

$$
\begin{pmatrix} x_1 \\ x_2 \\ x_3 \end{pmatrix} \otimes \begin{pmatrix} y_1 \\ y_2 \end{pmatrix} = \begin{pmatrix} y_1 \\ y_2 \end{pmatrix} \otimes \begin{pmatrix} x_1 \\ x_2 \\ x_3 \end{pmatrix} \Rightarrow \begin{pmatrix} x_1 y_1 \\ x_1 y_2 \\ x_2 y_1 \\ x_2 y_2 \\ x_3 y_1 \\ x_3 y_2 \end{pmatrix} = \begin{pmatrix} x_1 y_1 \\ x_2 y_1 \\ x_3 y_1 \\ x_1 y_2 \\ x_2 y_2 \\ x_3 y_2 \end{pmatrix}
$$

we obtain the four equations

$$
x_1 y_2 = y_1 x_2, \quad x_2 y_1 = y_1 x_3, \quad x_2 y_2 = y_2 x_1, \quad x_3 y_1 = y_2 x_2
$$

with the six unknowns x_1, x_2, x_3, y_1, y_2. Do we have a manifold embedded in $\mathbb{R}^6$?

Solution 55.   If $\mathbf{y} = \mathbf{0}$, then $\mathbf{x}$ is arbitrary. If $y_1 \neq 0$, $y_2 = 0$, then $x_2 = x_3 = 0$ and

$$
\begin{pmatrix} a \\ 0 \\ 0 \end{pmatrix} \otimes \begin{pmatrix} b \\ 0 \end{pmatrix} = \begin{pmatrix} b \\ 0 \end{pmatrix} \otimes \begin{pmatrix} a \\ 0 \\ 0 \end{pmatrix}.
$$

If $y_1 = 0$, $y_2 \neq 0$, then $x_1 = x_2 = 0$ and

$$
\begin{pmatrix} 0 \\ 0 \\ a \end{pmatrix} \otimes \begin{pmatrix} 0 \\ b \end{pmatrix} = \begin{pmatrix} 0 \\ b \end{pmatrix} \otimes \begin{pmatrix} 0 \\ 0 \\ a \end{pmatrix}.
$$

If $y_1 \neq 0$, $y_2 \neq 0$ we obtain $x_1 = x_2 = x_3$ and $y_1 = y_2$ and hence

$$
\begin{pmatrix} a \\ a \\ a \end{pmatrix} \otimes \begin{pmatrix} b \\ b \end{pmatrix} = \begin{pmatrix} b \\ b \end{pmatrix} \otimes \begin{pmatrix} a \\ a \\ a \end{pmatrix}.
$$

Problem 56.   Let $A = (a_{jk})_{j,k=1}^n$ be an $n \times n$ symmetric matrix over $\mathbb{R}$. The eigenvalue problem is given by $A\mathbf{x} = \lambda \mathbf{x}$ ($\mathbf{x} \neq \mathbf{0}$), where we assume that the eigenvectors are normalized, i.e. $\mathbf{x}^T \mathbf{x} = 1$. Thus

$$
\sum_{k=1}^n a_{jk} x_k = \lambda x_j, \quad j = 1, \ldots, n.
$$

Consider the function $f : \mathbb{R}^n \to \mathbb{R}$

$$f(\mathbf{x}) = \sum_{j,k=1}^{n} a_{jk} x_j x_k \equiv \mathbf{x}^T A \mathbf{x}.$$

To obtain its maximum and minimum of f subject to the constraint (compact manifold)

$$\sum_{j=1}^{n} x_j^2 = 1$$

we consider the *Lagrange function*

$$L(\mathbf{x}) = f(\mathbf{x}) - \lambda \sum_{j=1}^{n} x_j^2.$$

Show that the largest and smallest eigenvalues of A are the maximum and minimum of the function f subject to the constraint $\sum_{j=1}^{n} x_j^2 = 1$.

Solution 56. We obtain

$$\frac{\partial L}{\partial x_j} = 0 \Rightarrow \sum_{k=1}^{n} a_{jk} x_k - \lambda x_j = 0, \quad j = 1, \dots, n.$$

Therefore

$$\sum_{k=1}^{n} a_{jk} x_{k0} = \lambda x_{j0}, \quad j = 1, \dots, n.$$

Hence $\mathbf{x}_0$ is a constrained critical point of the function f subject to $\sum_{j=1}^{n} x_j^2 = 1$ iff $A\mathbf{x}_0 = \lambda \mathbf{x}_0$ for some λ, i.e. λ is an eigenvalue and $\mathbf{x}_0$ is a normalized eigenvector. It follows that

$$f(\mathbf{x}_0) = \sum_{j=1}^{n} \left(\sum_{k=1}^{n} a_{jk} x_{k0} \right) x_{j0} = \sum_{j=1}^{n} (\lambda x_{j0}) x_{j0} = \lambda \sum_{j=1}^{n} x_{j0}^2 = \lambda.$$

Consequently the largest and smallest eigenvalues of A are the maximum and minimum of f subject to the constraint (manifold) $\sum_{j=1}^{n} x_{j0}^2 = 1$.

Problem 57. Consider the Lie group $GL(2, \mathbb{R})$ and $V \in \mathbb{R}^{2 \times 2}$, i.e. V is a 2×2 matrix over $\mathbb{R}$. Let $g \in GL(2, \mathbb{R})$. Show that the derivative of the function $f = \det : \mathbb{R}^{2 \times 2} \to \mathbb{R}$ is given by

$$df(g)V = \det(g) \operatorname{tr}(g^{-1}V). \tag{1}$$

Note that

$$g^{-1} = \frac{1}{\det(g)} \begin{pmatrix} g_{22} & -g_{12} \\ -g_{21} & g_{11} \end{pmatrix}.$$

Solution 57.   For the left-hand side of (1) we have

$$df(g)V \equiv \frac{d}{d\tau} f(g + \tau V)\Big|_{\tau=0} = \frac{d}{d\tau} \det(g + \tau V)\Big|_{\tau=0}$$

$$= \frac{d}{d\tau} \det \begin{pmatrix} g_{11} + \tau v_{11} & g_{12} + \tau v_{12} \\ g_{21} + \tau v_{21} & g_{22} + \tau v_{22} \end{pmatrix}\Big|_{\tau=0}$$

$$= \frac{d}{d\tau} ((g_{11} + \tau v_{11})(g_{22} + \tau v_{22}) - (g_{21} + \tau v_{21})(g_{12} + \tau v_{12})\Big|_{\tau=0}$$

$$= g_{11}v_{22} + g_{22}v_{11} - g_{12}v_{21} - g_{21}v_{12}.$$

For the right-hand side we have

$$\det(g)\mathrm{tr}(g^{-1}V) = \det(g)\det(g^{-1})\mathrm{tr}\begin{pmatrix} g_{22}v_{11} - g_{12}v_{21} & * \\ * & -g_{21}v_{12} + g_{11}v_{22} \end{pmatrix}$$

$$= g_{22}v_{11} + g_{11}v_{22} - g_{12}v_{21} - g_{21}v_{12}.$$

Equation (1) also holds for arbitrary n, i.e. $GL(n,\mathbb{R})$ and $V \in \mathbb{R}^{n \times n}$.

Problem 58.   Let $a > 0$.   Consider the transformation between *Minkowski coordinates* (x_0, x_3) and *Rindler coordinates* (ζ, η)

$$x_0(\zeta, \eta) = \frac{1}{a}\exp(a\zeta)\sinh(a\eta), \qquad x_3(\zeta, \eta) = \frac{1}{a}\exp(a\zeta)\cosh(a\eta).$$

Find the inverse transformation.

Solution 58.   Division x_0/x_3 yields

$$\frac{x_0}{x_3} = \frac{\sinh(a\eta)}{\cosh(a\eta)} = \tanh(a\eta) \;\Rightarrow\; \eta = \frac{1}{a}\mathrm{arctanh}(x_0/x_3).$$

Now we have

$$x_3\cosh(a\eta) - x_0\sinh(a\eta) = \frac{1}{a}\exp(a\eta) \Rightarrow x_3^2 ae^{-a\eta} - x_0^2 ae^{-a\eta} = \frac{1}{a}e^{a\eta}$$

and therefore $a^2(x_3^2 - x_0^2) = e^{2a\zeta}$. Finally

$$\zeta = \frac{1}{2a}\ln\left(a^2(x_3^2 - x_0^2)\right).$$

Programming Problems

Problem 59. Show that the curve in the plane $x_2^2 + x_1(x_1 - 1)^3 = 0$ is parametrized by

$$x_1(\tau) = \frac{1}{\tau^2 + 1}, \qquad x_2(\tau) = -\frac{\tau^3}{(\tau^2 + 1)^2}.$$

Note that $x_1(0) = 1$, $x_2(0) = 0$.

Solution 59. The following Maxima program will do the job

```
x1: 1/(tau*tau+1);
x2: -tau*tau*tau/((tau*tau+1)*(tau*tau+1));
r1: x2*x2+x1*(x1-1)*(x1-1)*(x1-1);
r1: ratsimp(r1);
```

Here `ratsimp()` provides the simplification of the rational expression. Is the curve regular?

Problem 60. Consider the curve in the plane

$$x_1(\tau) = 3\cos(\tau) - \cos(3\tau), \qquad x_2(\tau) = 3\sin(\tau) - \sin(3\tau).$$

Note that

$$(x_1(\tau = 0) = 2, x_2(\tau = 0) = 0), \quad (x_1(\tau = \pi) = -2, x_2(\tau = \pi) = 0).$$

Provide Gnuplot code to draw the curve. Note that in Gnuplot the dummy variable is t for curves. Give Tikz code to draw the curve.

Solution 60. For Gnuplot we have

```
> set parametric
> plot 3*cos(t)-cos(3*t),3*sin(t)-sin(3*t)
```

For Tikz we have

```
\begin{tikzpicture}
\draw[variable=\t,domain=0:360,samples=200]
      plot({3*cos(\t)-cos(3*\t)},{3*sin(\t)-sin(3*\t)});
\draw[->](-4,0)--(4,0);
\draw[->](0,-4)--(0,4);
\end{tikzpicture}
```

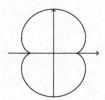

Problem 61.   Consider the curve in $\mathbb{R}^2$ given by

$$x_1(\tau) = \cos(\tau)(2\cos(\tau) - 1), \qquad x_2(\tau) = \sin(\tau)(2\cos(\tau) - 1)$$

where $\tau \in [0, 2\pi]$. Draw the curve with Gnuplot. Note that in Gnuplot the dummy variable is t for curves.

Solution 61.   Note that $x_1^2(t) + x_2^2(t) = (2\cos(t) - 1)^2$. The curve can be plotted with Gnuplot as follows

```
set parametric
set size square
set xrange [-pi:pi]
set yrange [-pi:pi]
plot [0:2*pi] cos(t)*(2.0*cos(t)-1.0),sin(t)*(2.0*cos(t)-1.0)
```

Find the longest distance between two points on the curve.

Problem 62.   Consider the curve in $\mathbb{R}^3$

$$\mathbf{x}(\tau) = \begin{pmatrix} x_1(\tau) \\ x_2(\tau) \\ x_3(\tau) \end{pmatrix} = \begin{pmatrix} 1 + \cos(\tau) \\ \sin(\tau) \\ \sin(\tau/2) \end{pmatrix}.$$

(i) Show that the curve is regular.
(ii) Calculate $(x_1 - 1)^2 + x_2^2$, $x_1^2 + x_2^2 + 4x_3^2$. Note that

$$\cos(\tau) \equiv 1 - 2\sin^2(\tau/2), \quad \sin(\tau) \equiv 2\sin(\tau/2)\cos(\tau/2).$$

Solution 62.   (i) We have

$$\frac{d\mathbf{x}(\tau)}{d\tau} = \begin{pmatrix} -\sin(\tau) \\ \cos(\tau) \\ \frac{1}{2}\cos(\tau/2) \end{pmatrix}.$$

Hence there is no τ so that $d\mathbf{x}/d\tau$ is the zero vector.
(ii) The Maxima program will do the job

```
/* curve.mac */
x1: 1+cos(tau); x2: sin(tau); x3: sin(tau/2);
r1: (x1-1)*(x1-1)+x2*x2; r1: trigsimp(r1);
r2: x1*x1+x2*x2+4*x3*x3; r2: trigsimp(r2);
r2: subst(1-2*sin(tau/2)*sin(tau/2),cos(tau),r2);
r2: trigsimp(r2);
```

where **trigsimp** provides simplifications for trigonometric expressions. We obtain $(x_1 - 1)^2 + x_2^2 = 1$, $x_1^2 + x_2^2 + 4x_3^2 = 4$. Thus the curve lies on the intersection of the two surfaces

$$(x_1 - 1)^2 + x_2^2 = 1, \qquad x_1^2 + x_2^2 + 4x_3^2 = 4.$$

Problem 63. Consider the closed regular curve in $\mathbb{R}^3$ (*trefoil knot*)

$$\mathbf{x}(\tau) = \begin{pmatrix} x_1(\tau) \\ x_2(\tau) \\ x_3(\tau) \end{pmatrix} = \begin{pmatrix} (\sqrt{2} + \cos(2\tau)) \cos(3\tau) \\ (\sqrt{2} + \cos(2\tau)) \sin(3\tau) \\ \sin(2\tau) \end{pmatrix}$$

for $0 \leq \tau \leq 2\pi$. Are the vectors $\mathbf{x}(0)$, $\mathbf{x}(\pi/2)$, $\mathbf{x}(\pi)$ linearly independent? Note that $\mathbf{x}(\tau) = \mathbf{x}(\tau + 2\pi)$. Apply a Maxima program.

Solution 63. The Maxima program

```
/* trefoil.mac */
x10: (sqrt(2)+cos(0))*cos(0);
x20: (sqrt(2)+cos(0))*sin(0);
x30: sin(0);
x1pih: (sqrt(2)+cos(2*%pi/2))*cos(3*%pi/2);
x2pih: (sqrt(2)+cos(2*%pi/2))*sin(3*%pi/2);
x3pih: sin(2*%pi/2);
x1pi: (sqrt(2)+cos(2*%pi))*cos(3*%pi);
x2pi: (sqrt(2)+cos(2*%pi))*sin(3*%pi);
x3pi: sin(2*%pi);
```

provides

$$\mathbf{x}(0) = \begin{pmatrix} \sqrt{2} + 1 \\ 0 \\ 0 \end{pmatrix}$$

$$\mathbf{x}(\pi/2) = \begin{pmatrix} (\sqrt{2} + \cos(\pi)) \cos(3\pi/2) \\ (\sqrt{2} + \cos(\pi)) \sin(3\pi/2) \\ \sin(\pi) \end{pmatrix} = \begin{pmatrix} 0 \\ 1 - \sqrt{2} \\ 0 \end{pmatrix}$$

$$\mathbf{x}(\pi) = \begin{pmatrix} (\sqrt{2} + \cos(2\pi)) \cos(3\pi) \\ (\sqrt{2} + \cos(2\pi)) \sin(3\pi) \\ \sin(2\pi) \end{pmatrix} = \begin{pmatrix} -(1 + \sqrt{2}) \\ 0 \\ 0 \end{pmatrix}.$$

Thus the vectors are linearly dependent.

Problem 64.   Let $a > 0$, $b > 0$. Consider the *circular helix* (a regular curve in $\mathbb{R}^3$ since $d\mathbf{x}/d\tau \neq \mathbf{0}$ for all $\tau \in \mathbb{R}$)

$$\mathbf{x}(\tau) = \begin{pmatrix} x_1(\tau) \\ x_2(\tau) \\ x_3(\tau) \end{pmatrix} = \begin{pmatrix} a\cos(\tau) \\ a\sin(\tau) \\ b\tau \end{pmatrix}.$$

Find the *curvature* given by

$$\kappa(\tau) = \frac{\left| \frac{d\mathbf{x}(\tau)}{d\tau} \times \frac{d^2\mathbf{x}(\tau)}{d\tau^2} \right|}{\left| \frac{d\mathbf{x}(\tau)}{d\tau} \right|^3}.$$

Find the *torsion* given by

$$T(\tau) = \frac{\det\left(\frac{d\mathbf{x}(\tau)}{d\tau} \quad \frac{d^2\mathbf{x}(\tau)}{d\tau^2} \quad \frac{d^3\mathbf{x}(\tau)}{d\tau^3} \right)}{\left| \frac{d\mathbf{x}(\tau)}{d\tau} \times \frac{d^2\mathbf{x}(\tau)}{d\tau^2} \right|^2}$$

where det denotes the determinant (of the 3×3 matrix). Provide a Maxima program.

Solution 64.   Since

$$\frac{d\mathbf{x}}{d\tau} = \begin{pmatrix} -a\sin(\tau) \\ a\cos(\tau) \\ b \end{pmatrix}, \quad \frac{d^2\mathbf{x}}{d\tau^2} = \begin{pmatrix} -a\cos(\tau) \\ -a\sin(\tau) \\ 0 \end{pmatrix}, \quad \frac{d^3\mathbf{x}}{d\tau^3} = \begin{pmatrix} a\sin(\tau) \\ -a\cos(\tau) \\ 0 \end{pmatrix}$$

we find for the curvature and torsion, respectively

$$\kappa(\tau) = \frac{a}{a^2 + b^2}, \qquad T(\tau) = \frac{b}{a^2 + b^2}.$$

Both are independent of τ. The following Maxima program will calculate the torsion. The command `trigsimp` will do the simplification $\sin^2(\alpha) + \cos^2(\alpha) = 1$.

```
/* Torsion.mac */
x: [a*cos(tau),a*sin(tau),b*tau];
```

```
x[1]; x[2]; x[3];
xd: [diff(x[1],tau),diff(x[2],tau),diff(x[3],tau)];
xd[1]; xd[2]; xd[3];
xdd: [diff(xd[1],tau),diff(xd[2],tau),diff(xd[3],tau)];
xdd[1]; xdd[2]; xdd[3];
xddd: [diff(xdd[1],tau),diff(xdd[2],tau),diff(xdd[3],tau)];
xddd[1]; xddd[2]; xddd[3];
C: matrix([xd[1],xdd[1],xddd[1]],[xd[2],xdd[2],xddd[2]],
          [xd[3],xdd[3],xddd[3]]);
D: determinant(C); D: trigsimp(D);
vpxdxdd: [xd[2]*xdd[3]-xd[3]*xdd[2],xd[3]*xdd[1]-xd[1]*xdd[3],
          xd[1]*xdd[2]-xd[2]*xdd[1]];
vpxdxdd: trigsimp(vpxdxdd);
vpxdxdd[1]; vpxdxdd[2]; vpxdxdd[3];
square: vpxdxdd[1]*vpxdxdd[1]+vpxdxdd[2]*vpxdxdd[2]
          +vpxdxdd[3]*vpxdxdd[3];
square: trigsimp(square);
Torsion: D/square;
Torsion: ratsimp(Torsion);
```

Problem 65.   Given a smooth surface in the Euclidean space $\mathbb{E}^3$ described by

$$\mathbf{x}(u,v) = \begin{pmatrix} x_1(u,v) \\ x_2(u,v) \\ x_3(u,v) \end{pmatrix}.$$

The *Gaussian curvature* is calculated as follows. First we calculate $E(u,v)$, $F(u,v)$, $G(u,v)$ of the *first fundamental form*

$$E(u,v) := \frac{\partial \mathbf{x}}{\partial u} \cdot \frac{\partial \mathbf{x}}{\partial u}, \qquad F(u,v) := \frac{\partial \mathbf{x}}{\partial u} \cdot \frac{\partial \mathbf{x}}{\partial v}, \qquad G(u,v) := \frac{\partial \mathbf{x}}{\partial v} \cdot \frac{\partial \mathbf{x}}{\partial v}$$

where $\cdot$ denotes the scalar product. Next we calculate the normal vector

$$\mathbf{n}(u,v) := \frac{\frac{\partial \mathbf{x}}{\partial u} \times \frac{\partial \mathbf{x}}{\partial v}}{\left| \frac{\partial \mathbf{x}}{\partial u} \times \frac{\partial \mathbf{x}}{\partial v} \right|}.$$

Using $\mathbf{n}$ we calculate $L(u,v)$, $M(u,v)$, $N(u,v)$ of the *second fundamental form*

$$L(u,v) := \mathbf{n} \cdot \frac{\partial^2 \mathbf{x}}{\partial u^2}, \qquad M(u,v) := \mathbf{n} \cdot \frac{\partial^2 \mathbf{x}}{\partial u \partial v}, \qquad N(u,v) := \mathbf{n} \cdot \frac{\partial^2 \mathbf{x}}{\partial v^2}.$$

Then the *Gaussian curvature* $K(u,v)$ is given by

$$K := \frac{LN - M^2}{EG - F^2}.$$

Write a SymbolicC++ program that calculates K and apply it to the *Möbius band* given by

$$\mathbf{x}(u,v) = \begin{pmatrix} (2 - v\sin(u/2))\sin(u) \\ (2 - v\sin(u/2))\cos(u) \\ v\cos(u/2) \end{pmatrix}$$

where $-1/2 < v < 1/2$ and $u \in [0, 2\pi)$.

Solution 65.   In the code % denotes the vector product and | denotes the scalar product.

```cpp
// curvature.cpp
#include <iostream>
#include "symbolicc++.h"
using namespace std;

Symbolic curvature(const Symbolic &x,const Symbolic &u,
                const Symbolic &v,const Equations &rules)
{
 Symbolic dxdu = df(x,u); Symbolic dxdv = df(x,v);
 Symbolic d2xdu2  = df(dxdu,u); Symbolic d2xdudv = df(dxdu,v);
 Symbolic d2xdv2  = df(dxdv,v);
 Symbolic E = (dxdu | dxdu).subst(rules);
 Symbolic F = (dxdu | dxdv).subst(rules);
 Symbolic G = (dxdv | dxdv).subst(rules);
 Symbolic n = (dxdu % dxdv).subst(rules);
 n = n/sqrt((n | n).subst(rules));
 Symbolic L = (n | d2xdu2).subst(rules);
 Symbolic M = (n | d2xdudv).subst(rules);
 Symbolic N = (n | d2xdv2).subst(rules);
 return (L*N-M*M)/(E*G-F*F);
}

int main(void)
{
 Symbolic x("x",3), u("u"), v("v");
 Equations rules = (sin(u/2)*sin(u/2)==1-cos(u/2)*cos(u/2),
                 cos(u/2)*cos(u/2)==(cos(u)+1)/2,
                 sin(u)*sin(u)==1-cos(u)*cos(u));
 x(0) = (2-v*sin(u/2))*sin(u); x(1) = (2-v*sin(u/2))*cos(u);
 x(2) = v*cos(u/2);
 cout << "x = " << x << endl;
 cout << "K = " << curvature(x,u,v,rules) << endl;
 return 0;
}
```

The output is

```
K = -(3/4*v^(2)-4*v*sin(1/2*u)-1/2*cos(u)*v^(2)+4)^(-2)
```

With $\sin^2(\alpha/2) \equiv \frac{1}{2} - \frac{1}{2}\cos(\alpha)$ we can also write

$$K = -\frac{1}{(v^2/4 + (2 - v\sin(u/2))^2)^2}.$$

1.3   Supplementary Problems

Problem 1.   Consider the curve in the plane

$$x_1(\tau) = -2c_1\tau, \qquad x_2(\tau) = c_2 + c_1\tau^2$$

where c_1, c_2 are positive constants. Show that

$$f(x_1, x_2; c_1, c_2) \equiv c_2 + \frac{1}{4c_1}x_1^2 - x_2 = 0.$$

Problem 2.   (i) Study the curve in the plane

$$x_1(\tau) = \cos(\tau) + \sin(\tau) + 1, \quad x_2(\tau) = \frac{1}{\cos(\tau) + \sin(\tau) + 1}.$$

Note that $x_1(0) = 2$, $x_2(0) = 1/2$.
(ii) Study the curve in the plane

$$x_1(\tau) = \cosh(\tau) + \sinh(\tau), \quad x_2(\tau) = \frac{1}{\cosh(\tau) + \sinh(\tau)}.$$

Note that $x_1(0) = x_2(0) = 1$.

Problem 3.   Consider the subset of $\mathbb{R}^2$

$$S := \{ (x_1, x_2) : x_1 x_2 = 1 \}.$$

It actually consists of two non-intersecting curves. Find the shortest distance between these two curves. Find the normal vectors and tangent vectors at $(1, 1)$ and $(-1, -1)$.

Problem 4.   (i) Study the curve in the plane $\sin(x) = \sinh(y)$. Note that $(0, 0)$ satisfies this equation and $y = \text{arcsinh}(\sin(x))$.

(ii) Study the curve in the plane $\sin(x + y) = \sin(xy)$. Note that $(0,0)$ and $(2,2)$ satisfy this equation.

Problem 5.   Describe the curve in the plane $f(x, y) = x^2 - y^3 = 0$. Note that $f(0,0) = 0$ and $\nabla f(0,0) = 0$. Show it has a single branch at the *cusp point* $(0,0)$.

Problem 6.   Draw the curve in the plane $x + y = xy$. Note that $(0,0)$ and $(2,2)$ are solutions and that the equation is invariant under $x \mapsto y$, $y \mapsto x$. For a given x we can solve for y as

$$y = \frac{x}{x-1}.$$

For $0 \leq x < 1$ we have $\lim_{x \to 1} y = -\infty$.

Problem 7.   Study the curve in the plane

$$x_1(t) = \cos\left(c_0 t + \frac{c_1}{\omega}\sin(\omega t)\right), \quad x_2(t) = -\sin\left(c_0 t + \frac{c_1}{\omega}\sin(\omega t)\right)$$

with $c_0, c_1, \omega > 0$, where c_0, c_1, ω have the dimension of a frequency and t is the time.

Problem 8.   (i) Consider the map $\mathbf{f} : \mathbb{R} \to \mathbb{R}^2$ defined by

$$\mathbf{f}(x) = (2\cos(x - \pi/2), \sin(2(x - \pi/2))).$$

Show that $(\mathbf{f}, \mathbb{R})$ is an immersed submanifold of the manifold $\mathbb{R}^2$, but not an embedded submanifold. Note that $\mathbf{f}(0) = (0,0)$.
(ii) Show that the map $\mathbf{g} : (\pi/4, 7\pi/4) \to \mathbb{R}^2$

$$\mathbf{g}(\theta) = \begin{pmatrix} \sin(\theta)\cos(2\theta) \\ \cos(\theta)\cos(2\theta) \end{pmatrix}$$

is an injective immersion. Note that $\mathbf{g}(\pi/2) = (-1 \quad 0)^T$. Show that the image of $\mathbf{g}$ is an injectively immersed submanifold.

Problem 9.   Consider the set (manifold) $M := \{\, w \; : \; w_0^2 - w_1^2 - w_2^2 = 1 \,\}$. Show that the map

$$f : M \ni w \mapsto g = \begin{pmatrix} w_0 + w_1 & w_2 \\ w_2 & w_0 - w_1 \end{pmatrix} \in SL(2, \mathbb{R})$$

establishes a $1 : 1$ correspondence between the manifold M and a subset of symmetric matrices in the Lie group $SL(2, \mathbb{R})$.

Problem 10. Let $z \in \mathbb{C}$. Study the curve in the plane defined by

$$z + \bar{z} = z\bar{z}.$$

Set $z = x + iy$ with $x, y \in \mathbb{R}$. Then $\bar{z} = x - iy$. An alternative would be setting $z = re^{i\phi}$ with $r \geq 0$ and $\bar{z} = re^{-i\phi}$.

Problem 11. Study the curve $\sin(x_1 + x_2) - \cos(x_1 x_2) + 1 = 0$ in the Euclidean plane. Note that $(0,0)$ satisfies the equation. With $f(x_1, x_2) = \sin(x_1 + x_2) - \cos(x_1 x_2) - 1$ we have

$$df = \cos(x_1+x_2)dx_1 + \cos(x_1+x_2)dx_2 + x_2\sin(x_1 x_2)dx_1 + x_1\sin(x_1 x_2)dx_2.$$

Hence $df(0,0) = dx_1 + dx_2$ and $dx_1 = -dx_2$. There are infinite many disconnected closed curves.

Problem 12. Let $a > 0$. The equation for the *hypocycloid* with four *cusps* is given by $x^{2/3} + y^{2/3} = a^{2/3}$.
(i) Show that the parametric form is given by

$$x(\tau) = a\cos^3(\tau), \quad y(\tau) = a\sin^3(\tau).$$

(ii) Let $a = 1$. Find the normal line and tangent line at the point

$$x_0 = y_0 = 1/(2\sqrt{2}).$$

Problem 13. Consider the Euclidean space $\mathbb{E}^2$. The equation for the *lemniscate* is given by the curve $r^2(\theta) = \cos(2\theta)$. Consider the point $(3/2, 0)$. Find the shortest distance from this point to the lemniscate. Apply the Lagrange multiplier method. Note that $\cos(2\theta) \equiv \cos^2(\theta) - \sin^2(\theta)$.

Problem 14. Consider the curve in $\mathbb{R}^3$

$$\mathbf{x}(\tau) = \begin{pmatrix} x_1(\tau) \\ x_2(\tau) \\ x_3(\tau) \end{pmatrix} = \begin{pmatrix} (\sqrt{2} + \cos(2\tau))\cos(3\tau) \\ (\sqrt{2} + \cos(2\tau))\sin(3\tau) \\ \sin(2\tau) \end{pmatrix}.$$

Show that the curve is regular.

Problem 15. The *three-leaved rose* in the plane is described by

$$r(\theta) = \cos(\theta)$$

or $x(\theta) = \cos(3\theta)\sin(\theta)$, $y(\theta) = \cos(3\theta)\cos(\theta)$. Find the tangent line and normal line at the point $x_0 = \sqrt{3}/2$, $y_0 = -1/2$ i.e. $\theta_0 = 2\pi/3$ with $\cos(3\theta_0) = 1$.

Problem 16. (i) The *stereographic projection* is the map $\phi : \mathbb{S}^2 \setminus N \to \mathbb{C}$ (N north pole) defined by

$$\phi(x_1, x_2, x_3) = \frac{x_1}{1 - x_3} + i\frac{x_2}{1 - x_3}.$$

Show that the inverse of the stereographic projection takes a complex number $u + iv$ $(u, v \in \mathbb{R})$

$$\left(\frac{2u}{1 + u^2 + v^2}, \frac{2v}{1 + u^2 + v^2}, \frac{1 - u^2 - v^2}{1 + u^2 + v^2}\right)$$

to the unit sphere.
(ii) Show that the *stereographic projection* of the two-dimensional sphere

$$\mathbb{S}^2 := \{\, \mathbf{v} \in \mathbb{R}^3 \,:\, \mathbf{v}^2 = v_0^2 + v_1^2 + v_2^2 = 1 \,\}$$

is given by $\mathbf{v} = (\cos(2|\zeta|), \sin(2|\zeta|)\cos(\phi), \sin(2|\zeta|)\sin(\phi))$.

Problem 17. Consider the manifold (unit ball)

$$\mathbb{S}^2 = \{\, (x_1, x_2, x_3) \,:\, x_1^2 + x_2^2 + x_3^2 = 1 \,\}.$$

Find the plane $a_1 x_1 + a_2 x_2 + a_3 x_3 - a_4 = 0$ that touches (tangent plane) the unit ball at the point $x_1 = x_2 = x_3 = 1/\sqrt{3}$.

Problem 18. Let $\mathbf{d} = (d_0, d_1, \ldots, d_n)$ be an $(n+1)$-tuple of integers $d_j > 1$. We define

$$V(\mathbf{d}) := \{\, \mathbf{z} = (z_0, z_1, \ldots, z_n) \in \mathbb{C}^{n+1} \,:\, f(\mathbf{z}) := z_0^{d_0} + z_1^{d_1} + \cdots + z_n^{d_n} = 0 \,\}.$$

Let $\mathbb{S}^{2n+1}$ denote the unit sphere in $\mathbb{C}^{n+1}$, i.e.

$$z_0\bar{z}_0 + z_1\bar{z}_1 + \cdots + z_n\bar{z}_n = 2.$$

We define

$$\Sigma(\mathbf{d}) := V(\mathbf{d}) \cap \mathbb{S}^{2n+1}.$$

Show that $\Sigma(\mathbf{d})$ is a smooth manifold of dimension $2n - 1$. The manifolds $\Sigma(\mathbf{d})$ are called *Brieskorn manifolds*.

Problem 19. (i) Let A be a nonzero symmetric $n \times n$ matrix over $\mathbb{R}$. Let $0 \neq b \in \mathbb{R}$. Show that the surface

$$M = \{ \mathbf{x} \in \mathbb{R}^n \: : \: \mathbf{x}^T A \mathbf{x} = b \}$$

is a smooth $(n - 1)$-dimensional submanifold of the manifold $\mathbb{R}^n$.
(ii) Let $M(n, \mathbb{R})$ be the vector space of all $n \times n$ matrices over $\mathbb{R}$. Show that

$$M = \{ A \in M(n, \mathbb{R}) \: : \: \mathrm{tr}(A) = 0 \}$$

is a manifold.

Problem 20. Let $x_1, x_2 \in \mathbb{R}$. Show that the eigenvalues of the 2×2 matrix

$$\begin{pmatrix} 1 + x_1^2 & -x_1 x_2 \\ -x_1 x_2 & 1 + x_2^2 \end{pmatrix}$$

are given by $\lambda_1 = 1 + x_1^2 + x_2^2$ and $\lambda_2 = 1$. What curve in the plane is described by

$$\det \begin{pmatrix} 1 + x_1^2 & -x_1 x_2 \\ -x_1 x_2 & 1 + x_2^2 \end{pmatrix} = 0?$$

Problem 21. Describe the set

$$S = \{ (x, y) \in \mathbb{R} \: : \: \sin(y) \cosh(x) = 1 \}.$$

Then study the complex numbers given by $z = x + iy$ with $x, y \in S$.

Problem 22. Let $\mathbb{E}^3$ be the three-dimensional Euclidean space. Show that

$$M = \{ \mathbf{x} \in \mathbb{E}^3 \: : \: (\sqrt{x_1^2 + x_2^2} - 2)^2 + x_3^2 = 1 \}$$

is a differentiable manifold. Show that a parameter representation is

$$\mathbf{x}(u_1, u_2) = \begin{pmatrix} x_1(u_1, u_2) \\ x_2(u_1, u_2) \\ x_3(u_1, u_2) \end{pmatrix} = \begin{pmatrix} \cos(u_1)(2 + \cos(u_2)) \\ \sin(u_1)(2 + \cos(u_2)) \\ \sin(u_2) \end{pmatrix}.$$

Let $u_1 = \pi/4$. Study the resulting curve in $\mathbb{E}^3$.

Problem 23. Draw the curve (*three-cusped hypocycloid*) in the plane

$$x_1(\theta) = \frac{1}{3}(\cos(2\theta) - 2\sin(\theta)), \quad x_2(\theta) = \frac{1}{3}(\sin(2\theta) - 2\cos(\theta)).$$

Problem 24. Consider the two curves in the Euclidean plane $y_1 = e^{x_1}$ and $y_2 = \ln(x_2)$ $(x_2 > 0)$. Find the shortest distance between the curves. Note that $\exp(0) = 1$ and $\ln(1) = 0$. Describe the distance vector.

Problem 25. Let $\mathbf{z} = (\begin{array}{ccccc} z_1 & z_2 & z_3 & z_4 & z_5 \end{array})$ and $\mathbf{z}^T$ be the transpose and $\overline{\mathbf{z}}$ the complex conjugate. Consider the complex *Lie ball*

$$D^5 = \{\, \mathbf{z} \in \mathbb{C}^5 \,:\, 1 + |\mathbf{zz}^T|^2 - 2\overline{\mathbf{z}}\mathbf{z}^T > 0 \,,\, |\mathbf{zz}^T| < 1 \,\}.$$

Show that the boundary of D^5 is given by

$$Q^5 = \{\, \mathbf{w} = \mathbf{x}e^{i\theta} \,,\, \mathbf{x} \in \mathbb{R}^5 \,,\, \mathbf{xx}^T = 1 \,\}, \;\; 0 < \theta < \pi.$$

Show that the Lie ball is included in the complex unit ball

$$C^5 = \{\, \mathbf{z} \in \mathbb{C}^5 \,:\, |\mathbf{zz}^T| < 1 \,\}$$

and contains the real unit ball $B^5 = \{\, \mathbf{x} \in \mathbb{R}^5 \,:\, \mathbf{xx}^T < 1 \,\}$.

Problem 26. Let $\mathbf{x}, \mathbf{y} \in \mathbb{R}^3$ and $\times$ be the *vector product*

$$\mathbf{x} \times \mathbf{y} = \begin{pmatrix} x_2 y_3 - x_3 y_2 \\ x_3 y_1 - x_1 y_3 \\ x_1 y_2 - x_2 y_1 \end{pmatrix}.$$

Hence the vector product is a map from $\mathbb{R}^3 \times \mathbb{R}^3$ into $\mathbb{R}^3$. Show that the map is differentiable.

Problem 27. Find the asymptotic curves of the *catenoid*

$$\mathbf{x}(u_1, u_2) = \begin{pmatrix} x_1(u_1, u_2) \\ x_2(u_1, u_2) \\ x_3(u_1, u_2) \end{pmatrix} = \begin{pmatrix} \cosh(u_2)\cos(u_1) \\ \cosh(u_2)\sin(u_1) \\ u_2 \end{pmatrix}.$$

Problem 28. The Lie group of unitary 2×2 matrices with determinant $+1$ is denoted by $SU(2)$. Let $\mathbb{S}^3$ be the unit sphere in $\mathbb{R}^4$ centered at the origin $(0, 0, 0, 0)$. Let $\mathbf{x} \in \mathbb{S}^3$ and we define the 2×2 matrix over $\mathbb{C}$

$$U_{\mathbf{x}} = \begin{pmatrix} x_1 + ix_2 & -x_3 + ix_4 \\ x_3 + ix_4 & x_1 - ix_2 \end{pmatrix}$$

with $\det(U_{\mathbf{x}}) = 1$. Show that the map $\mathbf{f} : \mathbf{x} \to U_{\mathbf{x}}$ is a *bijection* from $\mathbb{S}^3$ onto $SU(2)$. We can view $\mathbb{S}^3$ as a smooth submanifold of $\mathbb{R}^4$.

Problem 29. Consider the two circles in $\mathbb{R}^3$

$$S_1 = \{ (x_1, x_2, x_3) : x_1^2 + x_2^2 = 1, x_3 = 0 \}$$

$$S_2 = \{ (x_1, x_2, x_3) : (x_1 + 1)^2 + x_3^2 = 1, x_2 = 0 \}.$$

Is the point $\mathbf{p} = (1, 1, 1) \in \mathbb{R}^3$ closer to circle 1 or circle 2?

Problem 30. Let

$$B^n := \{ (x_1, \dots, x_n) \in \mathbb{R}^n : x_1^2 + \cdots + x_n^2 \leq 1 \}.$$

(i) Show that B^n is a smooth n-manifold with boundary.
(ii) Show that its boundary is the sphere $\mathbb{S}^{n-1}$.

Problem 31. (i) Show that the subset of $\mathbb{R}^6$

$$M = \{ (x_1^2, x_2^2, x_3^2, x_2 x_3, x_3 x_1, x_1 x_2) : x_1, x_2, x_3 \in \mathbb{R}, x_1^2 + x_2^2 + x_3^2 = 1 \}$$

is a differentiable two-manifold in $\mathbb{R}^6$.
(ii) Is the subset of $\mathbb{R}^3$

$$S = \{ (x_1, x_2, x_2) : x_1, x_2, x_3 \in \mathbb{R}^3, x_1^2 + x_2^2 + x_3^2 = 1 \}$$

a submanifold of $\mathbb{R}^3$?

Problem 32. Consider the map $\mathbf{f} : \mathbb{R}^3 \to \mathbb{R}^4$ given by

$$\mathbf{f}(u_1, u_2, u_3) = \begin{pmatrix} f_1(u_1, u_2, u_3) \\ f_2(u_1, u_2, u_3) \\ f_3(u_1, u_2, u_3) \\ f_4(u_1, u_2, u_3) \end{pmatrix} = \begin{pmatrix} u_1^2 - u_2^2 \\ u_1 u_2 \\ u_1 u_3 \\ u_2 u_3 \end{pmatrix}.$$

Note that $\mathbf{f}(0, 0, u_3) = (0, 0, 0, 0)$. The map is invariant under $u_1 \to -u_1$, $u_2 \to -u_2$, $u_3 \to -u_3$. Note that

$$\begin{pmatrix} \partial f_1/\partial u_1 & \partial f_1/\partial u_2 & \partial f_1/\partial u_3 \\ \partial f_2/\partial u_1 & \partial f_2/\partial u_2 & \partial f_2/\partial u_3 \\ \partial f_3/\partial u_1 & \partial f_3/\partial u_2 & \partial f_3/\partial u_3 \\ \partial f_4/\partial u_1 & \partial f_4/\partial u_2 & \partial f_4/\partial u_3 \end{pmatrix} = \begin{pmatrix} 2u_1 & -2u_2 & 0 \\ u_2 & u_1 & 0 \\ u_3 & 0 & u_1 \\ 0 & u_3 & u_2 \end{pmatrix}.$$

If $u_3 \neq 0$ the maximal rank is 3. If $u_1 = u_2 = 0$, $u_3 \neq 0$ the rank is 2. Show that one has a non-orientable closed surface in $\mathbb{R}^4$. Study the subcase that $u_1^2 + u_2^2 + u_3^2 = 1$, i.e. we have the map $\mathbf{g} : \mathbb{S}^2 \to \mathbb{E}^4$.

Problem 33.   Let $n \geq 1$ and A be an $(n+1) \times (n+1)$ hermitian matrix. Hence all eigenvalues are real. Consider the real-valued map $f : \mathbb{C}^{n+1} \setminus \{\mathbf{0}\} \to \mathbb{R}$ defined by

$$f(\mathbf{v}) = \frac{\mathbf{v}^* A \mathbf{v}}{\mathbf{v}^* \mathbf{v}}$$

for $\mathbf{v} = (z_1, \ldots, z_{n+1})^T$. Show that if $z \in \mathbb{C} \setminus \{0\}$, then $f(z\mathbf{v}) = f(\mathbf{v})$. The map f induces the map

$$[f] = (\mathbb{C}^{n+1} \setminus \{\mathbf{0}\}) / (\mathbb{C} \setminus \{0\}) = \mathbb{CP}^n \to \mathbb{R}$$

by $[f]([\mathbf{v}]) = f(\mathbf{v})$. The *critical points* of the maps f and $[f]$ are points where the derivative vanishes. Show that the equations for critical points of $[f]$ are the lines $[\mathbf{v}]$ for

$$A\mathbf{v} = \lambda \mathbf{v}$$

where $\lambda = f(\mathbf{v})$.

Problem 34.   Let $GL(n, \mathbb{R})$ be the group (under matrix multiplication) of all $n \times n$ matrices, i.e. $\det(M) \neq 0$ for all $M \in GL(n, \mathbb{R})$. Show that $GL(n, \mathbb{R})$ is an n^2-dimensional manifold. Apply the vec-operator which maps the $n \times n$ matrix into an element of $\mathbb{R}^{n^2}$. A distance measure between two $n \times n$ matrices A and B is given by

$$d(A, B) = \sqrt{\sum_{j,k=1}^{n} (a_{jk} - b_{jk})^2}.$$

Problem 35.   Let M be a differentiable manifold. Suppose that $f : M \to M$ is a diffeomorphism with $N_m(f) < \infty$, $m = 1, 2, \ldots$. Here $N_m(f)$ is the number of fixed points of the m-th iterate of f, i.e. $f^{(m)}$. One defines the *zeta function* of f as the formal power series

$$\zeta_f(t) := \exp\left(\sum_{m=1}^{\infty} \frac{1}{m} N_m(f) t^m \right).$$

(i) Show that $\zeta_f(t)$ is an invariant of the topological conjugacy class of f.
(ii) Find $N_m(f)$ for the map $f : \mathbb{R} \to \mathbb{R}$ and $f(x) = \sinh(x)$.

Problem 36.   The transformation between the orthogonal *ellipsoidal coordinates* (ρ, μ, ν) and the Cartesian coordinates (x_1, x_2, x_3) is

$$x_1^2 = \frac{\rho^2 \mu^2 \nu^2}{h^2 k^2}$$

$$x_2^2 = \frac{(\rho^2 - \mu^2)(\mu^2 - h^2)(h^2 - \nu^2)}{h^2(k^2 - h^2)}$$

$$x_3^2 = \frac{(\rho^2 - k^2)(k^2 - \mu^2)(k^2 - \nu^2)}{k^2(k^2 - h^2)}$$

where $k^2 = a_1^2 - a_3^2$, $h^2 = a_1^2 - a_2^2$ and $a_1 > a_2 > a_3$ denote the three semi-axes of the ellipsoid. The three surfaces in $\mathbb{R}^3$, $\rho = \text{constant}$, $(k \le \rho \le \infty)$, $\mu = \text{constant}$, $(h \le \mu \le k)$ and $\nu = \text{constant}$, $(0 \le \nu \le h)$, represent ellipsoids and hyperboloids of one and two sheets, respectively. Find the inverse transformation.

Problem 37. (i) Let $R > 0$. Study the manifold

$$\frac{x_1^2}{R^2 e^{-\epsilon}} + \frac{x_2^2}{R^2 e^{-\epsilon}} + \frac{x_3^2}{R^2 e^{2\epsilon}} = 1$$

where ϵ is a deformation parameter.
(ii) Show that the volume V of the spheroid is given by

$$V = \frac{4\pi}{3} R^3.$$

Problem 38. Let $\alpha, \theta, \phi, \omega \in \mathbb{R}$. Consider the vector in $\mathbb{R}^5$

$$\mathbf{x}(\alpha, \theta, \phi, \omega) = \begin{pmatrix} \cosh(\alpha)\sin(\theta)\cos(\phi) \\ \cosh(\alpha)\sin(\theta)\sin(\phi) \\ \cosh(\alpha)\cos(\theta) \\ \sinh(\alpha)\cos(\omega) \\ \sinh(\alpha)\sin(\omega) \end{pmatrix}.$$

Show that $x_1^2 + x_2^2 + x_3^2 - x_4^2 - x_5^2 = 1$. This vector plays a role for the Lie group $SO(3,2)$.

Problem 39. (i) Let C be the topological space given by the boundary of

$$[0,1]^n := [0,1] \times \cdots \times [0,1].$$

This means C is the surface of the n-dimensional unit cube. Show that C can be endowed with the structure of a differential manifold.
(ii) Show that the surface ∂C of the unit cube

$$C = \{(x_1, x_2, x_3) : 0 \le x_1 \le 1,\ 0 \le x_2 \le 1,\ 0 \le x_3 \le 1\}$$

can be made into a differentiable manifold.

Problem 40.   Let $a > 0$. Consider

$$x_1(u,v) = a\frac{1-v^2}{1+v^2}\cos(u), \quad x_2(u,v) = a\frac{1-v^2}{1+v^2}\sin(u),$$

$$x_3(u,v) = a\frac{2v}{1+v^2}.$$

Show that

$$x_1^2(u,v) + x_2^2(u,v) + x_3^2(u,v) = a^2.$$

Calculate the vector product

$$\frac{\partial \mathbf{x}}{\partial u} \times \frac{\partial \mathbf{x}}{\partial v}.$$

Problem 41.   (i) Let M be a manifold and $f : M \to M$, $g : M \to M$. Assume that f is invertible. Then we say that the map f is a *symmetry* of the map g if

$$f \circ g \circ f^{-1} = g.$$

Let $M = \mathbb{R}$ and $f(x) = \sinh(x)$. Find all g such that

$$f \circ g \circ f^{-1} = g.$$

(ii) Let f and g be invertible maps. We say that g has a *reversing symmetry* f if

$$f \circ g \circ f^{-1} = g^{-1}.$$

Let $M = \mathbb{R}$ and $f(x) = \sinh(x)$. Find all g that satisfy this equation.
(iii) Let M be a smooth manifold. A *diffeomorphism* is a one-to-one smooth map $f : M \to M$ such that $f^{-1} : M \to M$ is also smooth. Show that the set of diffeomorphism form a group under function composition. Let $M = \mathbb{R}$. Show that the function $f(x) = \sinh(x)$ is a diffeomorphism.

Problem 42.   Consider the unit circle around $(0,0,0)$ in the $x_1 - x_2$ plane

$$\mathbf{r}_1(\tau) = \begin{pmatrix} x_{1,1}(\tau) \\ x_{1,2}(\tau) \\ x_{1,3}(\tau) \end{pmatrix} = \begin{pmatrix} \cos(\tau) \\ \sin(\tau) \\ 0 \end{pmatrix}, \quad \tau \in [0, 2\pi]$$

and the unit circle around $(0,1,0)$ in the $x_2 - x_3$ plane

$$\mathbf{r}_2(s) = \begin{pmatrix} x_{2,1}(s) \\ x_{2,2}(s) \\ x_{2,3}(s) \end{pmatrix} = \begin{pmatrix} 0 \\ 1 + \cos(s) \\ \sin(s) \end{pmatrix}, \quad s \in [0, 2\pi].$$

Then the derivatives are

$$\frac{d\mathbf{r}_1(\tau)}{d\tau} = \begin{pmatrix} -\sin(\tau) \\ \cos(\tau) \\ 0 \end{pmatrix}, \qquad \frac{d\mathbf{r}_2(s)}{ds} = \begin{pmatrix} 0 \\ -\sin(s) \\ \cos(s) \end{pmatrix}.$$

Calculate (*Gauss formula*)

$$\frac{1}{4\pi} \oint \oint d\tau ds \left(\frac{d\mathbf{r}_1(\tau)}{d\tau} \times \frac{d\mathbf{r}_2(s)}{ds} \right) \cdot \frac{\mathbf{r}_1(\tau) - \mathbf{r}_2(s)}{|\mathbf{r}_1(\tau) - \mathbf{r}_2(s)|^3}$$

where $\times$ denotes the vector product, $\cdot$ denotes the scalar product and contour integrations run from 0 to 2π.

Problem 43. Consider the two *hyperplanes* $(n \geq 1)$

$$x_1 + x_2 + \cdots + x_n = 2, \qquad x_1 + x_2 + \cdots + x_n = -2.$$

The hyperplanes do not intersect. Find the shortest distance between the hyperplanes. What happens if $n \to \infty$? Discuss first the cases $n = 1$, $n = 2$ and $n = 3$.

Problem 44. Consider the *space cardioid*

$$\mathbf{x}(\tau) = \begin{pmatrix} x_1(\tau) \\ x_2(\tau) \\ x_3(\tau) \end{pmatrix} = \begin{pmatrix} (1 - \cos(\tau))\cos(\tau) \\ (1 - \cos(\tau))\sin(\tau) \\ \sin(\tau) \end{pmatrix}.$$

Find the torsion. The *torsion* is given by

$$\frac{((d\mathbf{x}/d\tau) \times (d^2\mathbf{x}/d\tau^2)) \cdot d^3\mathbf{x}/d\tau^3}{\|d\mathbf{x}/d\tau \times d^2\mathbf{x}/d\tau^2\|}.$$

Apply computer algebra.

Problem 45. Let $\omega_1, \omega_2 > 0$ (frequencies) and $R > r > 0$. Consider the curve

$$x_1(t) = (R + r\cos(\omega_1 t))\sin(\omega_2 t)$$
$$x_2(t) = (R + r\cos(\omega_1 t))\cos(\omega_2 t)$$
$$x_3(t) = r\sin(\omega_1 t).$$

Find the curvature and torsion. Discuss the dependence on the frequencies.

Problem 46. Let $x_1 > 0$, $x_2 > 0$, $x_3 > 0$. Consider the surface

$$x_1 x_2 x_3 = 2$$

in $\mathbb{R}^3$. Find the shortest distance from the origin $(0,0,0)$ the surface. Apply the Lagrange multiplier method. Apply differential forms and symmetry considerations (permutation group).

Problem 47. The unit circle

$$\mathbb{S}^1 := \{(x_1, x_2) : x_1^2 + x_2^2 = 1\}$$

is a smooth manifold of dimension one. Show that a possible atlas is given by $\mathcal{A} = \{(U_1, \phi_1), (U_2, \phi_2)\}$ with

$U_1 = \mathbb{S}^1 \backslash \{(-1, 0)\}$ $\phi_1(x_1, x_2) = \arctan(x_2/x_1)$ for $-\pi < \phi_1(x_1, x_2) < \pi$

$U_2 = \mathbb{S}^1 \backslash \{(1, 0)\}$ $\phi_2(x_1, x_2) = \arctan(x_2/x_1)$ for $0 < \phi_2(x_1, x_2) < 2\pi$.

Note that

$$\arctan(0) = 0, \quad \arctan(1) = \pi/4, \quad \arctan(-1) = -\pi/4.$$

Problem 48. Show that the two-dimensional surface

$$\mathbb{T}^2 = \left\{ (x_1, x_2, x_3) \in \mathbb{R}^3 : (\sqrt{x_1^2 + x_2^2} - 1)^2 + x_3^2 = \frac{1}{4} \right\}$$

in $\mathbb{R}^3$ is a differentiable manifold (*torus*).

Problem 49. Consider the triaxial ellipsoid

$$Q : \frac{x_1^2}{a_1^2} + \frac{x_2^2}{a_2^2} + \frac{x_3^2}{a_3^2} = 1$$

with $0 < a_3 < a_2 < a_1$. Show that any billard trajectory inside Q has as caustics two elements of the family of confocal quadrics

$$Q_\epsilon = \left\{ (x_1, x_2, x_3) \in \mathbb{R}^3 : \frac{x_1^2}{a_1 - \epsilon} + \frac{x_2^2}{a_2 - \epsilon} + \frac{x_3^2}{a_3 - \epsilon} = 1 \right\}.$$

Problem 50. Let n be a positive integer. Consider the manifold

$$C_n := \left\{ (x, y) \in \mathbb{R}^2 : \left(x - \frac{1}{n} \right)^2 + y^2 = \frac{1}{n^2} \right\}.$$

We have a circle in the plane with radius $1/n$ and centre $(1/n, 0)$. Find the area of the circle.

Problem 51. The manifold $\mathbb{S}^3$ is given by

$$x_1^2 + x_2^2 + x_3^2 + x_4^2 = 1$$

with $x_1, x_2, x_3, x_4 \in \mathbb{R}$. The manifold $\mathbb{S}^2$ is given by

$$y_1^2 + y_2^2 + y_3^2 = 1$$

with $y_1, y_2, y_3 \in \mathbb{R}$. The *Hopf map* $f : \mathbb{S}^3 \to \mathbb{S}^2$ is given by

$$f : \quad y_1 = 2(x_1 x_3 + x_2 x_4), \quad y_2 = 2(x_2 x_3 - x_1 x_4), \quad y_3 = x_1^2 + x_2^2 - x_3^2 - x_4^2.$$

Show that

$$y_1^2 + y_2^2 + y_3^2 = (x_1^2 + x_2^2 + x_3^2 - x_4^2)^2.$$

Note that $(1/2, 1/2, 1/2, 1/2) \mapsto (1, 0, 0)$.

Problem 52. The matrix

$$A(\tau) = \begin{pmatrix} \cos(\tau) & -\sin(\tau) \\ \sin(\tau) & \cos(\tau) \end{pmatrix}$$

is an element of the Lie group $SO(2)$. Applying the vec-operator we obtain the curve in $\mathbb{R}^4$

$$\mathbf{x}(\tau) = \begin{pmatrix} x_1(\tau) \\ x_2(\tau) \\ x_3(\tau) \\ x_4(\tau) \end{pmatrix} = \begin{pmatrix} \cos(\tau) \\ \sin(\tau) \\ -\sin(\tau) \\ \cos(\tau) \end{pmatrix}.$$

Can this vector in $\mathbb{R}^4$ be written as the Kronecker product of two vectors from $\mathbb{R}^2$? Find the curvature and torsion.

Problem 53. Consider the Hilbert space $L_2[0, \pi]$, the curve in the plane expressed in polar coordinates $r(\theta) = 1 + \cos(2\theta)$ and

$$r(\theta) = \sum_{n=-\infty}^{+\infty} C_n \exp(in\theta), \quad C_n = \frac{1}{\pi} \int_0^\pi r(\theta) \exp(-in\theta) d\theta.$$

Find the coefficents C_n.

Problem 54. Let $a_1, a_2, a_3 > 0$. Consider the analytic function $f : \mathbb{R}^3 \to \mathbb{R}$

$$f(x_1, x_2, x_3) = x_1 x_2 + x_2 x_3 + x_3 x_1.$$

Show that the extrema value of f subject to the constraint (compact manifold)

$$\frac{x_1^2}{a_1^2} + \frac{x_2^2}{a_2^2} + \frac{x_3^2}{a_3^2} = 1$$

are given by the largest and smallest eigenvalues of the 3×3 matrix

$$\begin{pmatrix} 0 & a_1^2 & a_1^2 \\ a_2^2 & 0 & a_2^2 \\ a_3^2 & 0 & a_3^2 \end{pmatrix}.$$

Problem 55.   Let $\alpha_1, \alpha_2, \alpha_3 \in \mathbb{R}$. Consider the vector in $\mathbb{R}^3$

$$\mathbf{v}(\alpha_1, \alpha_2, \alpha_3) = \begin{pmatrix} \sin(\alpha_2 - \alpha_3) \\ \sin(\alpha_3 - \alpha_1) \\ \sin(\alpha_1 - \alpha_2) \end{pmatrix}.$$

Can one find α_1, α_2, α_3 such that $\|\mathbf{v}(\alpha_1, \alpha_2, \alpha_3)\| = 1$?

Problem 56.   Let $(u_1, u_2) \in \mathbb{R}^2$. Describe the smooth surface in $\mathbb{R}^3$ given by

$$\mathbf{x}(u_1, u_2) = \begin{pmatrix} x_1(u_1, u_2) \\ x_2(u_1, u_2) \\ x_3(u_1, u_2) \end{pmatrix} = \begin{pmatrix} u_1 - u_2 \\ u_1 + u_2 \\ 2(u_1^2 + u_2^2) \end{pmatrix}.$$

Note that $\mathbf{x}(0,0) = \mathbf{0}$ and if $u_1 = u_2 = u$ we have the curve in $\mathbb{R}^3$

$$\mathbf{x}(u) = \begin{pmatrix} 0 & 2u & 4u^2 \end{pmatrix}^T.$$

Problem 57.   Consider the subset of $\mathbb{R}^3$ given by

$$S = \{ (x_1, x_2, x_3) : x_1^2 + x_2^2 - x_3^2 = 0 \}$$

(*double cone*). Show that S is not a two-manifold. Note that with

$$f(x_1, x_2, x_3) = x_1^2 + x_2^2 - x_3^2$$

we have $df = 2x_1 + 2x_2 - 2x_3$, $df|_{x_1 = x_2 = x_3 = 0} = 0$.

Problem 58.   Consider the two smooth surfaces in $\mathbb{R}^3$

$$3x_1^2 x_2 + x_2^2 x_3 + 2 = 0, \quad 2x_1 x_3 - x_1^2 x_2 - 3 = 0.$$

The surfaces intersect and form a curve in $\mathbb{R}^3$. For example the point $(1, -1, 1)$ satisfies these equations. The command in Maxima

```
load(grobner);
R: poly_reduced_grobner([3*x1*x1*x2+x2*x2*x3+2,2*x1*x3-x1*x1*x2-3],
                        [x1,x2,x3]);
```

provides a generator for the ideal generated by $\{3x_1^2x_2 + x_2^2x_3 + 2 = 0, 2x_1x_3 - x_1^2x_2 - 3 = 0\}$. One finds

$$-x_2^2x_3 - 6x_1x_3 + 7 = 0$$
$$12x_2^2x_3^2 + x_2^5x_3 + 24x_3 - 7x_2^3 + 42x_1x_2 = 0$$
$$12x_2^2x_3^3 + x_2^5x_3^2 + 24x_3^2 - 14x_2^3x_3 + 49x_2 = 0.$$

Discuss.

Problem 59. Consider the curve in the plane

$$f(x_1, x_2) = x_2^5 - 4x_2^4 + 4x_2^3 + 2x_1^2x_2^2 - x_1x_2^2 + 2x_1x_2 + x_1^4 + x_1^3 = 0.$$

Show that $f(0,0) = 0$, $f(-1,0) = 0$, $f(0,2) = 0$. Show that $df(0,0) = 0$ and the implicit function theorem cannot be applied at $(0,0)$. Show that the curve is self-intersecting at $(0,0)$. Show that the implicit function theorem can be applied for the point at $(-1,0)$. Show that the curve around $(0,0)$ admits the expansions

$$x_1(\tau) = \tau, \quad x_2(\tau) = -\frac{1}{2}\tau^2 + \frac{1}{8}\tau^4 - \frac{1}{8}\tau^5 + \cdots,$$

and

$$x_1(\tau) = -2\tau^2, \quad x_2(\tau) = \tau + \frac{1}{4}\tau^2 - \frac{27}{32}\tau^3 + \cdots$$

Problem 60. Let $f : \mathbb{R}^n \to \mathbb{R}$ be a smooth function. Show that the regular points of a *level surface*

$$M := \{\, \mathbf{x} \in \mathbb{R}^n \, : \, f(\mathbf{x}) = c, \; df(\mathbf{x}) \neq 0 \,\}$$

form a $(n - 1)$-dimensional submanifold endowed with the topology induced from $\mathbb{R}^n$.

Problem 61. (i) Find the symmetries of the curve in the plane given by

$$f(x, y) = (x^2 + y^2)^3 - 4x^2y^2 = 0.$$

Find the area enclosed by the curve for $x \geq 0$ and $y \geq 0$.
(ii) Find the symmetries of the curve in the plane given by

$$f(x, y) = (x^2 + y^2)^2 + 3x^2y - y = 0.$$

Find the area enclosed by the curve for $y \geq 0$.

Problem 62. Show that the equation of a *hyperplane* passing through the n points $\mathbf{x}_1, \mathbf{x}_2, \ldots, \mathbf{x}_n$ in the Euclidean space $\mathbb{E}^n$ is given by

$$\det \begin{pmatrix} 1 & 1 & 1 & \cdots & 1 \\ \mathbf{x} & \mathbf{x}_1 & \mathbf{x}_2 & \cdots & \mathbf{x}_n \end{pmatrix} = 0.$$

Problem 63. Show that *Scherk's surface*

$$u(x_1, x_2) = \ln \left(\frac{\cos(x_1)}{\cos(x_2)} \right) \equiv \ln(\cos(x_1)) + \ln(\cos(x_2))$$

$(x_1, x_2 \in (-\pi/2, +\pi/2))$ is a *minimal surface*, i.e. it satisfies the partial differential equation

$$\frac{\partial^2 u}{\partial x_1^2} \left(1 + \left(\frac{\partial u}{\partial x_2} \right)^2 \right) - 2 \frac{\partial u}{\partial x_1} \frac{\partial u}{\partial x_2} \frac{\partial^2 u}{\partial x_1 \partial x_2} + \frac{\partial^2 u}{\partial x_2^2} \left(1 + \left(\frac{\partial u}{\partial x_1} \right)^2 \right) = 0.$$

Problem 64. Given the surface in $\mathbb{R}^3$

$$\mathbf{f}(\tau, \theta) = \begin{pmatrix} (1 + \tau \sin(\theta/2)) \cos(\theta) \\ (1 + \tau \cos(\theta/2)) \sin(\theta) \\ \tau \sin(\theta/2) \end{pmatrix}$$

where $\tau \in (-1/2, 1/2)$, $\theta \in \mathbb{R}$.
(i) Build three models of this surface using paper, glue and a pair of scissors. Color the first model with the South African flag. For the second model keep τ fixed (say $\tau = 0$) and cut the second model along the θ parameter. For the third model keep θ fixed (say $\theta = 0$) and cut the model along the τ parameter.
(ii) Describe the curves with respect to τ for θ fixed. Describe the curve with respect to θ for τ fixed. For fixed τ the curve $(x_1(\theta), x_2(\theta), x_3(\theta))$ can be considered as a solution of a differential equation. Find this differential equation. Then τ plays the role of a bifurcation parameter.

Chapter 2

Vector Fields, Lie Series and Lie Algebras

2.1 Notations and Definitions

Consider the autonomous system of first order ordinary differential equations in $\mathbb{R}^n$

$$\frac{du_j}{d\tau} = f_j(\mathbf{u}), \quad \mathbf{u} = (u_1, \ldots, u_n)^T, \; j = 1, \ldots, n$$

where $f_j : \mathbb{R}^n \to \mathbb{R}$ are analytic functions with the initial value $\mathbf{u}(\tau = 0) = \mathbf{u}_0$ at $\tau = 0$. Then, for sufficiently small τ the solution of the initial value problem is given by the *Lie series*

$$u_j(\tau) = \exp(\tau V) u_j|_{\mathbf{u}=\mathbf{u}_0}, \quad j = 1, \ldots, n$$

with the *vector field*

$$V = \sum_{j=1}^{n} f_j(\mathbf{u}) \frac{\partial}{\partial u_j}.$$

We have

$$V(c_1 g_1(\mathbf{u}) + c_2 g_2(\mathbf{u})) = c_1 V(g_1(\mathbf{u})) + c_2 V(g_2(\mathbf{u}))$$

$$V(g_1(\mathbf{u})g_2(\mathbf{u})) = (V(g_1(\mathbf{u}))g_2(\mathbf{u}) + g_1(\mathbf{u})V(g_2(\mathbf{u}))$$
$$e^{\tau V}\mathbf{g}(\mathbf{u}) = \mathbf{g}(e^{\tau V}\mathbf{u}).$$

Expanding the exponential function yields

$$u_j(\tau) = u_j(0) + \tau V(u_j)|_{\mathbf{u}=\mathbf{u}(0)} + \frac{\tau^2}{2}V(V(u_j))|_{\mathbf{u}=\mathbf{u}(0)} + \cdots.$$

This method is called the *Lie series technique*.

Given two analytic vector fields

$$V = \sum_{j=1}^{n} f_j(\mathbf{u})\frac{\partial}{\partial u_j}, \qquad W = \sum_{k=1}^{n} g_k(\mathbf{u})\frac{\partial}{\partial u_k}$$

then the commutator of the two vector fields is given by the analytic vector field

$$[V, W] = \sum_{j=1}^{n}\left(\sum_{k=1}^{n} f_k\frac{\partial g_j}{\partial u_k} - g_k\frac{\partial f_j}{\partial u_k}\right)\frac{\partial}{\partial u_j}.$$

The smooth vector fields in $\mathbb{R}^n$ form a Lie algebra with the *Jacobi identity* given by

$$[V, [W, X]] + [X, [V, W]] + [W, [X, V]] = 0$$

for all vector fields V, W, X.

Let $g : \mathbb{R}^n \to \mathbb{R}$ be a smooth function. Then g is called a *first integral* of the dynamical system given by the vector field V if

$$V(g) \equiv L_V(g) = \sum_{j=1}^{n} V_j\frac{\partial g}{\partial x_j} = 0$$

where $L_V(.)$ denotes the Lie derivative. Two first integrals for a given dynamical system are independent if there is at least one point $x_0 \in \mathbb{R}$ such that their gradients are linearly independent. If we have r first integrals $I_1, \ldots, I_r$ then the condition is

$$\text{rank}(\nabla I_1(x_0), \nabla I_2(x_0), \ldots, \nabla I_r(x_0)) = r.$$

The *fixed points* of the autonomous system of first order differential equations are the solutions of the equations

$$f_j(\mathbf{u}) = 0, \quad j = 1, \ldots, n.$$

2.2   Solved Problems

Problem 1.   Let $c \in \mathbb{R}$. Calculate

$$\exp(cd/du)u, \quad \exp(cd/du)u^2, \quad \exp(cd/du)f(u)$$

where $f : \mathbb{R} \to \mathbb{R}$ is an analytic function.

Solution 1.   We have

$$\exp(cd/du)u = \left(1 + \frac{c}{1!}\frac{d}{du} + \frac{c^2}{2!}\frac{d^2}{du^2} + \frac{c^3}{3!}\frac{d^3}{du^3} + \cdots\right)u = u + c$$

and

$$\exp(cd/du)u^2 = (e^{cd/du}u)^2 = (u + c)^2 = u^2 + 2cu + c^2$$
$$\exp(cd/du)f(u) = f(e^{cd/du}u) = f(u + c).$$

Problem 2.   Let $c \in \mathbb{R}$. Calculate

$$\exp(cud/du)u, \quad \exp(cud/du)u^2, \quad \exp(cud/du)f(u)$$

where $f : \mathbb{R} \to \mathbb{R}$ is an analytic function.

Solution 2.   We have

$$\exp(cud/du)u = \left(1 + \frac{c}{1!}u\frac{d}{du} + \frac{c^2}{2!}u\frac{d}{du}u\frac{d}{du} + \cdots\right)u = e^c u$$

and thus

$$\exp(cud/du)u^2 = (\exp(cud/du)u)(\exp(cud/du)u) = e^c u e^c u = e^{2c}u^2$$
$$\exp(cud/du)f(u) = f(\exp(cud/du)u) = f(e^c u).$$

Problem 3.   Consider the first order linear ordinary differential equation

$$\frac{du}{d\tau} = u + 1$$

with the corresponding vector field

$$V = (u + 1)\frac{d}{du}.$$

Calculate the map
$$u \mapsto \exp(\tau V)u.$$

Solve the initial value problem of the differential equation and compare. Find the *fixed points* of the differential equation, i.e. the solutions of $u + 1 = 0$.

Solution 3.   We have the expansion

$$e^{\tau V} u = \left(1 + \frac{\tau}{1!}V + \frac{\tau^2}{2!}V^2 + \frac{\tau^3}{3!}V^3 + \cdots\right)u$$

$$= \left(1 + \tau(u+1)\frac{d}{du} + \frac{1}{2!}\tau^2(u+1)\frac{d}{du}(u+1)\frac{d}{du} + \cdots\right)u$$

$$= u + \tau(u+1) + \frac{1}{2!}\tau^2(u+1) + \cdots$$

$$= u\left(1 + \tau + \frac{\tau^2}{2!} + \cdots\right) + \left(\tau + \frac{\tau^2}{2!} + \cdots\right)$$

$$= ue^{\tau} + e^{\tau} - 1$$

$$= (u+1)e^{\tau} - 1.$$

The solution of the linear differential equation is given by $u(\tau) = Ce^{\tau} - 1$, where C is the constant of integration. With $u(\tau = 0) = C - 1$ we have $C = u(0) + 1$. Thus the solution of the initial value problem is

$$u(\tau) = (u(0) + 1)e^{\tau} - 1$$

which is the map given above. The fixed points are the solutions of $u^* + 1 = 0$. Thus there is only one fixed point given by $u^* = -1$.

Problem 4.   Consider the three vector fields in $\mathbb{R}$

$$V_1 = \frac{d}{du}, \quad V_2 = u\frac{d}{du}, \quad V_3 = u^2\frac{d}{du}.$$

(i) Calculate the commutators $[V_1, V_2]$, $[V_1, V_3]$, $[V_2, V_3]$ and thus show that the vector fields form a basis of a Lie algebra.
(ii) Calculate
$$e^{\tau V_1} f(u), \quad e^{\tau V_2} f(u), \quad e^{\tau V_3} f(u)$$
where $f : \mathbb{R} \to \mathbb{R}$ is an analytic function.

Solution 4.   (i) We find the nonzero commutators

$$[V_1, V_2] = \left[\frac{d}{du}, u\frac{d}{du}\right] = \frac{d}{du} = V_1$$

$$[V_1, V_3] = \left[\frac{d}{du}, u^2 \frac{d}{du}\right] = 2u \frac{d}{du} = 2V_2$$

$$[V_2, V_3] = \left[u \frac{d}{du}, u^2 \frac{d}{du}\right] = u^2 \frac{d}{du} = V_3.$$

Thus the set is closed under the commutator. Since all (analytic) vector fields form a Lie algebra we have a basis for a Lie subalgebra.
(ii) We obtain

$$e^{\tau V_1} f(u) = f(e^{\tau V_1} u) = f(u + \tau)$$
$$e^{\tau V_2} f(u) = f(e^{\tau V_2} u) = f(e^\tau u)$$
$$e^{\tau V_3} f(u) = f(e^{\tau V_3} u) = f(u/(1 - \tau u)).$$

Problem 5. Consider the nonlinear differential equation

$$\frac{du}{d\tau} = \sin(u)$$

with the initial value $u(\tau = 0) = u_0 = \pi/2$.
(i) Find the fixed points.
(ii) Solve the differential equation by direct integration. Hint.

$$\int \frac{du}{\sin(cu)} = \frac{1}{c} \ln\left(\tan\left(\frac{cu}{2}\right)\right).$$

What happens if $\tau \to \infty$?
(iii) Find the solution of the initial value problem using the *Lie series expansion*

$$u(\tau) = \exp\left(\tau \sin(u) \frac{d}{du}\right) u\Big|_{u=u_0}.$$

Solution 5. (i) From the equation $\sin(u^*) = 0$ it follows that the fixed points are given by $u^* = n\pi$, where $n \in \mathbb{Z}$.
(ii) Integrating the differential equation yields

$$\tan\left(\frac{u(\tau)}{2}\right) = e^\tau e^C.$$

For $\tau = 0$ we have $e^C = 1$ since $\tan(\pi/4) = 1$. Thus $u(\tau) = 2 \arctan(e^\tau)$. For $\tau \to \infty$ we obtain $u(\infty) = \pi$ which is a fixed point of the differential equation.

(iii) We have

$$\sin(u)\frac{d}{du}u = \sin(u)$$

$$\sin(u)\frac{d}{du}\sin(u)\frac{d}{du}u = \sin(u)\cos(u)$$

$$\sin(u)\frac{d}{du}\sin(u)\frac{d}{du}\sin(u)\frac{d}{du}u = \sin(u)\cos(2u)$$

where we used the identity $\cos^2(u) - \sin^2(u) \equiv \cos(2u)$. Thus we obtain the expansion

$$u(\tau) = u_0 + \tau\sin(u_0) + \frac{\tau^2}{2}\sin(u_0)\cos(u_0) + \frac{\tau^3}{3!}\sin(u_0)\cos(2u_0) + \cdots$$

With $u_0 = \pi/2$ we arrive at (since $\cos(\pi/2) = 0$)

$$u(\tau) = \frac{\pi}{2} + \tau - \frac{\tau^3}{3!} + \cdots$$

Note that for $u \geq 1$ we have the expansion

$$\arctan(u) = \frac{\pi}{2} - \frac{1}{u} + \frac{1}{3u^3} - \frac{1}{5u^5} + \cdots$$

Problem 6. Consider the vector field d/dx and the differential form dx in $\mathbb{R}^1$. Note that $d/dx \rfloor dx = 1$, where $\rfloor$ denotes the *interior product*. Find the vector field $f_*(d/dx)$ and the differential form $f^*(dx)$ with $f(x) = \sinh(x)$.

Solution 6. Let $u'(x'(x)) = u(x)$ and $x'(x) = \sinh(x)$. Then

$$\frac{du'}{dx} = \frac{du'}{dx'}\frac{dx'}{dx} = \frac{du}{dx} \Rightarrow \cosh(x)\frac{du'}{dx'} = \frac{du}{dx}.$$

Thus $f_*(d/dx) = \frac{1}{\cosh(x)}d/dx$. We have

$$f^*(dx) = df = \cosh(x)dx, \qquad (f_*(d/dx))\rfloor f^*(dx) = 1.$$

Problem 7. Consider the nonlinear differential equation

$$\frac{du}{d\tau} = -u^2 \tag{1}$$

with the initial condition $u(\tau = 0) \equiv u_0 = 1$ and $\tau \in [0, \infty)$.
(i) Show that the exact solution of the initial value problem is given by

$$u(\tau) = \frac{1}{1 + \tau}. \tag{2}$$

(ii) Solve the initial value problem of the differential equation (1) with the help of Lie series.
(iii) What are the fixed points of the differential equation?

Solution 7.   (i) We have

$$\frac{du}{u^2} = -d\tau \quad \Rightarrow \quad -\frac{1}{u} = -\tau + C$$

and with $u(0) = 1$ we obtain $u(\tau) = 1/(1 + \tau)$.
(ii) The right hand side of (1) corresponds to the vector field

$$V = -u^2 \frac{d}{du}.$$

For sufficiently small τ the solution of (1) is given by

$$u(\tau) = \exp(\tau V) u|_{u \to u_0}.$$

We expand $\exp(\tau V)$ with respect to τ and obtain

$$\exp(\tau V) u = (1 + \tau V + \frac{\tau^2}{2!} V^2 + \frac{\tau^3}{3!} V^3 + \cdots) u$$

where $V^2(u) := V(V(u))$, $V^3(u) := V(V(V(u)))$ and so on. We find

$$Vu = -u^2, \quad V^2 u = V(Vu) = V(-u^2) = 2u^3.$$

In general we have $V^n u = (n!)(-1)^n u^{n+1}$. Thus

$$\exp(\tau V) u = \sum_{n=1}^{\infty} (-1)^{n+1} \tau^{n-1} u^n.$$

Therefore

$$\exp(\tau V) u|_{u \to 1} = \sum_{n=1}^{\infty} (-1)^{n+1} \tau^{n-1}.$$

Now

$$\frac{1}{1 + \tau} = 1 - \tau + \tau^2 - \tau^3 + \cdots \qquad -1 < \tau < 1.$$

(iii) From $-u^2 = 0$ we obtain the only fixed point $u^* = 0$. The solution given above tends to this fixed point for $\tau \to \infty$.

Problem 8. Consider the initial value problem of the nonlinear differential equation

$$\frac{du}{d\tau} = u - u^2, \qquad u(\tau = 0) = u_0 > 0.$$

The fixed points are given by $u^* = 0$ and $u^* = 1$.
(i) Solve the differential equation by direct integration. Find $u(\tau)$ for $\tau \to \infty$.
(ii) Solve the differential equation using the Lie series

$$u(\tau) = e^{\tau V} u\big|_{u \to u_0}$$

where V is the *vector field* V associated with the differential equation

$$V = (u - u^2)\frac{d}{du}.$$

Solution 8. (i) We have

$$\int_{u_0}^{u} \frac{du}{u(1 - u)} = \int_0^{\tau} d\tau$$

$$(\ln(u) - \ln(u - 1))|_{u_0}^{u} = \tau$$

$$\ln\left(\frac{u}{u - 1}\right) - \ln\left(\frac{u_0}{u_0 - 1}\right) = \tau$$

$$\ln\left(\frac{u(u_0 - 1)}{u_0(u - 1)}\right) = \tau$$

$$\frac{u}{u - 1} = \frac{u_0}{u_0 - 1}e^{\tau}.$$

Thus

$$u(\tau) = \frac{u_0 e^{\tau}}{u_0 e^{\tau} - u_0 + 1}.$$

Since $u_0 > 0$ we obtain $u(\tau \to \infty) = 1$ which is a fixed point of the differential equation.
(ii) We have

$$(u - u^2)\frac{d}{du}u = u - u^2$$

$$(u - u^2)\frac{d}{du}(u - u^2)\frac{d}{du}u = u - 3u^2 + 2u^3$$

$$u(\tau) = u_0 + \tau(u_0 - u_0^2) + \frac{\tau^2}{2!}(u_0 - 3u_0^2 + 2u_0^3) + \cdots$$

$$= u_0(1 + \tau(1 - u_0) + \frac{\tau^2}{2!}(1 - 3u_0 + 2u_0^2) + \cdots).$$

Problem 9.   (i) Solve the initial value problem $(u(\tau = 0) = u_0 > 0)$ of the differential equation

$$\frac{du}{d\tau} = u^3$$

using the Lie series technique. The only fixed point is $u^* = 0$.
(ii) Solve the initial value problem by direct integration of the differential equation.

Solution 9.   (i) We have

$$\exp\left(\tau u^3 \frac{d}{du}\right) u = u + \tau u^3 + \frac{3}{2!}\tau^2 u^5 + \frac{5 \cdot 3}{3!}\tau^3 u^7 + \cdots$$

$$= u(1 + \tau u^2 + \frac{3}{2!}\tau^2 u^4 + \frac{5 \cdot 3}{3!}\tau^3 u^6 + \cdots)$$

$$= u(1 + (\tau u^2) + \frac{3}{2!}(\tau u^2)^2 + \frac{5 \cdot 3}{3!}(\tau u^2)^3 + \cdots)$$

$$= u(1 - 2(\tau u^2))^{-1/2} = \frac{u}{\sqrt{1 - 2\tau u^2}}$$

where we used the Taylor expansion of $(1 - 2x)^{-1/2}$ for $|2x| < 1$

$$(1 - 2x)^{-1/2} = 1 + x + \frac{3x^2}{2!} + \frac{15x^3}{3!} + \cdots.$$

Hence the solution of the initial value problem is given by

$$u(\tau) = \exp\left(\tau u^3 \frac{d}{du}\right) u \Big|_{u=u_0} = \frac{u_0}{\sqrt{1 - 2\tau u_0^2}}.$$

(ii) Direct integration yields

$$\frac{du}{d\tau} = u^3 \Rightarrow \int u^{-3} du = \int d\tau \Rightarrow -\frac{u^{-2}}{2} = \tau + C.$$

Then

$$\tau = 0 \Rightarrow u = u_0 \Rightarrow C = -\frac{u_0^{-2}}{2}.$$

Thus $u^{-2} = -2\tau + u_0^{-2}$ and finally

$$u(\tau) = \frac{u_0}{\sqrt{1 - 2\tau u_0^2}}.$$

Problem 10.   Consider the nonlinear differential equations

$$\frac{du}{d\tau} = u^2 - u, \qquad \frac{du}{d\tau} = -\sin(u)$$

with the corresponding vector fields

$$V = (u^2 - u)\frac{d}{du}, \qquad W = -\sin(u)\frac{d}{du}.$$

(i) Show that both differential equations admit the fixed point $u^* = 0$.
(ii) Consider the vector field given by the commutator of the two vector fields V and W, i.e. $[V, W]$. Show that the corresponding differential equation of this vector field also admits the fixed point $u^* = 0$.

Solution 10.   (i) We have to solve $u^* - (u^*)^2 = 0$ and $\sin(u^*) = 0$. Both equations admit $u^* = 0$ as solution.
(ii) The commutator yields the vector field

$$[V, W] = ((-u^2 + u)\cos(u) + (2u - 1)\sin(u))\frac{d}{du}$$

with the corresponding differential equation

$$\frac{du}{d\tau} = (-u^2 + u)\cos(u) + (2u - 1)\sin(u).$$

Obviously

$$(-(u^*)^2 + u^*)\cos(u^*) + (2u^* - 1)\sin(u^*) = 0$$

also admits $u^* = 0$ as a solution. Study the stability of the fixed point.

Problem 11.   Consider the initial value problem of the differential equation

$$\frac{du}{d\tau} = \frac{1}{2u}, \qquad u(\tau = 0) = u_0 = 1.$$

Use the Lie series technique to solve this differential equation.

Solution 11.   With the vector field

$$V = \frac{1}{2u}\frac{d}{du}$$

we have

$$e^{\tau V}u = (1 + \tau V + \frac{\tau^2}{2!}V^2 + \frac{\tau^3}{3!}V^3 + \cdots)u = u + \frac{\tau}{2}\frac{1}{u} - \frac{\tau^2}{8}\left(\frac{1}{u^3}\right) + \cdots$$

it follows that

$$u(\tau) = u_0 + \frac{\tau}{2}\frac{1}{u_0} - \frac{\tau^2}{8}\left(\frac{1}{u_0^3}\right) + \cdots$$

With $u_0 = 1$ we arrive at

$$u(\tau) = 1 + \frac{1}{2}\tau - \frac{1}{8}\tau^2 + \cdots .$$

This is the *Taylor expansion* of $\sqrt{1+\tau}$ around 1.

Problem 12.   Let $a > 0$, $b > 0$. Consider the differential equation

$$\frac{du}{d\tau} = b(u^2 + a^2)$$

with the vector field

$$V = b(u^2 + a^2)\frac{d}{du}.$$

The differential equation admits no fixed points. Solve the initial value problem with $u(\tau = 0) = 0$. The Lie series is given by

$$u(\tau) = e^{\tau V}u\big|_{u \to 0}.$$

Solution 12.   Direct integration provides

$$u(\tau) = a\tan(an\tau + \arctan(u(0)/a))$$

and with $u(0) = 0$ (since $\arctan(0) = 0$) we arrive at $u(\tau) = a\tan(ab\tau)$. Note that

$$\tan(y) = y + \frac{y^3}{3} + \frac{2y^5}{15} + \frac{17y^7}{315} + \cdots + \frac{2^{2n}(2^{2n} - 1)B_n y^{2n-1}}{(2n)!} + \cdots$$

where $|y| < \pi/2$ and B_n are the *Bernoulli numbers*. The Lie series is

$$u(\tau) = e^{\tau V}u\big|_{u \to 0} = a\tan(ab\tau).$$

Problem 13.   Consider the initial value problem of the ordinary differential equation

$$\frac{du}{dx} = 1 + x - u.$$

We set $u(x) = u_1(x)$ and $u_2(x) = x$ with $u_2(0) = u_{20} = 0$. Then we obtain the autonomous system of differential equations

$$\frac{du_1}{dx} = 1 + u_2 - u_1, \qquad \frac{du_2}{dx} = 1$$

with the vector field

$$V = (1 + u_2 - u_1)\frac{\partial}{\partial u_1} + \frac{\partial}{\partial u_2}$$

and $u_1(0) = u_{10}$, $u_2(x = 0) = 0$. Find

$$\begin{pmatrix} u_1(x) \\ u_2(x) \end{pmatrix} = e^{xV}\begin{pmatrix} u_1 \\ u_2 \end{pmatrix}\Bigg|_{u_1 \mapsto u_{10}, u_2 \mapsto 0}.$$

Solution 13.   We find $u_1(x) = x + u_{10}e^{-x}$ and $u_2(x) = x$.

Problem 14.   Consider the autonomous system of first order differential equations

$$\frac{du_1}{d\tau} = -u_1 + u_2 + 2, \qquad \frac{du_2}{d\tau} = -u_1 - u_2$$

with the vector field

$$V = (-u_1 + u_2 + 2)\frac{\partial}{\partial u_1} + (-u_1 - u_2)\frac{\partial}{\partial u_2}.$$

(i) Find the *fixed points*.
(ii) Let $u_1(\tau = 0) = u_{1,0} = 2$, $u_2(\tau = 0) = u_{2,0} = 0$ be the initial conditions. Find the curve $(u_1(\tau), u_2(\tau))$ in the plane.
(iii) Find $\lim_{\tau \to \infty} u_1(\tau)$ and $\lim_{\tau \to \infty} u_2(\tau)$.

Solution 14.   (i) The fixed points are the solutions of

$$-u_1^* + u_2^* + 2 = 0, \qquad -u_1^* - u_2^* = 0.$$

The only solution is $u_1^* = 1$, $u_2^* = -1$.
(ii) The solution of the initial value problem is the curve

$$u_1(\tau) = 1 + e^{-\tau}(\cos(\tau) + \sin(\tau)), \qquad u_2(\tau) = -1 + e^{-\tau}(\cos(\tau) - \sin(\tau))$$

which can be obtained from the Lie series

$$\begin{pmatrix} u_1(\tau) \\ u_2(\tau) \end{pmatrix} = e^{\tau V} \begin{pmatrix} u_1 \\ u_2 \end{pmatrix}\Bigg|_{u_1 = u_{1,0}, u_2 = u_{2,0}}.$$

(iii) We find

$$\lim_{\tau \to \infty} u_1(\tau) = 1, \qquad \lim_{\tau \to \infty} u_2(\tau) = -1$$

i.e. the curve tends to the fixed point $(1, -1)$.

Problem 15. Let u be a differentiable function of x_1, x_2, x_3 which satisfies the partial differential equation

$$(x_2 - x_3)\frac{\partial u}{\partial x_1} + (x_3 - x_1)\frac{\partial u}{\partial x_2} + (x_1 - x_2)\frac{\partial u}{\partial x_3} = 0$$

i.e. we have the vector field in $\mathbb{R}^3$

$$V = (x_2 - x_3)\frac{\partial}{\partial x_1} + (x_3 - x_1)\frac{\partial}{\partial x_2} + (x_1 - x_2)\frac{\partial}{\partial x_3}.$$

Show that u contains x_1, x_2, x_3 only in combinations $x_1 + x_2 + x_3$ and $x_1^2 + x_2^2 + x_3^2$.

Solution 15. The *auxiliary equations* of the partial differential equation are

$$\frac{dx_1}{x_2 - x_3} = \frac{dx_2}{x_3 - x_1} = \frac{dx_3}{x_1 - x_2} = \frac{du}{0}.$$

They are equivalent to the three differential relations

$$du = 0, \qquad dx_1 + dx_2 + dx_3 = 0, \qquad x_1 dx_1 + x_2 dx_2 + x_3 dx_3 = 0.$$

Thus the integrals are $u = c_1$, $x_1 + x_2 + x_3 = c_2$, $x_1^2 + x_2^2 + x_3^2 = c_3$, where c_1, c_2, c_3 are constants. Therefore the general solution is given by

$$u(x_1, x_2, x_3) = f(x_1 + x_2 + x_3, x_1^2 + x_2^2 + x_3^2)$$

where f is an arbitrary differentiable function.

Problem 16. (i) Solve the first order partial differential equation

$$F\left(u, \frac{\partial u}{\partial t}, \frac{\partial u}{\partial x}\right) \equiv \frac{\partial u}{\partial t} + u\frac{\partial u}{\partial x} = 0 \tag{1}$$

with the initial condition

$$u(x, t = 0) = x. \tag{2}$$

(ii) Show that the nonlinear partial differential equation

$$\frac{\partial u}{\partial t} + (\alpha + \beta u)\frac{\partial u}{\partial x} = 0 \tag{3}$$

admits the general solution

$$u(x,t) = f(x - (\alpha + \beta u(x,t))t) \tag{4}$$

to the initial value problem, where $u(x, t = 0) = f(x)$ and $\alpha, \beta \in \mathbb{R}$. Simplify case (ii) to case (i).

Solution 16.   (i) From (1) we obtain the surface

$$F(u, p, q) \equiv p + uq = 0.$$

Then we obtain the autonomous system of first-order ordinary differential equations

$$\frac{dt}{ds} = \frac{\partial F}{\partial p} = 1$$

$$\frac{dx}{ds} = \frac{\partial F}{\partial q} = u$$

$$\frac{du}{ds} = p\frac{\partial F}{\partial p} + q\frac{\partial F}{\partial q} = p + uq = 0$$

where we have used $F(u, p, q) \equiv p + uq$. The corresponding vector field is

$$V = \frac{\partial}{\partial t} + u\frac{\partial}{\partial x} + 0\frac{\partial}{\partial u} = \frac{\partial}{\partial t} + u\frac{\partial}{\partial x}.$$

This autonomous system of differential equations determines the *characteristic strip*. Applying Lie series we find the solution of the initial value problem of this autonomous system of differential equations

$$t(s) = s + t_0, \quad x(s) = u_0 s + x_0, \quad u(s) = u_0.$$

To impose the initial condition (2) we have to set

$$x_0(\alpha) = \alpha, \quad t_0(\alpha) = 0, \quad u_0(\alpha) = \alpha.$$

Thus we obtain as solution of the autonomous system

$$x(s, \alpha) = \alpha s + \alpha, \quad t(s, \alpha) = s, \quad u(s, \alpha) = \alpha.$$

Since

$$D := \det \begin{pmatrix} \partial t/\partial s & \partial t/\partial \alpha \\ \partial x/\partial s & \partial x/\partial \alpha \end{pmatrix} = \det \begin{pmatrix} 1 & 0 \\ \alpha & s+1 \end{pmatrix} = 1 + s$$

we can solve these equations with respect to s and α if $D \neq 0$. We find

$$s(x,t) = t, \quad \alpha(x,t) = \frac{x}{1+t}.$$

Inserting s and α into $u(s,\alpha) = \alpha$ gives the solution $u(x,t) = x/(1+t)$ of the initial value problem.

(ii) Since

$$\frac{\partial u}{\partial t} = \left(-\alpha - \beta u - \beta t \frac{\partial u}{\partial t} \right) f', \qquad \frac{\partial u}{\partial x} = \left(1 - \beta t \frac{\partial u}{\partial x} \right) f'$$

where f' is the derivative of f with respect to the argument we obtain

$$\left(1 - \beta t \frac{\partial u}{\partial x} \right) \frac{\partial u}{\partial t} = \left(-\alpha - \beta u - \beta t \frac{\partial u}{\partial t} \right) \frac{\partial u}{\partial x}$$

or

$$\frac{\partial u}{\partial t} = (-\alpha - \beta u) \frac{\partial u}{\partial x}$$

which is (3). Let us now simplify (3) to case (i). Let $\alpha = 0$ and $\beta = 1$. Then we obtain from (4) that

$$u(x,t) = f(x - u(x,t)t).$$

Since $u(x, t = 0) = f(x) = x$ it follows that

$$u(x,t) = f(x - u(x,t)t) = x - u(x,t)t.$$

From this solution we see that the solution $u(x,t) = x/(1+t)$ follows.

Problem 17. Consider the vector fields

$$V = x_1 \frac{\partial}{\partial x_1} + x_2 \frac{\partial}{\partial x_2}, \qquad W = x_1 \frac{\partial}{\partial x_2} - x_2 \frac{\partial}{\partial x_1}$$

defined on $\mathbb{R}^2$.

(i) Do the vector fields V, W form a basis of a Lie algebra? If so, what type of Lie algebra do we have.

(ii) Express the two vector fields in *polar coordinates* $x_1(r,\theta) = r\cos(\theta)$, $x_2(r,\theta) = r\sin(\theta)$.

(iii) Calculate the commutator of the two vector fields expressed in polar coordinates. Compare with the result of (i).

Solution 17.   (i) Calculating the commutator we find $[V, W] = 0$. Thus we have a basis of a commutative Lie algebra.
(ii) Applying the *chain rule* to $f(x_1(r, \theta), x_2(r, \theta))$ we obtain

$$\frac{\partial f}{\partial r} = \frac{\partial f}{\partial x_1}\frac{\partial x_1}{\partial r} + \frac{\partial f}{\partial x_2}\frac{\partial x_2}{\partial r} = \frac{\partial f}{\partial x_1}\cos(\theta) + \frac{\partial f}{\partial x_2}\sin(\theta)$$

$$\frac{\partial f}{\partial \theta} = \frac{\partial f}{\partial x_1}\frac{\partial x_1}{\partial \theta} + \frac{\partial f}{\partial x_2}\frac{\partial x_2}{\partial \theta} = -\frac{\partial f}{\partial x_1}r\sin(\theta) + \frac{\partial f}{\partial x_2}r\cos(\theta).$$

Thus

$$\frac{\partial f}{\partial r}r\sin(\theta) = \frac{\partial f}{\partial x}r\cos(\theta)\sin(\theta) + \frac{\partial f}{\partial x_2}r\sin^2(\theta)$$

$$\frac{\partial f}{\partial \theta}\cos(\theta) = -\frac{\partial f}{\partial x_1}r\sin(\theta)\cos(\theta) + \frac{\partial f}{\partial x_2}r\cos^2(\theta).$$

Therefore

$$\frac{\partial}{\partial x_2} = \sin(\theta)\frac{\partial}{\partial r} + \frac{\cos(\theta)}{r}\frac{\partial}{\partial \theta}.$$

Analogously

$$-\frac{\partial f}{\partial r}r\cos(\theta) = -\frac{\partial f}{\partial x_1}r\cos^2(\theta) - \frac{\partial f}{\partial x_2}r\sin(\theta)\cos(\theta)$$

$$\frac{\partial f}{\partial \theta}\sin(\theta) = -\frac{\partial f}{\partial x_1}r\sin^2(\theta) + \frac{\partial f}{\partial x_2}r\cos(\theta)\sin(\theta).$$

Hence

$$\frac{\partial}{\partial x_1} = \cos(\theta)\frac{\partial}{\partial r} - \frac{\sin(\theta)}{r}\frac{\partial}{\partial \theta}.$$

Thus the vector field V takes the form $r\partial/\partial r$ and the vector field W takes the form $\partial/\partial\theta$.
(iii) For the commutator we find $[r\partial/\partial r, \partial/\partial\theta] = 0$. Thus under the transformation to polar coordinates the commutator is preserved.

Problem 18.   (i) Consider the vector field $V = x_1\partial/\partial x_1 + x_2\partial/\partial x_2$ in $\mathbb{R}^2$ and

$$V(u(x_1, x_2)) = x_1\frac{\partial u}{\partial x_1} + x_2\frac{\partial u}{\partial x_2}.$$

Apply the transformation

$$\tilde{u}(r(x_1, x_2), \phi(x_1, x_2)) = u(x_1, x_2), \quad x_1(r, \phi) = r\cos(\phi), \quad x_2(r, \phi) = r\sin(\phi)$$

and thus find how the vector field V transforms.

(ii) Consider the vector field $W = x_1 \partial/\partial x_2 - x_2 \partial/\partial x_1$ in $\mathbb{R}^2$ and

$$W(u(x_1, x_2)) = x_1 \frac{\partial u}{\partial x_2} - x_2 \frac{\partial u}{\partial x_1}.$$

Apply the transformation

$$\widetilde{u}(r(x_1, x_2), \phi(x_1, x_2)) = u(x_1, x_2), \quad x_1(r, \phi) = r \cos(\phi), \quad x_2(r, \phi) = r \sin(\phi)$$

and thus find how the vector field W transforms. Utilize the calculations from (i).

(iii) Find the commutator $[V, W]$ and then the commutator of the transformed vector fields.

Solution 18.   (i) Applying the chain rule provides

$$\frac{\partial \widetilde{u}}{\partial x_1} = \frac{\partial \widetilde{u}}{\partial r} \frac{\partial r}{\partial x_1} + \frac{\partial \widetilde{u}}{\partial \phi} \frac{\partial \phi}{\partial x_1} = \frac{\partial u}{\partial x_1}$$

$$\frac{\partial \widetilde{u}}{\partial x_2} = \frac{\partial \widetilde{u}}{\partial r} \frac{\partial r}{\partial x_2} + \frac{\partial \widetilde{u}}{\partial \phi} \frac{\partial \phi}{\partial x_2} = \frac{\partial u}{\partial x_2}.$$

With $r^2(x_1, x_2) = x_1^2 + x_2^2$ and $\phi(x_1, x_2) = \arctan(x_2/x_1)$ we obtain

$$\frac{\partial r}{\partial x_1} = \frac{x_1}{r}, \quad \frac{\partial r}{\partial x_2} = \frac{x_2}{r}$$

$$\frac{\partial \phi}{\partial x_1} = -\frac{x_2}{x_1^2 + x_2^2} = \frac{-\sin(\phi)}{r}, \quad \frac{\partial \phi}{\partial x_2} = \frac{x_1}{x_1^2 + x_2^2} = \frac{\cos(\phi)}{r}.$$

It follows that

$$\frac{\partial u}{\partial x_1} = \frac{\partial \widetilde{u}}{\partial r} \frac{x_1}{r} + \frac{\partial \widetilde{u}}{\partial \phi} \frac{(-\sin(\phi))}{r}, \quad \frac{\partial u}{\partial x_2} = \frac{\partial \widetilde{u}}{\partial r} \frac{x_2}{r} + \frac{\partial \widetilde{u}}{\partial \phi} \frac{(\cos(\phi))}{r}$$

and $V(u)$ takes the form $r \partial \widetilde{u}/\partial r$. Thus the vector field $V = x_1 \partial/\partial x_1 + x_2 \partial/\partial x_2$ transform into $r \partial/\partial r$.

(ii) Utilizing the results from (i) we obtain that $W(u)$ takes the form $\partial \widetilde{u}/\partial \phi$ and the vector field $W = x_1 \partial/\partial x_2 - x_2 \partial x_1$ transform to $\partial/\partial \phi$.

(iii) We have $[V, W] = 0$, $[r \partial/\partial r, \partial/\partial \phi] = 0$. We could also consider the transformation

$$u'(x_1(r, \phi), x_2(r, \phi)) = u(r, \phi), \quad x_1(r, \phi) = r \cos(\phi), \quad x_2(r, \phi) = r \sin(\phi)$$

and do the calculations as follows. We obtain the two equations

$$\frac{\partial u'}{\partial r} = \frac{\partial u'}{\partial x_1} \frac{\partial x_1}{\partial r} + \frac{\partial u'}{\partial x_2} \frac{\partial x_2}{\partial r} = \cos(\phi) \frac{\partial u'}{\partial x_1} + \sin(\phi) \frac{\partial u'}{\partial x_2} = \frac{\partial u}{\partial r}$$

$$\frac{\partial u'}{\partial \phi} = \frac{\partial u'}{\partial x_1}\frac{\partial x_1}{\partial \phi} + \frac{\partial u'}{\partial x_2}\frac{\partial x_2}{\partial \phi} = -r\sin(\phi)\frac{\partial u'}{\partial x_1} + r\cos(\phi)\frac{\partial u'}{\partial x_2} = \frac{\partial u}{\partial \phi}.$$

Multiplying the first equation with $r\cos(\phi)$ and the second equation with $-\sin(\phi)$ and adding yields

$$\frac{\partial u'}{\partial x_1} = \cos(\phi)\frac{\partial u}{\partial r} - \frac{1}{r}\sin(\phi)\frac{\partial u}{\partial \phi}.$$

Multiplying the first equation with $r\sin(\phi)$ and the second equation with $\cos(\phi)$ and adding yields

$$\frac{\partial u'}{\partial x_2} = \sin(\phi)\frac{\partial u}{\partial r} + \frac{1}{r}\cos(\phi)\frac{\partial u}{\partial \phi}.$$

Consequently

$$x_1\frac{\partial}{\partial x_2} - x_2\frac{\partial}{\partial x_1} \mapsto \frac{\partial}{\partial \phi}.$$

Analogously we have

$$x_1\frac{\partial}{\partial x_1} + x_2\frac{\partial}{\partial x_2} \mapsto r\frac{\partial}{\partial r}.$$

Note that $[\partial/\partial\phi, r\partial/\partial r] = 0$.

Problem 19.   Consider the vector fields (differential operators)

$$E = x\frac{\partial}{\partial y}, \quad F = y\frac{\partial}{\partial x}, \quad H = x\frac{\partial}{\partial x} - y\frac{\partial}{\partial y}.$$

Show that these vector fields form a basis of a Lie algebra, i.e. calculate the commutators. Consider the basis for $n \in \mathbb{Z}$

$$\{\, x^j y^k \,:\, j, k \in \mathbb{Z}, \, j + k = n \,\}.$$

Find $E(x^j y^k)$, $F(x^j y^k)$, $H(x^j y^k)$.

Solution 19.   We find $[E, F] = H$, $[H, E] = 2E$, $[H, F] = -2F$. We obtain

$$E(x^j y^k) = k x^{j+1} y^{k-1}, \quad F(x^j y^k) = j x^{j-1} y^{k+1}, \quad H(x^j y^k) = (j-k) x^j y^k.$$

Problem 20.   Consider the Lie algebra $o(3,2)$. The ten vector fields

$$V_1 = \frac{\partial}{\partial t}, \quad V_2 = t\frac{\partial}{\partial t} + \frac{1}{2}x\frac{\partial}{\partial x}, \quad V_3 = t^2\frac{\partial}{\partial t} + tx\frac{\partial}{\partial x} + \frac{1}{4}x^2\frac{\partial}{\partial u},$$

$$V_4 = \frac{\partial}{\partial x}, \quad V_5 = t\frac{\partial}{\partial x} + \frac{1}{2}x\frac{\partial}{\partial u}, \quad V_6 = \frac{\partial}{\partial u}$$

$$V_7 = \frac{1}{2}x\frac{\partial}{\partial x} + u\frac{\partial}{\partial u}, \quad V_8 = \frac{1}{2}xt\frac{\partial}{\partial t} + (tu + \frac{1}{4}x^2)\frac{\partial}{\partial x} + \frac{1}{2}xu\frac{\partial}{\partial u}$$

$$V_9 = \frac{1}{4}x^2\frac{\partial}{\partial t} + ux\frac{\partial}{\partial x} + u^2\frac{\partial}{\partial u}, \quad V_{10} = \frac{1}{2}x\frac{\partial}{\partial t} + u\frac{\partial}{\partial x}$$

form a basis of this Lie algebra. Show that the vector fields $V_1, \ldots, V_7$ form a Lie subalgebra.

Solution 20.   For the commutators between the vector fields $V_1, \ldots, V_7$ we find

$$[V_1, V_2] = V_1, \quad [V_1, V_3] = 2V_2, \quad [V_1, V_4] = 0, \quad [V_1, V_5] = V_4, \quad [V_1, V_6] = 0,$$

$$[V_1, V_7] = 0, \quad [V_2, V_3] = V_3, \quad [V_2, V_4] = -\frac{1}{2}V_4, \quad [V_2, V_5] = 2V_2,$$

$$[V_2, V_6] = 0, \quad [V_2, V_7] = 0,$$

$$[V_3, V_4] = V_3, \quad [V_3, V_5] = 0, \quad [V_3, V_6] = 0, \quad [V_3, V_7] = 0.$$

Thus the vector fields $V_1, \ldots, V_7$ form a Lie subalgebra.

Problem 21.   Let $X_1, X_2, \ldots, X_r$ be the basis of a Lie algebra with the commutator

$$[X_i, X_j] = \sum_{k=1}^{r} C_{ij}^k X_k$$

where the C_{ij}^k are the *structure constants*. The structure constants satisfy (*third fundamental theorem*)

$$C_{ij}^k = -C_{ji}^k$$

$$\sum_{m=1}^{r} \left(C_{ij}^m C_{mk}^\ell + C_{jk}^m C_{mi}^\ell + C_{ki}^m C_{mj}^\ell \right) = 0.$$

We replace the X_i's by c-number differential operators (linear vector fields)

$$X_i \mapsto V_i = \sum_{\ell=1}^{r} \sum_{k=1}^{r} x_k C_{i\ell}^k \frac{\partial}{\partial x_\ell}, \quad i = 1, \ldots, r$$

which preserve the commutators.

Consider the Lie algebra with $r = 3$ and the generators X_1, X_2, X_3 and the commutators $[X_1, X_3] = X_1$, $[X_2, X_3] = -X_2$. All other commutators are 0. The Lie algebra is *solvable*. Find the corresponding linear vector fields. Find the smooth function $f : \mathbb{R}^3 \to \mathbb{R}$ such that

$$V_j f(\mathbf{x}) = 0 \qquad \text{for } j = 1, 2, 3.$$

Solution 21.   From the commutation relation we find the structure constants

$$C_{13}^1 = 1, \quad C_{31}^1 = -1, \quad C_{23}^2 = -1, \quad C_{32}^2 = 1.$$

All other structure constants are 0. Thus

$$V_1 = \sum_{\ell=1}^{3} \sum_{k=1}^{3} x_k C_{1\ell}^k \frac{\partial}{\partial x_\ell} = x_1 \frac{\partial}{\partial x_3}$$

$$V_2 = \sum_{\ell=1}^{3} \sum_{k=1}^{3} x_k C_{2\ell}^k \frac{\partial}{\partial x_\ell} = -x_2 \frac{\partial}{\partial x_3}$$

$$V_3 = \sum_{\ell=1}^{3} \sum_{k=1}^{3} x_k C_{3\ell}^k \frac{\partial}{\partial x_\ell} = x_2 \frac{\partial}{\partial x_2} - x_1 \frac{\partial}{\partial x_1}.$$

The function f that is invariant is $f(x_1, x_2, x_3) = x_1 x_2$ since

$$V_1 f(\mathbf{x}) = V_2 f(\mathbf{x}) = V_3 f(\mathbf{x}) = 0.$$

Problem 22.   Consider the smooth vector fields in $\mathbb{R}^2$

$$V_1 = \frac{\partial}{\partial x_1}, \quad V_2 = 2x_1 \frac{\partial}{\partial x_1} + x_2 \frac{\partial}{\partial x_2}, \quad V_3 = x_1^2 \frac{\partial}{\partial x_1} + x_1 x_2 \frac{\partial}{\partial x_2}.$$

Find the commutators and thus show that we have a basis of a Lie algebra.

Solution 22.   We obtain $[V_1, V_2] = 2V_1$, $[V_1, V_3] = V_2$, $[V_2, V_3] = 2V_3$. Thus we have a basis of a Lie algebra, the Lie algebra $s\ell(2, \mathbb{R})$.

Problem 23.   Let V, W be two smooth vector fields

$$V = f_1 \frac{\partial}{\partial u_1} + f_2 \frac{\partial}{\partial u_2} + f_3 \frac{\partial}{\partial u_3}, \quad W = g_1 \frac{\partial}{\partial u_1} + g_2 \frac{\partial}{\partial u_2} + g_3 \frac{\partial}{\partial u_3}$$

defined on $\mathbb{R}^3$. Let $du/d\tau = \mathbf{f}(\mathbf{u})$ and $du/d\tau = \mathbf{g}(\mathbf{u})$ be the corresponding autonomous system of first order differential equations. The fixed points of V are defined by the solutions of the equations $f_j(u_1^*, u_2^*, u_3^*) = 0$ ($j = 1, 2, 3$) and the fixed points of W are defined as the solutions of the equations $g_j(u_1^*, u_2^*, u_3^*) = 0$ ($j = 1, 2, 3$). What can be said about the fixed points of $[V, W]$?

Solution 23.   For the commutator we have

$$[V, W] = \sum_{i=1}^{3} \sum_{j=1}^{3} \left(f_i \frac{\partial g_j}{\partial u_i} \frac{\partial}{\partial u_j} - g_j \frac{\partial f_i}{\partial u_j} \frac{\partial}{\partial u_i} \right).$$

Thus we can write

$$[V, W] = \sum_{n=1}^{3} \sum_{m=1}^{3} \left(f_m \frac{\partial g_n}{\partial u_m} - g_n \frac{\partial f_n}{\partial u_m} \right) \frac{\partial}{\partial u_n}.$$

It follows that the determining equations for the fixed points are

$$f_1 \frac{\partial g_1}{\partial u_1} + f_2 \frac{\partial g_1}{\partial u_2} + f_3 \frac{\partial g_1}{\partial u_3} - g_1 \frac{\partial f_1}{\partial u_1} - g_2 \frac{\partial f_1}{\partial u_2} - g_3 \frac{\partial f_1}{\partial u_3} = 0$$

$$f_1 \frac{\partial g_2}{\partial u_1} + f_2 \frac{\partial g_2}{\partial u_2} + f_3 \frac{\partial g_2}{\partial u_3} - g_1 \frac{\partial f_2}{\partial u_1} - g_2 \frac{\partial f_2}{\partial u_2} - g_3 \frac{\partial f_2}{\partial u_3} = 0$$

$$f_1 \frac{\partial g_3}{\partial u_1} + f_2 \frac{\partial g_3}{\partial u_2} + f_3 \frac{\partial g_3}{\partial u_3} - g_1 \frac{\partial f_3}{\partial u_1} - g_2 \frac{\partial f_3}{\partial u_2} - g_3 \frac{\partial f_3}{\partial u_3} = 0.$$

Thus if V and W have a common fixed point, then the commutator also has this fixed point.

Problem 24.   Consider the vector fields

$$V_1 = (u_2 + u_1 u_3) \frac{\partial}{\partial u_1} + (-u_1 + u_2 u_3) \frac{\partial}{\partial u_2} + (1 + u_3^2) \frac{\partial}{\partial u_3}$$

$$V_2 = (1 + u_1^2) \frac{\partial}{\partial u_1} + (u_1 u_2 + u_3) \frac{\partial}{\partial u_2} + (-u_2 + u_1 u_3) \frac{\partial}{\partial u_3}$$

$$V_3 = (u_1 u_2 - u_3) \frac{\partial}{\partial u_1} + (1 + u_2^2) \frac{\partial}{\partial u_2} + (u_1 + u_2 u_3) \frac{\partial}{\partial u_3}.$$

Find the commutators $[V_1, V_2]$, $[V_2, V_3]$, $[V_3, V_1]$ and thus show that we have a basis of the Lie algebra $so(3, \mathbb{R})$.

Solution 24.   We find $[V_1, V_2] = V_3$, $[V_2, V_3] = V_1$, $[V_3, V_1] = V_2$.

Problem 25.   Let $\{ , \}$ denote the *Poisson bracket*. Consider the functions

$$S_1 = \frac{1}{4}(x_1^2 + p_1^2 - x_2^2 - p_2^2), \quad S_2 = \frac{1}{2}(p_1 p_2 + x_1 x_2), \quad S_3 = \frac{1}{2}(x_1 p_2 - x_2 p_1).$$

Calculate $\{S_1, S_2\}$, $\{S_2, S_3\}$, $\{S_3, S_1\}$ and thus establish that we have a basis of a Lie algebra. Classify the Lie algebra.

Solution 25.   We find $\{S_1, S_2\} = S_3$, $\{S_2, S_3\} = S_1$, $\{S_3, S_1\} = S_2$. Thus we have a representation of the simple Lie algebra $so(3, \mathbb{R})$.

Problem 26.   Consider in $\mathbb{R}^3$ the vector fields

$$V_{12} = x_2 \frac{\partial}{\partial x_1} - x_1 \frac{\partial}{\partial x_2}, \quad V_{23} = x_3 \frac{\partial}{\partial x_2} - x_2 \frac{\partial}{\partial x_3}, \quad V_{31} = x_1 \frac{\partial}{\partial x_3} - x_3 \frac{\partial}{\partial x_1}$$

with the commutators

$$[V_{12}, V_{23}] = V_{31}, \quad [V_{23}, V_{31}] = V_{12}, \quad [V_{31}, V_{12}] = V_{23}.$$

Thus we have a basis of the simple Lie algebra $so(3, \mathbb{R})$.
(i) Find the curl of these vector fields.
(ii) Let $\Omega = dx_1 \wedge dx_2 \wedge dx_3$ be the *volume form* in $\mathbb{R}^3$. Find the differential two-forms

$$V_{12} \rfloor \Omega, \quad V_{23} \rfloor \Omega, \quad V_{31} \rfloor \Omega.$$

(iii) Let $\star$ be the Hodge duality operator. Find the differential one-forms

$$\star(V_{12} \rfloor \Omega), \quad \star(V_{23} \rfloor \Omega), \quad \star(V_{31} \rfloor \Omega).$$

Solution 26.   (i) We find the vector fields

$$2 \frac{\partial}{\partial x_3}, \quad 2 \frac{\partial}{\partial x_1}, \quad 2 \frac{\partial}{\partial x_2}.$$

They form a basis of an abelian Lie algebra.
(ii) We have

$$V_{12} \rfloor (dx_1 \wedge dx_2 \wedge dx_3) = x_2 dx_2 \wedge dx_3 - x_1 dx_3 \wedge dx_1$$
$$V_{23} \rfloor (dx_1 \wedge dx_2 \wedge dx_3) = x_3 dx_3 \wedge dx_1 - x_2 dx_1 \wedge dx_2$$
$$V_{31} \rfloor (dx_1 \wedge dx_2 \wedge dx_3) = x_1 dx_1 \wedge dx_2 - x_3 dx_2 \wedge dx_3.$$

(iii) We obtain the differential one-forms

$$\star(V_{12} \rfloor \Omega) = x_2 dx_1 - x_1 dx_2$$
$$\star(V_{23} \rfloor \Omega) = x_3 dx_2 - x_2 dx_3$$
$$\star(V_{31} \rfloor \Omega) = x_1 dx_3 - x_3 dx_1.$$

Problem 27. Give four different representations of the simple Lie algebra $s\ell(2,\mathbb{R})$ using vector fields V_1, V_2, V_3 which have to satisfy

$$[V_1, V_2] = V_1, \quad [V_2, V_3] = V_3, \quad [V_1, V_3] = 2V_2.$$

Solution 27. A representation in $\mathbb{R}$ is

$$\frac{d}{dx}, \quad x\frac{d}{dx}, \quad x^2\frac{d}{dx}.$$

A representation in $\mathbb{R}^2$ is

$$\frac{\partial}{\partial x_2}, \quad x_1\frac{\partial}{\partial x_1} + x_2\frac{\partial}{\partial x_2}, \quad 2x_1x_2\frac{\partial}{\partial x_1} + x_2^2\frac{\partial}{\partial x_2}.$$

Another representation in $\mathbb{R}^2$ is

$$\frac{\partial}{\partial x_1}, \quad x_1\frac{\partial}{\partial x_1} + x_2\frac{\partial}{\partial x_2}, \quad 2x_1x_2\frac{\partial}{\partial x_1} + (-x_1^2 + x_2^2)\frac{\partial}{\partial x_2}.$$

Finally a fourth representation in $\mathbb{R}^2$ is

$$\frac{\partial}{\partial x_2}, \quad x_1\frac{\partial}{\partial x_1} + x_2\frac{\partial}{\partial x_2}, \quad 2x_1x_2\frac{\partial}{\partial x_1} + (x_1^2 + x_2^2)\frac{\partial}{\partial x_2}.$$

Problem 28. Consider the vector fields

$$V_1 = \frac{\partial}{\partial x_0}, \quad V_2 = x_0\frac{\partial}{\partial x_0} + \frac{2}{3}\left(x_1\frac{\partial}{\partial x_1} + x_2\frac{\partial}{\partial x_2} + x_3\frac{\partial}{\partial x_3}\right),$$

$$V_3 = x_3\frac{\partial}{\partial x_2} - x_2\frac{\partial}{\partial x_3}, \quad V_4 = x_1\frac{\partial}{\partial x_3} - x_3\frac{\partial}{\partial x_1}, \quad V_5 = x_2\frac{\partial}{\partial x_1} - x_1\frac{\partial}{\partial x_2}.$$

Find the commutators and thus show we have a basis of a Lie algebra.

Solution 28. The nonzero commutators are

$$[V_1, V_2] = V_1, \quad [V_3, V_4] = V_5, \quad [V_4, V_5] = V_3, \quad [V_5, V_3] = V_4.$$

The Lie algebra is the direct sum $a_2 \oplus so(3, \mathbb{R})$.

Problem 29. (i) Consider the vector field $V_1(x_1, x_2) = x_2\frac{\partial}{\partial x_1}$ with the corresponding autonomous system of differential equations

$$\frac{dx_1}{d\tau} = x_2, \quad \frac{dx_2}{d\tau} = 0.$$

Find the solution of the initial value problem. Discuss.

(ii) Consider the vector field $V_2(x_1, x_2) = \frac{x_1^2}{2}\frac{\partial}{\partial x_2}$ with the corresponding autonomous system of differential equations

$$\frac{dx_1}{d\tau} = 0, \qquad \frac{dx_2}{d\tau} = \frac{1}{2}x_1^2.$$

Find the solution of the initial value problem. Discuss.

(iii) Find the vector field $V_3 = [V_1, V_2]$. Write down the corresponding autonomous system of differential equations and solve the initial value problem. Discuss.

(iv) Find the vector field $V_4 = V_1 + V_2$ and write down the corresponding autonomous system of differential equations and solve the initial value problem. Discuss.

Solution 29.   (i) The solution of the initial value problem is

$$x_1(\tau) = x_{2,0}\tau + x_{1,0}, \qquad x_2(\tau) = x_{2,0}.$$

Thus we have a complete vector field, i.e. $\tau \in \mathbb{R}$.

(ii) The solution of the initial value problem is

$$x_1(\tau) = x_{1,0}, \qquad x_2(\tau) = \frac{1}{2}x_{1,0}^2\tau + x_{2,0}.$$

Thus we have a complete vector field, i.e. $\tau \in \mathbb{R}$.

(iii) The commutator yields

$$V_3 = [V_1, V_2] = -x_1^2\frac{\partial}{\partial x_1} + x_1 x_2 \frac{\partial}{\partial x_2}$$

with the corresponding system of differential equations

$$\frac{dx_1}{d\tau} = -\frac{1}{2}x_1^2, \qquad \frac{dx_2}{d\tau} = x_1 x_2.$$

The solution of the initial value problem is

$$x_1(\tau) = \frac{2x_{1,0}}{2 + x_{1,0}\tau}, \qquad x_2(\tau) = \frac{x_{2,0}}{4}(x_{1,0}\tau + 2)^2.$$

Thus we do not have a *global flow*. The conclusion is that the set of complete vector fields on $\mathbb{R}^2$ is neither closed under the commutator nor a vector space.

(iv) We obtain

$$V_4 = x_2\frac{\partial}{\partial x_1} + \frac{1}{2}x_1^2\frac{\partial}{\partial x_2}$$

with the corresponding system of differential equations

$$\frac{dx_1}{d\tau} = x_2, \qquad \frac{dx_2}{d\tau} = \frac{1}{2}x_1^2.$$

The first integral is $x_1^3 - 3x_2^2$.

Problem 30.   Consider the *Lorenz model*

$$\frac{dx_1}{dt} = -\sigma x_1 + \sigma x_2 = V_1(x_1, x_2, x_3)$$

$$\frac{dx_2}{dt} = -x_1 x_3 + rx_1 - x_2 = V_2(x_1, x_2, x_3)$$

$$\frac{dx_3}{dt} = x_1 x_2 - bx_3 = V_3(x_1, x_2, x_3)$$

with the vector field

$$V = V_1(x_1, x_2, x_3)\frac{\partial}{\partial x_1} + V_2(x_1, x_2, x_3)\frac{\partial}{\partial x_2} + V_3(x_1, x_2, x_3)\frac{\partial}{\partial x_3}.$$

(i) Find curl(V).
(ii) Show that curl(curl(V)) = **0**.
(iii) Since curl(curl(V)) = **0** we can find a smooth function ϕ such that

$$\text{curl}(V) = \text{grad}(\phi).$$

Find ϕ.

Solution 30.   (i) We have

$$\text{curl}(V) := \begin{pmatrix} \partial V_3/\partial x_2 - \partial V_2/\partial x_3 \\ \partial V_1/\partial x_3 - \partial V_3/\partial x_1 \\ \partial V_2/\partial x_1 - \partial V_1/\partial x_2 \end{pmatrix} \Rightarrow \text{curl}(V) = \begin{pmatrix} 2x_1 \\ -x_2 \\ -x_3 + r - \sigma \end{pmatrix}.$$

(ii) From (i) we find curl(curl(V)) = **0**.
(iii) The function ϕ is given by

$$\phi(x_1, x_2, x_3) = x_1^2 - \frac{1}{2}x_2^2 - \frac{1}{2}x_3^2 + (r - \sigma)x_3.$$

Problem 31.   Consider the three 2×2 matrices

$$T_1 = \begin{pmatrix} 0 & 1 \\ 1 & 0 \end{pmatrix}, \quad T_2 = \begin{pmatrix} 0 & 1 \\ -1 & 0 \end{pmatrix}, \quad T_3 = \begin{pmatrix} 1 & 0 \\ 0 & -1 \end{pmatrix}$$

with the commutation relations $[T_1, T_2] = -2T_3$, $[T_2, T_3] = -2T_1$, $[T_3, T_1] = 2T_2$. Thus we have a basis of a simple Lie algebra. Consider the vector fields

$$V_1 = (x_1 \quad x_2) T_1 \begin{pmatrix} \partial/\partial x_1 \\ \partial/\partial x_2 \end{pmatrix} = x_1 \frac{\partial}{\partial x_2} + x_2 \frac{\partial}{\partial x_1}$$

$$V_2 = (x_1 \quad x_2) T_2 \begin{pmatrix} \partial/\partial x_1 \\ \partial/\partial x_2 \end{pmatrix} = x_1 \frac{\partial}{\partial x_2} - x_2 \frac{\partial}{\partial x_1}$$

$$V_3 = (x_1 \quad x_2) T_3 \begin{pmatrix} \partial/\partial x_1 \\ \partial/\partial x_2 \end{pmatrix} = x_1 \frac{\partial}{\partial x_1} - x_2 \frac{\partial}{\partial x_2}.$$

Find the commutators $[V_1, V_2]$, $[V_2, V_3]$, $[V_3, V_1]$. Discuss.

Solution 31.   We have $[V_1, V_2] = -2V_3$, $[V_2, V_3] = -2V_1$, $[V_3, V_1] = 2V_2$.

Problem 32.   Consider the autonomous system of first order ordinary differential equations

$$\frac{du_1}{d\tau} = -u_1 u_3, \quad \frac{du_2}{d\tau} = u_2 u_3, \quad \frac{du_3}{d\tau} = u_1^2 - u_2^2$$

with the corresponding vector field

$$V = -u_1 u_3 \frac{\partial}{\partial u_1} + u_2 u_3 \frac{\partial}{\partial u_2} + (u_1^2 - u_2^2) \frac{\partial}{\partial u_3}.$$

Show that $I = u_1 u_2$ is a first integral.

Solution 32.   We have

$$\frac{d}{d\tau} I = u_2 \frac{du_1}{d\tau} + u_1 \frac{du_2}{d\tau} = -u_1 u_3 u_2 + u_1 u_2 u_3 = 0.$$

Problem 33.   Consider the vector field in $\mathbb{R}^3$

$$V = x_1 \frac{\partial}{\partial x_2} - x_2 \frac{\partial}{\partial x_1} + \frac{\partial}{\partial x_3}.$$

Find

$$e^{\tau V} \begin{pmatrix} x_1 \\ x_2 \\ x_3 \end{pmatrix} \equiv \begin{pmatrix} e^{\tau V} x_1 \\ e^{\tau V} x_2 \\ e^{\tau V} x_3 \end{pmatrix}.$$

Solution 33.   Note that $[x_1 \partial/\partial x_2 - x_2 \partial/\partial x_1, \partial/\partial x_3] = 0$. Thus

$$e^{\tau V} = e^{\tau (x_1 \partial/\partial x_2 - x_2 \partial/\partial x_1)} e^{\tau \partial/\partial x_3}$$

and

$$e^{\tau V} \begin{pmatrix} x_1 \\ x_2 \\ x_3 \end{pmatrix} = \begin{pmatrix} x_1 \cos(\tau) - x_2 \sin(\tau) \\ x_2 \cos(\tau) + x_1 \sin(\tau) \\ x_3 + \tau \end{pmatrix}.$$

Problem 34. Consider the autonomous system of first order differential equations

$$\frac{du_j}{d\tau} = V_j(\mathbf{u}), \quad j = 1, \ldots, n \tag{1}$$

with the analytic vector field

$$V = \sum_{j=1}^{n} V_j(\mathbf{u}) \frac{\partial}{\partial u_j}.$$

The analytic function $g : \mathbb{R}^n \to \mathbb{R}$ is a first integral if $L_V(g) = 0$. To study explicitly τ-dependent first integrals we extend the autonomous system (1) to the autonomous system

$$\frac{du_j}{d\epsilon} = V_j(\mathbf{u}), \quad \frac{d\tau}{d\epsilon} = 1, \quad j = 1, \ldots, n$$

where $\tau(\epsilon = 0) = 0$ and the vector field

$$W = V + \frac{\partial}{\partial \tau}.$$

The $g(\mathbf{u}, \tau)$ is a first integral if $L_W g(\mathbf{u}, \tau) = 0$. Consider

$$\frac{du_1}{d\tau} = cu_1 + c_{23}u_2u_3, \quad \frac{du_2}{d\tau} = cu_2 + c_{13}u_1u_3, \quad \frac{du_3}{d\tau} = cu_3 + c_{12}u_1u_2.$$

Show that

$$I_1(\mathbf{u}, \tau) = \frac{1}{2}(c_{13}u_1^2 - c_{23}u_2^2)e^{-2c\tau}, \quad I_2(\mathbf{u}, \tau) = \frac{1}{2}(c_{12}u_1^2 - c_{23}u_3^2)e^{-2c\tau}$$

are explicitly τ dependent first integrals.

Solution 34. With the vector field

$$W = V + \frac{\partial}{\partial \tau}$$

we find $L_W(I_1) = 0$ and $L_W(I_2) = 0$.

Problem 35.   Consider an autonomous system of first order ordinary differential equations

$$\frac{d\mathbf{u}}{d\tau} = V(\mathbf{u})$$

where $\mathbf{u} = (u_1, u_2, \ldots, u_n)$ and V_j are smooth functions of $u_1, u_2, \ldots, u_n$. A smooth function $I(\mathbf{u})$ is called a *first integral* of the differential equations if

$$\frac{d}{d\tau} I(\mathbf{u}(\tau)) \equiv \sum_{j=1}^{n} \frac{\partial I}{\partial u_j} \frac{du_j}{d\tau} \equiv \sum_{j=1}^{n} \frac{\partial I}{\partial u_j} V_j(\mathbf{u}) = 0.$$

A smooth function $I(\mathbf{u}(\tau), \tau)$ is called an explicitly τ-dependent first integral of the differential equations if

$$\frac{d}{d\tau} I(\mathbf{u}(\tau), \tau) = \frac{\partial I}{\partial \tau} + \sum_{j=1}^{n} \frac{\partial I}{\partial u_j} V_j(\mathbf{u}(\tau), \tau) = 0.$$

Consider the system of differential equations

$$\frac{du_1}{d\tau} = u_2 u_3, \qquad \frac{du_2}{d\tau} = u_1 u_3, \qquad \frac{du_3}{d\tau} = u_1 u_2.$$

Show that

$$I_1(\mathbf{u}) = u_1^2 - u_2^2, \qquad I_2(\mathbf{u}) = u_3^2 - u_1^2, \qquad I_3(\mathbf{u}) = u_2^2 - u_3^2$$

are first integrals of the system.

Solution 35.   We have

$$\frac{d}{d\tau}(u_1^2 - u_2^2) = 2u_1 \frac{du_1}{d\tau} - 2u_2 \frac{du_2}{d\tau} = 2u_1(u_2 u_3) - 2u_2(u_1 u_3) = 0.$$

Analogously we find that I_2 and I_3 are first integrals.

Problem 36.   In *Nambu mechanics* the phase space is spanned by an n-tuple of dynamical variables u_i, for $\iota = 1, \ldots, n$. The equations of motion of Nambu mechanics (i.e. the autonomous system of first order ordinary differential equations) can be constructed as follows. Let $I_k : \mathbb{R}^n \to \mathbb{R}$, for $k = 1, \ldots, n-1$ be smooth functions, then

$$\frac{du_i}{dt} = \frac{\partial(u_i, I_1, \ldots, I_{n-1})}{\partial(u_1, u_2, \ldots, u_n)}$$

where $\partial(u_i, I_1, \ldots, I_{n-1})/\partial(u_1, u_2, \ldots, u_n)$ denotes the *Jacobian*. Consequently, the equations of motion can also be written as (summation convention)

$$\frac{du_i}{dt} = \epsilon_{ijk\ldots l} \partial_j I_1 \ldots \partial_l I_{n-1}$$

where $\epsilon_{ijk...l}$ is the generalized *Levi-Civita symbol* and $\partial_j \equiv \partial/\partial u_j$. The proof that $I_1, \ldots, I_{n-1}$ are first integrals of the system is as follows. Since (summation convention)

$$\frac{dI_i}{dt} = \frac{\partial I_i}{\partial u_j}\frac{du_j}{dt}$$

we have

$$\frac{dI_i}{dt} = (\partial_j I_i)\epsilon_{jl_1...l_{n-1}}(\partial_{l_1} I_1)\ldots(\partial_{l_{n-1}} I_{n-1}) = \frac{\partial(I_i, I_1, \ldots, I_{n-1})}{\partial(u_1, \ldots, u_n)} = 0.$$

Since the Jacobian matrix has two equal rows, it is singular. If the first integrals are polynomials, then the dynamical system is algebraically completely integrable. For the case $n = 3$, we obtain the equations of motion

$$\frac{du_1}{dt} = \frac{\partial I_1}{\partial u_2}\frac{\partial I_2}{\partial u_3} - \frac{\partial I_1}{\partial u_3}\frac{\partial I_2}{\partial u_2}, \quad \frac{du_2}{dt} = \frac{\partial I_1}{\partial u_3}\frac{\partial I_2}{\partial u_1} - \frac{\partial I_1}{\partial u_1}\frac{\partial I_2}{\partial u_3},$$

$$\frac{du_3}{dt} = \frac{\partial I_1}{\partial u_1}\frac{\partial I_2}{\partial u_2} - \frac{\partial I_1}{\partial u_2}\frac{\partial I_2}{\partial u_1}.$$

We consider the following first integrals

$$I_1 = u_1 + u_2 + u_3 \text{ and } I_2 = u_1 u_2 u_3.$$

The program uses the first equation to construct the 3×3 matrices and then calculate their determinants

$$\frac{du_i}{dt} = \det\begin{pmatrix} \dfrac{\partial u_i}{\partial u_1} & \dfrac{\partial I_1}{\partial u_1} & \dfrac{\partial I_2}{\partial u_1} \\ \dfrac{\partial u_i}{\partial u_2} & \dfrac{\partial I_1}{\partial u_2} & \dfrac{\partial I_2}{\partial u_2} \\ \dfrac{\partial u_i}{\partial u_3} & \dfrac{\partial I_1}{\partial u_3} & \dfrac{\partial I_2}{\partial u_3} \end{pmatrix}, \quad i = 1, 2, 3.$$

Solution 36.

```
// nambu.cpp
#include <iostream>
#include "symbolicc++.h"
using namespace std;

void nambu(const Symbolic &I,const Symbolic &u,int n)
{
 int i, j;
 Symbolic J("J",n,n);
```

```
  for(i=0;i<n;i++)
    for(j=1;j<n;j++) J(i,j) = df(I(j-1),u(i));
  for(i=0;i<n;i++)
  {
  for(j=0;j<n;j++) J(j,0) = df(u(i),u(j));
  cout << "du(" << i << ")/dt = " << det(J) << endl;
  }
}

int main(void)
{
Symbolic u("u",3), I("I",2);
I(0) = u(0)+u(1)+u(2); I(1) = u(0)*u(1)*u(2);
cout << "The equations are " << endl;
nambu(I,u,3);
return 0;
}
```

The output provides the autonomous system of first order differential equations

$$\frac{du_1}{dt} = u_1 u_2 - u_1 u_3, \quad \frac{du_2}{dt} = u_2 u_3 - u_1 u_2, \quad \frac{du_3}{dt} = u_1 u_3 - u_2 u_3.$$

2.3   Supplementary Problems

Problem 1. Let $\alpha \in \mathbb{R}$. Let $f : \mathbb{R} \to \mathbb{R}$ be an analytic function. Calculate

$$\cosh\left(\alpha\frac{d}{dx}\right) f(x), \qquad \sinh\left(\alpha\frac{d}{dx}\right) f(x).$$

Problem 2. Find two smooth vector fields V and W in $\mathbb{R}^n$ such that

$$[[W, V], V] = 0 \quad \text{but} \quad [W, V] \neq 0.$$

Find two $n \times n$ matrices A and B such that

$$[[B, A], A] = 0_n \quad \text{but} \quad [B, A] \neq 0_n.$$

Problem 3.   (i) Show that the vector fields

$$\left\{ \frac{\partial}{\partial x_k}, \; x_j \frac{\partial}{\partial x_k}, \; x_j \sum_{\ell=1}^{n} x_\ell \frac{\partial}{\partial x_\ell} \; : \; j,k = 1, \ldots, n \right\}$$

form a basis of a Lie algebra. This is the Lie algebra of the n-dimensional *projective group*. Is this group isomorphic to the special linear Lie algebra $s\ell(n+1, \mathbb{R})$?

(ii) Show that the 8 vector fields

$$\frac{\partial}{\partial x}, \quad x\frac{\partial}{\partial x}, \quad y\frac{\partial}{\partial x}, \quad xy\frac{\partial}{\partial x} + y^2\frac{\partial}{\partial y},$$

$$\frac{\partial}{\partial y}, \quad x\frac{\partial}{\partial y}, \quad y\frac{\partial}{\partial y}, \quad x^2\frac{\partial}{\partial x} + xy\frac{\partial}{\partial y}$$

form a basis of a Lie algebra. Find all sub Lie algebras.

(iii) Let $n \geq 2$. Show that the vector fields in $\mathbb{R}^2$

$$\frac{\partial}{\partial x_1}, \quad \frac{\partial}{\partial x_2}, \quad x_1\frac{\partial}{\partial x_1}, \quad x_1\frac{\partial}{\partial x_2}, \quad x_2\frac{\partial}{\partial x_2},$$

$$x_1^2\frac{\partial}{\partial x_1} + nx_1x_2\frac{\partial}{\partial x_2}, \quad x_1^2\frac{\partial}{\partial x_2}, \; \ldots, \; x_1^n\frac{\partial}{\partial x_2}$$

form a basis of a Lie algebra.

(iv) Show that the differential operators (vector fields)

$$\frac{\partial}{\partial y} + x\frac{\partial}{\partial z}, \quad y\frac{\partial}{\partial y} + z\frac{\partial}{\partial z}, \quad (xy - z)\frac{\partial}{\partial x} + y^2\frac{\partial}{\partial y} + yz\frac{\partial}{\partial z}$$

generate a finite-dimensional Lie algebra.

Problem 4.   Consider the vector fields

$$V_1 = \cos(\psi)\frac{\partial}{\partial\theta} + \frac{\sin(\psi)}{\sin(\theta)}\frac{\partial}{\partial\phi} - \cot(\theta)\sin(\psi)\frac{\partial}{\partial\psi}$$

$$V_2 = -\sin(\psi)\frac{\partial}{\partial\theta} + \frac{\cos(\psi)}{\sin(\theta)}\frac{\partial}{\partial\phi} - \cot(\theta)\cos(\psi)\frac{\partial}{\partial\psi}$$

$$V_3 = \frac{\partial}{\partial\psi}.$$

Calculate the commutators and show that V_1, V_2, V_3 form a basis of a Lie algebra.

Problem 5.   Let $n \geq 2$. Consider the compact Lie group $SO(n)$ and the corresponding Lie algebra $so(n)$.
(i) Show that every $A \in SO(n)$ can be written as $\exp(X)$ with $X \in so(n)$.
(ii) Let $A, B \in so(n)$. We form the vector fields

$$V_A = (x_1 \quad \cdots \quad x_n) A \begin{pmatrix} \partial/\partial x_1 \\ \vdots \\ \partial/\partial x_n \end{pmatrix}, \quad V_B = (x_1 \quad \cdots \quad x_n) B \begin{pmatrix} \partial/\partial x_1 \\ \vdots \\ \partial/\partial x_n \end{pmatrix}.$$

Find the commutator $[V_A, V_B]$. Is

$$[V_A, V_B] = (x_1 \quad \cdots \quad x_n) ([A, B]) \begin{pmatrix} \partial/\partial x_1 \\ \vdots \\ \partial/\partial x_n \end{pmatrix} ?$$

Problem 6.   Consider the autonomous system of differential equations

$$\frac{du_1}{d\tau} = u_1(u_2 - u_3), \quad \frac{du_2}{d\tau} = u_2(u_3 - u_1), \quad \frac{du_3}{d\tau} = u_3(u_1 - u_2)$$

with the vector field

$$V = u_1(u_2 - u_3)\frac{\partial}{\partial u_1} + u_2(u_3 - u_1)\frac{\partial}{\partial u_2} + u_3(u_1 - u_2)\frac{\partial}{\partial u_3}.$$

Find the solution by integrating

$$\frac{du_1}{u_1(u_2 - u_3)} = \frac{du_2}{u_2(u_3 - u_1)} = \frac{du_3}{u_3(u_1 - u_2)} = \frac{d\tau}{1}.$$

Problem 7.   Let V be smooth vector in $\mathbb{R}^n$ which can be written as $V = W + U$. Then for sufficiently small τ we have (*leapfrog scheme*)

$$e^{\tau V} = e^{\tau W/2} e^{\tau U} e^{\tau W/2} + O(\tau^3).$$

Let $n = 1$ and $V = W + U$ with

$$W = x\frac{d}{dx}, \quad U = \sin(x)\frac{d}{dx}.$$

Find $e^{\tau W/2}$, $e^{\tau U}$, $e^{\tau W/2} e^{\tau U} e^{\tau W/2}$.

Problem 8. (i) Consider the vector field

$$V = \frac{\partial}{\partial x_1} + \frac{\partial}{\partial x_2} + \frac{\partial}{\partial x_3}$$

in $\mathbb{R}^3$. Find all smooth vector fields W such that $[V, W] = 0$.
(ii) Consider the vector field

$$V = \sum_{k=1}^{n} x_k \frac{\partial}{\partial x_k}.$$

Find all smooth vector fields W such that $[V, W] = 0$. Find all smooth vector fields W such that $[V, W] = fV$, where f is a smooth function.

Problem 9. Consider the smooth vector fields in $\mathbb{R}^n$

$$V = \sum_{j,k=1}^{n} a_{jk} x_j \frac{\partial}{\partial x_k}, \qquad W = \sum_{j,k,\ell=1}^{n} c_{jk\ell} x_j x_k \frac{\partial}{\partial x_\ell}$$

where $a_{jk}, c_{jk\ell} \in \mathbb{R}$. Find the conditions on a_{jk} and $c_{jk\ell}$ such that $[V, W] = 0$.

Problem 10. (i) Give a vector field V in $\mathbb{R}^3$ such that $V \times \mathrm{curl}(V) \neq \mathbf{0}$.
(ii) Give a vector field V in $\mathbb{R}^3$ such that $V \times \mathrm{curl}(V) = \mathbf{0}$.

Problem 11. (i) Consider the four vector fields in $\mathbb{R}^2$

$$V_1 = \frac{\partial}{\partial x}, \quad V_2 = \frac{\partial}{\partial y}, \quad V_3 = x\frac{\partial}{\partial x} + y\frac{\partial}{\partial y}, \quad V_4 = (x^2 - y^2)\frac{\partial}{\partial x} + 2xy\frac{\partial}{\partial y}.$$

Find the fixed points of the corresponding autonomous systems of first order differential equations. Study their stability. Do the vector fields form a basis of a Lie algebra?
(ii) Show that the vector fields

$$V_1 = \frac{\partial}{\partial x}, \quad V_2 = x\frac{\partial}{\partial x} + y\frac{\partial}{\partial y}, \quad V_3 = (y^2 - x^2)\frac{\partial}{\partial x} - 2xy\frac{\partial}{\partial y}$$

form a basis for the Lie algebra $s\ell(2, \mathbb{R})$. Solve the initial value problem for the autonomous system

$$\frac{dx}{d\tau} = y^2 - x^2, \qquad \frac{dy}{d\tau} = -2xy.$$

Problem 12.   Consider the vector fields

$$\frac{\partial}{\partial u_{jk}}, \quad u_{jm}\frac{\partial}{\partial u_{jk}}, \quad u_{\ell k}\frac{\partial}{\partial u_{jk}}, \quad u_{jm}u_{\ell k}\frac{\partial}{\partial u_{jk}}$$

where $j = 1, 2, \ldots, p$; $k = 1, 2, \ldots, n$; $m = 1, 2, \ldots, n$; $\ell = 1, 2, \ldots, p$. Find the commutators. Show that the vector fields form a basis of the Lie algebra $g\ell(n + p, \mathbb{R})$. The dimension of the Lie algebra is $(n + p)^2$.

Problem 13.   (i) Find the Lie algebra generated by

$$V_1 = x_2\frac{\partial}{\partial x_2}, \quad V_2 = -x_2\frac{\partial}{\partial x_1}.$$

(ii) Let c be a constant. Find the Lie algebra generated by

$$V_1 = x_2\frac{\partial}{\partial x_2} + cx_3\frac{\partial}{\partial x_3}, \quad V_2 = -x_2\frac{\partial}{\partial x_1}, \quad V_3 = -cx_3\frac{\partial}{\partial x_1}.$$

Problem 14.   The *Poisson bracket* of two functions $f, g \in C^\infty(M)$ is the function $[f, g] \in C^\infty(M)$ defined by

$$[f, g] := V_f(g)$$

where V_f is the vector field

$$V_f := \sum_{j=1}^{m} \left(\frac{\partial f}{\partial p_j}\frac{\partial}{\partial q_j} - \frac{\partial f}{\partial q_j}\frac{\partial}{\partial p_j} \right).$$

Show that the Poisson bracket is antisymmetric, i.e. $[f, g] = -[g, f]$ and

$$[f, [g, h]] + [h, [f, g]] + [g, [h, f]] = 0.$$

Problem 15.   Consider the autonomous system of first order differential equations

$$\frac{du_1}{d\tau} = cu_1 + c_{234}u_2u_3u_4, \quad \frac{du_2}{d\tau} = cu_2 + c_{134}u_1u_3u_4,$$

$$\frac{du_3}{d\tau} = cu_3 + c_{124}u_1u_2u_4, \quad \frac{du_4}{d\tau} = cu_4 + c_{123}u_1u_2u_3.$$

Show that for $c = 0$ one has the three first integrals

$$c_{134}u_1^2 - c_{234}u_2^2, \quad c_{124}u_2^2 - c_{134}u_3^2, \quad c_{123}u_3^2 - c_{124}u_4^2.$$

Show that for $c > 0$ one has the three explicitly τ-dependent first integrals

$$(c_{134}u_1^2 - c_{234}u_2^2)e^{-c\tau}, \quad (c_{124}u_2^2 - c_{134}u_3^2)e^{-c\tau}, \quad (c_{123}u_3^2 - c_{124}u_4^2)e^{-c\tau}.$$

Problem 16. Consider the first order autonomous system of first order differential equation

$$\frac{du_1}{d\tau} = f_1(u_1, u_2), \qquad \frac{du_2}{d\tau} = f_2(u_1, u_2).$$

Find the conditions on f_1, f_2 such that $I(u_1, u_2) = u_1 u_2 (a + b u_1 + c u_2)$ is a first integral.

Problem 17. Consider the vector fields

$$V_1 = \frac{d}{dx}, \quad V_2 = x\frac{d}{dx}.$$

Then $[V_1, V_2] = V_1$. Consider $\widetilde{V}_1 = f_1(x) + V_1$, $\widetilde{V}_2 = f_2(x) + V_2$. Find the conditions on f_1, f_2 such that $[\widetilde{V}_1, \widetilde{V}_2] = \widetilde{V}_1$.

Problem 18. Let V, W be smooth vector fields with the corresponding flows f_{τ_1}, g_{τ_2}, respectively. Then

$$[V, W] = 0 \quad \Leftrightarrow \quad f_{\tau_1} \circ g_{\tau_2} = g_{\tau_2} \circ f_{\tau_1}$$

where $\circ$ denotes function composition. Apply it to the vector fields

$$V = x_1\frac{\partial}{\partial x_1} + x_2\frac{\partial}{\partial x_2}, \quad W = x_2\frac{\partial}{\partial x_1} + x_1\frac{\partial}{\partial x_2}$$

i.e. first show that $[V, W] = 0$ and then find f_{τ_1}, g_{τ_2} and show that $f_{\tau_1} \circ g_{\tau_2} = g_{\tau_2} \circ f_{\tau_1}$

Problem 19. Consider the smooth vector fields in $\mathbb{R}^3$

$$V = V_1(\mathbf{x})\frac{\partial}{\partial x_1} + V_2(\mathbf{x})\frac{\partial}{\partial x_2} + V_3(\mathbf{x})\frac{\partial}{\partial x_3}$$

$$W = W_1(\mathbf{x})\frac{\partial}{\partial x_1} + W_2(\mathbf{x})\frac{\partial}{\partial x_2} + W_3(\mathbf{x})\frac{\partial}{\partial x_3}.$$

Now

$$\text{curl}\begin{pmatrix} V_1 \\ V_2 \\ V_3 \end{pmatrix} = \begin{pmatrix} \frac{\partial V_3}{\partial x_2} - \frac{\partial V_2}{\partial x_3} \\ \frac{\partial V_1}{\partial x_3} - \frac{\partial V_3}{\partial x_1} \\ \frac{\partial V_2}{\partial x_1} - \frac{\partial V_1}{\partial x_2} \end{pmatrix}, \quad \text{curl}\begin{pmatrix} W_1 \\ W_2 \\ W_3 \end{pmatrix} = \begin{pmatrix} \frac{\partial W_3}{\partial x_2} - \frac{\partial W_2}{\partial x_3} \\ \frac{\partial W_1}{\partial x_3} - \frac{\partial W_3}{\partial x_1} \\ \frac{\partial W_2}{\partial x_1} - \frac{\partial W_1}{\partial x_2} \end{pmatrix}.$$

We consider the smooth vector fields

$$
V_c = \left(\frac{\partial V_3}{\partial x_2} - \frac{\partial V_2}{\partial x_3}\right)\frac{\partial}{\partial x_1} + \left(\frac{\partial V_1}{\partial x_3} - \frac{\partial V_3}{\partial x_1}\right)\frac{\partial}{\partial x_2} + \left(\frac{\partial V_2}{\partial x_1} - \frac{\partial V_1}{\partial x_2}\right)\frac{\partial}{\partial x_3}
$$

$$
W_c = \left(\frac{\partial W_3}{\partial x_2} - \frac{\partial W_2}{\partial x_3}\right)\frac{\partial}{\partial x_1} + \left(\frac{\partial W_1}{\partial x_3} - \frac{\partial W_3}{\partial x_1}\right)\frac{\partial}{\partial x_2} + \left(\frac{\partial W_2}{\partial x_1} - \frac{\partial W_1}{\partial x_2}\right)\frac{\partial}{\partial x_3}.
$$

Note that if $\alpha = V_1(\mathbf{x})dx_1 + V_2(\mathbf{x})dx_2 + V_3(\mathbf{x})dx_3$ then

$$
d\alpha = \left(\frac{\partial V_2}{\partial x_1} - \frac{\partial V_1}{\partial x_2}\right) dx_1 \wedge dx_2 + \left(\frac{\partial V_1}{\partial x_3} - \frac{\partial V_3}{\partial x_1}\right) dx_3 \wedge dx_1
$$
$$
+ \left(\frac{\partial V_3}{\partial x_2} - \frac{\partial V_2}{\partial x_3}\right) dx_2 \wedge dx_3.
$$

(i) Calculate the commutator $[V_c, W_c]$. Assume that $[V, W] = 0$. Can we conclude $[V_c, W_c] = 0$?
(ii) Assume that $[V, W] = R$. Can we conclude that $[V_c, W_c] = R_c$?

Problem 20.   Let $f_1 : \mathbb{R}^2 \to \mathbb{R}$, $f_2 : \mathbb{R}^2 \to \mathbb{R}$ be analytic function. Consider the analytic vector fields

$$
V = f_1(x_1, x_2)\frac{\partial}{\partial x_1} + \frac{\partial}{\partial x_2}, \qquad W = \frac{\partial}{\partial x_1} + f_2(x_1, x_2)\frac{\partial}{\partial x_2}
$$

in $\mathbb{R}^2$.
(i) Find the conditions on f_1 and f_2 such that $[V, W] = 0$.
(ii) Find the conditions on f_1 and f_2 such that $[V, W] = V + W$.

Problem 21.   (i) Consider the autonomous system of first order ordinary differential equations

$$
\frac{du_1}{d\tau} = -u_1 u_3, \qquad \frac{du_2}{d\tau} = u_2 u_3, \qquad \frac{du_3}{d\tau} = u_1^2 - u_2^2
$$

with the vector field

$$
V = -u_1 u_3 \frac{\partial}{\partial u_1} + u_2 u_3 \frac{\partial}{\partial u_2} + (u_1^2 - u_2^2)\frac{\partial}{\partial u_3}.
$$

Show that $I_1 = \frac{1}{2}(u_1 + u_2 + u_3)$, $I_2 = u_1 u_2$ are first integrals.
(ii) Consider the autonomous system of first order ordinary differential equations

$$
\frac{du_1}{d\tau} = u_2 - u_3, \qquad \frac{du_2}{d\tau} = -u_1 + u_1 u_3, \qquad \frac{du_3}{d\tau} = u_1 - u_1 u_2
$$

with the corresponding vector field

$$V = (u_2 - u_3)\frac{\partial}{\partial u_1} + (-u_1 + u_1 u_3)\frac{\partial}{\partial u_2} + (u_1 - u_1 u_2)\frac{\partial}{\partial u_3}.$$

Show that

$$I_1 = \frac{1}{2}(u_1^2 + u_2^2 + u_3^2), \quad I_2 = u_2 - \frac{1}{2}u_2^2 + u_3 - \frac{1}{2}u_3^2$$

are first integrals.

Problem 22. Consider the *Darboux-Halphen system*

$$\frac{dx_1}{d\tau} = x_2 x_3 - x_1 x_2 - x_3 x_1,$$

$$\frac{dx_2}{d\tau} = x_3 x_1 - x_1 x_2 - x_2 x_3,$$

$$\frac{dx_3}{d\tau} = x_1 x_2 - x_3 x_1 - x_2 x_3$$

with the corresponding vector field. Is the autonomous system of differential equations invariant under the transformation $(\alpha\delta - \beta\gamma \neq 0)$

$$(\tau, x_j) \mapsto \left(\frac{\alpha\tau + \beta}{\gamma\tau + \delta}, 2\gamma\frac{\gamma\tau + \delta}{\alpha\delta - \gamma\beta} + \frac{(\gamma\tau + \delta)^2}{\alpha\delta - \gamma\beta}x_j \right)$$

with $j = 1, 2, 3$?

Problem 23. The *geodesic flow* of an n-dimensional (pseudo-) Riemannian manifold (M, g) is locally described by an autonomous system of second order ordinary differential equations

$$\frac{d^2 x^j}{d\tau^2} + \sum_{k,\ell=1}^{n} \Gamma_{k\ell}^j \frac{dx^k}{d\tau}\frac{dx^\ell}{d\tau} = 0, \quad j = 1, \ldots, n$$

where $\Gamma_{k\ell}^j$ are the *Christoffel symbols*. This system of differential equations can be rewritten as a Hamilton system

$$\frac{dx^j}{d\tau} = \frac{\partial H}{\partial p_j}, \quad \frac{dp_j}{d\tau} = -\frac{\partial H}{\partial x^j}, \quad j = 1, \ldots, n$$

with Hamilton function

$$H(\mathbf{x}, \mathbf{p}) = \frac{1}{2}\sum_{j,k=1}^{n} g^{jk}(\mathbf{x})p_j p_k$$

and corresponding vector field

$$V = \sum_{j=1}^{n} \left(\frac{\partial H}{\partial p_j} \cdot \frac{\partial}{\partial x^j} - \frac{\partial H}{\partial x^j} \cdot \frac{\partial}{\partial p_j} \right).$$

Find the solution of the initial value problem with $n = 2$ and

$$(g_{jk}) = \begin{pmatrix} 0 & 1 \\ 1 & 0 \end{pmatrix}.$$

Problem 24.   Consider the *Weber equation*

$$\frac{d^2u}{dx^2} + \left(n + \frac{1}{2} - \frac{1}{4}x^2 \right) u = 0.$$

With the transformation $u(x) = \exp(-\frac{1}{4}x^2)v(x)$ we obtain

$$\frac{d^2v}{dx^2} - x\frac{dv}{dx} + nv = 0.$$

Setting $v_1 = v$, $v_2 = dv_1/dx$, $v_3(x) = x$, $dv_3/dx = 1$ we obtain

$$\frac{dv_1}{dx} = v_2, \quad \frac{dv_2}{dx} = v_3v_2 - nv_1, \quad \frac{dv_3}{dx} = 1$$

with $v_3(0) = 0$. The corresponding vector field is

$$V = v_2 \frac{\partial}{\partial v_1} + (v_2v_3 - nv_1) \frac{\partial}{\partial v_2} + \frac{\partial}{\partial v_3}.$$

Find

$$\exp(xV) \begin{pmatrix} v_1 \\ v_2 \\ v_3 \end{pmatrix} \Bigg|_{v_1=v_1(0),v_2=v_2(0),v_3=v_3(0)}$$

for the cases (i) $v_1(0) = 1$, $v_2(0) = 0$; (ii) $v_1(0) = 0$, $v_2(0) = 1$. Note that

$$Vv_1 = v_2, \quad Vv_2 = v_2v_3 - nv_1, \quad Vv_3 = 1$$

$$V(Vv_1) = v_2v_3 - nv_1, \quad V(V(Vv_1)) = -nv_2 + v_3(v_2v_3 - nv_1) + v_2.$$

Chapter 3

Metric Tensor Fields

3.1 Notations and Definitions

Let $n \geq 1$ and $\mathbb{E}^n$ be the n-dimensional *Euclidean space* with the metric tensor field

$$g = \sum_{j=1}^{n} (dx_j \otimes dx_j).$$

The distance of $\mathbf{x}, \mathbf{y} \in \mathbb{E}^n$ is given by

$$\|\mathbf{x} - \mathbf{y}\| = \sqrt{\sum_{j=1}^{n} (x_j - y_j)^2}.$$

Let $\mathbf{x} : M \mapsto \mathbb{E}^3$ be a smooth surface in $\mathbb{E}^3$. Choosing local coordinates $\mathbf{u} = (u_1, u_2)$ in a coordinate neighbourhood $U = M$, the surface $\mathbf{x}$ can be expressed as

$$\mathbf{x}(\mathbf{u}) = \begin{pmatrix} x_1(u_1, u_2) \\ x_2(u_1, u_2) \\ x_3(u_1, u_2) \end{pmatrix}.$$

The rank of the 2×3 matrix

$$\begin{pmatrix} \partial x_1/\partial u_1 & \partial x_2/\partial u_1 & \partial x_3/\partial u_1 \\ \partial x_1/\partial u_2 & \partial x_2/\partial u_2 & \partial x_3/\partial u_2 \end{pmatrix}$$

is 2. Then (*first fundamental form*)

$$dx_1(\mathbf{u}) \otimes dx_1(\mathbf{u}) + dx_2(\mathbf{u}) \otimes dx_2(\mathbf{u}) + dx_3(\mathbf{u}) \otimes dx_3(\mathbf{u})$$

$$= E(\mathbf{u})du_1 \otimes du_1 + F(\mathbf{u})du_1 \otimes du_2 + F(\mathbf{u})du_2 \otimes du_1 + G(\mathbf{u})du_2 \otimes du_2$$

with

$$E(\mathbf{u}) := \frac{\partial \mathbf{x}}{\partial u_1} \cdot \frac{\partial \mathbf{x}}{\partial u_1}, \quad F(\mathbf{u}) := \frac{\partial \mathbf{x}}{\partial u_1} \cdot \frac{\partial \mathbf{x}}{\partial u_2}, \quad G(\mathbf{u}) := \frac{\partial \mathbf{x}}{\partial u_2} \cdot \frac{\partial \mathbf{x}}{\partial u_2}$$

where $\cdot$ denotes the scalar product of the two vectors in $\mathbb{E}^3$. The *normal vector* $\mathbf{n}(u_1, u_2)$ to the surface of M is given by

$$\mathbf{n}(u_1, u_2) = \frac{(\partial \mathbf{x}/\partial u_1) \times (\partial \mathbf{x}/\partial u_2)}{\|(\partial \mathbf{x}/\partial u_1) \times (\partial \mathbf{x}/\partial u_2)\|}$$

where $\times$ denotes the vector product. Then one defines (coefficients of the *second fundamental form*)

$$L(\mathbf{u}) := \mathbf{n}(\mathbf{u}) \cdot \frac{\partial^2 \mathbf{x}}{\partial u_1^2}, \quad M(\mathbf{u}) := \mathbf{n}(\mathbf{u}) \cdot \frac{\partial^2 \mathbf{x}}{\partial u_1 \partial u_2}, \quad N(\mathbf{u}) := \mathbf{n}(\mathbf{u}) \cdot \frac{\partial^2 \mathbf{x}}{\partial u_2^2}.$$

The *Gaussian curvature* is given by

$$\frac{LN - M^2}{EG - F^2}$$

and the *mean curvature* is given by

$$\frac{NE - 2MF + LG}{EG - F^2}.$$

Let M be a smooth N-dimensional manifold. For each $p \in M$, the map $g_p : T_pM \times T_pM \to \mathbb{R}$ defines a positive definite inner product on T_pM and $p \to g_p$ is smooth as a section of $T^*M \otimes T^*M$. Hence in *local coordinates* $x_1, \ldots, x_N$ on a coordinate patch U we have the metric tensor field

$$g = \sum_{j,k=1}^{N} g_{jk}(\mathbf{x})dx_j \otimes dx_k$$

with $g_{jk} \in C^\infty(U)$. Let M be a Riemannian manifold or pseudo Riemannian manifold with metric tensor field

$$g = \sum_{j,k=1}^{N} g_{jk}(\mathbf{x})dx_j \otimes dx_k.$$

The inverse of the $N \times N$ matrix (g_{jk}) is denoted by (g^{jk}). The *Christoffel symbols* of the second kind are defined by

$$\Gamma^a_{mn} := \frac{1}{2} \sum_{b=1}^{N} g^{ab}(g_{bm,n} + g_{bn,m} - g_{mn,b})$$

where $g_{bm,n}$ denotes the partial derivative of g_{bm} with respect to the coordinates x_n, i.e.

$$g_{bm,n} := \frac{\partial g_{mn}}{\partial x_n}.$$

Note that

$$\Gamma^k_{ij} = \sum_{m=1}^{N} g^{km}\Gamma_{ijm}, \quad \Gamma_{ijm} = \frac{1}{2}(g_{jm,i} + g_{mi,j} - g_{ij,m})$$

where Γ_{ijm} are the Christoffel symbols of the first kind. From the Christoffel symbols of the second kind we find the *curvature tensor*

$$R^b_{mnq} = \Gamma^b_{mq,s} - \Gamma^b_{ms,q} + \sum_{n=1}^{N} \Gamma^b_{ns}\Gamma^n_{mq} - \sum_{n=1}^{N} \Gamma^b_{nq}\Gamma^n_{ms}.$$

The *Ricci tensor* R_{mq} is given by

$$R_{mq} = \sum_{s=1}^{N} R^s_{msq}, \quad R^m_q = \sum_{n=1}^{N} g^{mn} R_{nq}$$

and the *curvature scalar* is

$$R = \sum_{m=1}^{N} R^m_m.$$

A four-dimensional differentiable manifold M endowed with a non-degenerate, symmetric rank $(0,2)$ metric tensor field

$$g = \sum_{\mu=0,\nu=0}^{3} g_{\mu\nu}(\mathbf{x})dx_\mu \otimes dx_\nu$$

on M whose *signature* is

$$(-1, +1, +1, +1)$$

(or $(+1, -1, -1, -1)$) is called a *Lorentzian manifold*. One sets $x_0 = ct$. The *Einstein field equation* are given by

$$R_{\mu\nu} - \frac{1}{2}Rg_{\mu\nu} = \frac{8\pi G}{c^4}T_{\mu\nu}$$

with $\mu, \nu = 0, 1, 2, 3$ and $T = (T_{\mu\nu})$ is the energy-momentum tensor of any matter fields present with dimension $kg.meter^{-1}.sec^{-2}$. Here G is the gravitational constant. Furthermore

$$R = \sum_{\mu=0,\nu=0}^{3} R_{\mu\nu}g^{\mu\nu}.$$

Here $R_{\mu\nu}$ is the Ricci curvature tensor. Note that $(g^{\mu\nu})$ is the inverse matrix of $(g_{\mu\nu})$. In *local coordinates* one can write

$$\sum_{\mu,\nu=0}^{3} R_{\mu\nu}dx_\mu \otimes dx_\nu - \frac{1}{2}R \sum_{\mu,\nu=0}^{3} g_{\mu\nu}dx_\mu \otimes dx_\nu = \frac{8\pi G}{c^4} \sum_{\mu,\nu=0}^{3} T_{\mu\nu}dx_\mu \otimes dx_\nu.$$

The dimension of the curvature tensor is $meter^{-2}$. The *Hilbert action* is

$$S = \frac{1}{16\pi Gc^{-4}} \int R\sqrt{-g}dx_0dx_1dx_2dx_3.$$

The vacuum Einstein field equations are

$$R_{\mu\nu} - \frac{1}{2}Rg_{\mu\nu} = 0$$

and the *Einstein tensor* is

$$G_{\mu\nu} := R_{\mu\nu} - \frac{1}{2}Rg_{\mu\nu}.$$

They admit the *Schwarzschild solution*

$$g = \left(1 - \frac{r_s}{r}\right)dx_0 \otimes dx_0 - \left(1 - \frac{r_s}{r}\right)^{-1}dr \otimes dr$$
$$- r^2(d\theta \otimes d\theta + \sin(\theta)d\phi \otimes d\phi)$$

where

$$r_s = 2GM/c^2$$

is the *Schwarzschild radius*. This is the *Schwarzschild metric tensor field* for a static field of a non-rotating spherically symmetric black hole of mass M.

3.2 Solved Problems

Problem 1. Consider the metric tensor field

$$g = dx_1 \otimes dx_1 + dx_2 \otimes dx_2$$

for the two-dimensional Euclidean space $\mathbb{E}^2$ and the differential two-form $dx_1 \wedge dx_2$. Express the metric tensor field and the differential two-form in *polar coordinates* $x_1(r, \phi) = r\cos(\phi)$, $x_2(r, \phi) = r\sin(\phi)$.

Solution 1. We have

$$dx_1 = \cos(\phi)dr - r\sin(\phi)d\phi, \qquad dx_2 = \sin(\phi)dr + r\cos(\phi)d\phi.$$

Thus

$$
\begin{aligned}
dx_1 \otimes dx_1 = {}& \cos^2(\phi)dr \otimes dr + r\sin^2(\phi)d\phi \otimes d\phi \\
& - r\cos(\phi)\sin(\phi)(dr \otimes d\phi + d\phi \otimes dr) \\
dx_2 \otimes dx_2 = {}& \sin^2(\phi)dr \otimes dr + r\cos^2(\phi)d\phi \otimes d\phi \\
& + r\cos(\phi)\sin(\phi)(dr \otimes d\phi + d\phi \otimes dr).
\end{aligned}
$$

With $\cos^2(\phi) + \sin^2(\phi) = 1$ we obtain

$$dx_1 \otimes dx_1 + dx_2 \otimes dx_2 = dr \otimes dr + r^2 d\phi \otimes d\phi.$$

With $dr \wedge dr = d\phi \wedge d\phi = 0$ and $dr \wedge d\phi = -d\phi \wedge dr$ we find

$$dx_1 \wedge dx_2 = r\cos^2(\phi)dr \wedge d\phi - r\sin^2(\phi)d\phi \wedge dr = rdr \wedge d\phi.$$

Problem 2. Consider the metric tensor field

$$g = dx_1 \otimes dx_1 + dx_2 \otimes dx_2$$

of $\mathbb{E}^2$ and the differential two-form $dx_1 \wedge dx_2$.
(i) Express the metric tensor field in *parabolic coordinates*

$$x_1(u, v) = \frac{1}{2}(u^2 - v^2), \quad x_2(u, v) = uv.$$

(ii) Find the differential two-form $dx_1 \wedge dx_2$ in parabolic coordinates.
(iii) Find the *fixed points* of the map. The fixed points are the solutions of the system of equations

$$\frac{1}{2}(u^2 - v^2) = u, \qquad uv = v.$$

Solution 2. (i) We have $dx_1 = udu - vdv$, $dx_2 = vdu + udv$. Then

$$dx_1 \otimes dx_1 = u^2 du \otimes du + v^2 dv \otimes dv - uvdu \otimes dv - uvdv \otimes du$$
$$dx_2 \otimes dx_2 = v^2 du \otimes du + u^2 dv \otimes dv + uvdu \otimes dv + uvdv \otimes du$$

and it follows that

$$dx_1 \otimes dx_1 + dx_2 \otimes dx_2 = (u^2 + v^2)du \otimes du + (u^2 + v^2)dv \otimes dv.$$

(ii) With $du \wedge du = dv \wedge dv = 0$ and $du \wedge dv = -dv \wedge du$ we have

$$dx_1 \wedge dx_2 = u^2 du \wedge dv - v^2 dv \wedge du = (u^2 + v^2)du \wedge dv.$$

(iii) If $v = 0$, then we obtain $u^2 = 2u$ and find the solutions $u = 0$ and $u = 2$. Thus we have the fixed points $(0,0)$ and $(2,0)$. If $v \neq 0$, then $u = 1$ from the second equation and the first equation is $v^2 = -1$. Hence we have the two fixed points $(0,0)$ and $(2,0)$.

Problem 3. Consider the metric tensor field

$$g = dx_0 \otimes dx_0 - dx_1 \otimes dx_1.$$

Show that g is invariant under the transformation ($\alpha \in \mathbb{R}$)

$$\begin{pmatrix} x_0 \\ x_1 \end{pmatrix} \mapsto \begin{pmatrix} \cosh(\alpha) & \sinh(\alpha) \\ \sinh(\alpha) & \cosh(\alpha) \end{pmatrix} \begin{pmatrix} x_0 \\ x_1 \end{pmatrix}.$$

Solution 3. Since

$$dx_0 \mapsto \cosh(\alpha)dx_0 + \sinh(\alpha)dx_1, \quad dx_1 \mapsto \sinh(\alpha)dx_0 + \cosh(\alpha)dx_1$$

we obtain

$$dx_0 \otimes dx_0 \mapsto \cosh^2(\alpha)dx_0 \otimes dx_0 + \sinh^2(\alpha)dx_1 \otimes dx_1$$
$$+ \sinh(\alpha)\cosh(\alpha)dx_0 \otimes dx_1 + \sinh(\alpha)\cosh(\alpha)dx_1 \otimes dx_0$$

$$dx_1 \otimes dx_1 \mapsto \sinh^2(\alpha)dx_0 \otimes dx_0 + \cosh^2(\alpha)dx_1 \otimes dx_1$$
$$+ \sinh(\alpha)\cosh(\alpha)dx_0 \otimes dx_1 + \sinh(\alpha)\cosh(\alpha)dx_1 \otimes dx_0.$$

Then utilizing $\cosh^2(\alpha) - \sinh^2(\alpha) = 1$ we obtain

$$dx_0 \otimes dx_0 - dx_1 \otimes dx_1 \mapsto dx_0 \otimes dx_0 - dx_1 \otimes dx_1.$$

Problem 4. Let $a > 0$ and $b > 0$. Consider the *Hénon map* $\mathbf{f} : \mathbb{R}^2 \to \mathbb{R}^2$

$$f_1(x_1, x_2) = 1 + x_2 - ax_1^2, \quad f_2(x_1, x_2) = bx_1$$

with the differential one-forms $df_1 = dx_2 - 2ax_1 dx_1$, $df_2 = bdx_1$.
(i) Let $\Omega = dx_1 \wedge dx_2$. Find $\mathbf{f}^*(\Omega) = \mathbf{f}^*(dx_1 \wedge dx_2)$. Thus show that the Hénon map is invertible. Find the inverse.
(ii) Let $\alpha = x_1 dx_2 - x_2 dx_1$. Find $\mathbf{f}^*(\alpha)$.
(iii) Consider the *metric tensor field* of $\mathbb{E}^2$, $g = dx_1 \otimes dx_1 + dx_2 \otimes dx_2$. Find $\mathbf{f}^*(g)$.

Solution 4. (i) With $dx_1 \wedge dx_1 = 0$, $dx_2 \wedge dx_2 = 0$, $dx_1 \wedge dx_2 = -dx_2 \wedge dx_1$ we obtain

$$\begin{aligned} \mathbf{f}^*(dx_1 \wedge dx_2) &= df_1(x_1, x_2) \wedge df_2(x_1, x_2) = (d(1 + x_2 - ax_1^2)) \wedge (d(bx_1)) \\ &= (dx_2 - 2ax_1 dx_1) \wedge (bdx_1) \\ &= -bdx_1 \wedge dx_2. \end{aligned}$$

Thus the determinant of the functional matrix is given by $-b$. Since $b \neq 0$ the Hénon map is invertible. The inverse map is given by

$$x_{1,t} = \frac{1}{b} x_{2,t+1}, \quad x_{2,t} = x_{1,t+1} - 1 + \frac{a}{b} x_{2,t+1}^2.$$

(ii) We have

$$\begin{aligned} \mathbf{f}^*(\alpha) &= f_1 df_2 - f_2 df_1 = (1 + x_2 - ax_1^2)bdx_1 - bx_1(dx_2 - 2ax_1 dx_1) \\ &= b(1 + x_2 + ax_1^2)dx_1 - bx_1 dx_2. \end{aligned}$$

(iii) We have

$$\mathbf{f}^*(g) = (b^2 + 4a^2 x_1^2)dx_1 \otimes dx_1 - 2ax_1 dx_1 \otimes dx_2 - 2ax_1 dx_2 \otimes dx_1 + dx_2 \otimes dx_2$$

or written in matrix form

$$\begin{pmatrix} b^2 + 4a^2 x_1^2 & -2ax_1 \\ -2ax_1 & 1 \end{pmatrix}.$$

The determinant of this matrix is equal to b^2.

Problem 5. Consider the two-dimensional Euclidean space and the metric tensor field in polar coordinates $g = dr \otimes dr + r^2 d\theta \otimes d\theta$. Let $u \in \mathbb{R}$ and $R > 0$. Consider the transformation $(r, \theta) \mapsto (e^{u/R}, \theta)$. Find the metric tensor field.

Solution 5.   Since $dr(R) = d(e^{u/R}) = \frac{1}{R}e^{u/R}du$ we obtain

$$\widetilde{g} = \frac{1}{R^2}e^{2u/R}(du \otimes du + R^2 d\theta \otimes d\theta)$$

which is a conformal factor multiplying the natural metric tensor field for the space $\mathbb{S}^1 \otimes \mathbb{R}$.

Problem 6.   Consider the metric tensor field

$$g = dx_1 \otimes dx_1 + dx_2 \otimes dx_2 + dx_3 \otimes dx_3$$

of the three-dimensional Euclidean space. Express the metric tensor field in *cylindrical coordinates* $(\rho \geq 0)$

$$x_1(\rho, \phi) = \rho\cos(\phi), \quad x_2(\rho, \phi) = \rho\sin(\phi), \quad x_3 = x_3.$$

Solution 6.   We have

$$dx_1 = -\rho\sin(\phi)d\phi + \cos(\phi)d\rho$$
$$dx_2 = \rho\cos(\phi)d\phi + \sin(\phi)d\rho$$
$$dx_3 = dx_3.$$

Applying $\cos^2(\phi) + \sin^2(\phi) = 1$ we obtain the metric tensor field in cylindrical coordinates

$$\widetilde{g} = d\rho \otimes d\rho + \rho^2 d\phi \otimes d\phi + dx_3 \otimes dx_3.$$

Problem 7.   Consider *spherical coordinates* in $\mathbb{R}^3$

$$x_1(r, \theta, \phi) = r\sin(\theta)\cos(\phi)$$
$$x_2(r, \theta, \phi) = r\sin(\theta)\sin(\phi)$$
$$x_3(r, \theta, \phi) = r\cos(\theta)$$

$(r > 0, \theta \in [0, \pi], \phi \in [0, 2\pi))$ the metric tensor field

$$g = dx_1 \otimes dx_1 + dx_2 \otimes dx_2 + dx_3 \otimes dx_3$$

and the differential three-form $\Omega = dx_1 \wedge dx_2 \wedge dx_3$ (volume form in $\mathbb{R}^3$). Express the metric tensor field in spherical coordinates. Express the differential three-form in spherical coordinates.

Solution 7. We have

$$dx_1 = \sin(\theta)\cos(\phi)dr + r\cos(\theta)\cos(\phi)d\theta - r\sin(\theta)\sin(\phi)d\phi$$
$$dx_2 = \sin(\theta)\sin(\phi)dr + r\cos(\theta)\sin(\phi)d\theta + r\sin(\theta)\cos(\phi)d\phi$$
$$dx_3 = \cos(\theta)dr - r\sin(\theta)d\theta.$$

With $\cos^2(\theta) + \sin^2(\theta) = 1$ we find

$$\widetilde{g} = dr \otimes dr + r^2 d\theta \otimes d\theta + r^2 \sin^2(\theta)d\phi \otimes d\phi.$$

With $dr \wedge d\theta = -d\theta \wedge dr$, $dr \wedge d\phi = -d\phi \wedge dr$, $d\theta \wedge d\phi = -d\phi \wedge d\theta$ the volume form is given by

$$dx_1 \wedge dx_2 \wedge dx_3 = r^2 \sin(\theta)dr \wedge d\theta \wedge d\phi.$$

Note that (volume of the unit ball)

$$\int_{r=0}^{1} \int_{\theta=0}^{\pi} \int_{\phi=0}^{2\pi} r^2 \sin(\theta)drd\theta d\phi = \frac{4}{3}\pi.$$

Problem 8. Given the metric tensor field in $\mathbb{E}^3$. Express g in *quasi toroidal coordinates* (ρ, ϕ, ζ)

$$x_1(\rho, \phi, \zeta) = (R_0 + \rho\cos(\zeta))\cos(\phi)$$
$$x_2(\rho, \phi, \zeta) = (R_0 + \rho\cos(\zeta))\sin(\phi)$$
$$x_3(\rho, \phi, \zeta) = \rho\sin(\zeta)$$

where $R_0 > 0$. Note that $x_1^2 + x_2^2 = (R_0 + \rho\cos(\zeta))^2$.

Solution 8. We obtain

$$g_{QT} = d\rho \otimes d\rho + (R_0 + \rho\cos(\zeta))^2 d\phi \otimes d\phi + \rho^2 d\zeta \otimes d\zeta.$$

Problem 9. Consider the metric tensor field of $\mathbb{E}^3$. The *parabolic set* of unit-less coordinates (u, v, θ) is defined by a transformation of Cartesian coordinates ($0 \leq u \leq \infty, 0 \leq v \leq \infty$ and $0 \leq \phi \leq 2\pi$)

$$x_1(u, v, \phi) = auv\cos(\phi)$$
$$x_2(u, v, \phi) = auv\sin(\phi)$$
$$x_3(u, v, \phi) = \frac{1}{2}a(u^2 - v^2)$$

and $a > 0$ and a has the dimension of a length. Express g and the differential form $dx_1 \wedge dx_2 \wedge dx_3$ using *parabolic coordinates*.

Solution 9.   We have

$$dx_1 = a(v\cos(\phi)du + u\cos(\phi)dv - uv\sin(\phi)d\phi)$$
$$dx_2 = a(v\sin(\phi)du + u\sin(\phi)dv + uv\cos(\phi)d\phi)$$
$$dx_3 = audu - avdv.$$

Hence

$$\sum_{j=1}^{2} dx_j \otimes dx_j = a^2(v^2 du \otimes du + u^2 dv \otimes dv + u^2 v^2 d\phi \otimes d\phi + uv(du \otimes dv + dv \otimes du))$$

and

$$dx_3 \otimes dx_3 = a^2(u^2 du \otimes du + v^2 dv \otimes dv - uv(du \otimes dv + dv \otimes du)).$$

It follows that

$$\sum_{j=1}^{3}(dx_j \otimes dx_j) = a^2((u^2 + v^2)du \otimes du + (u^2 + v^2)dv \otimes dv + u^2 v^2 d\phi \otimes d\phi).$$

With $du \wedge dv = -dv \wedge du$, $du \wedge d\phi = -d\phi \wedge du$, $dv \wedge d\phi = -d\phi \wedge dv$ we obtain

$$dx_1 \wedge dx_2 = a^2(uv^2 du \wedge d\phi + u^2 v dv \wedge d\phi)$$

and

$$dx_1 \wedge dx_2 \wedge dx_3 = a^3(uv(u^2 + v^2)du \wedge dv \wedge d\phi).$$

Problem 10.   Let $\mathbb{E}^3$ be the three-dimensional Euclidean space. Express the metric tensor field in *oblate spheroidal coordinates* given by

$$x_1(\mu, \nu, \phi) = a\cosh(\mu)\cos(\nu)\cos(\phi)$$
$$x_2(\mu, \nu, \phi) = a\cosh(\mu)\cos(\nu)\sin(\phi)$$
$$x_3(\mu, \nu, \phi) = a\sinh(\mu)\sin(\nu)$$

where $\nu \in [-\pi/2, \pi/2]$, $\pi \in [-\pi, \pi]$, $\mu \geq 0$ and $a > 0$ with dimension length. Express the volume form $dx_1 \wedge dx_2 \wedge dx_3$ in $\mathbb{R}^3$ in oblate spheroidal coordinates. Note that

$$\frac{x_1^2 + x_2^2}{a^2 \cosh^2(\mu)} + \frac{x_3^2}{a^2 \sinh^2(\mu)} = 1.$$

Solution 10. With

$$dx_1 = a \sinh(\mu) \cos(\nu) \cos(\phi) d\mu - a \cosh(\mu) \sin(\nu) \cos(\phi) d\nu$$
$$- a \cosh(\mu) \cos(\nu) \sin(\phi) d\phi$$
$$dx_2 = a \sinh(\mu) \cos(\nu) \sin(\phi) d\mu - a \cosh(\mu) \sin(\nu) \sin(\phi) d\nu$$
$$+ a \cosh(\mu) \cos(\nu) \cos(\phi) d\phi$$
$$dx_3 = a \cosh(\mu) \sin(\nu) d\mu + a \sinh(\mu) \cos(\nu) d\nu$$

we obtain

$$g_O = a^2 (\sinh^2(\mu) \cos^2(\nu) + \cosh^2(\mu) \sin^2(\nu)) d\mu \otimes d\mu$$
$$+ a^2 (\sinh^2(\mu) \cos^2(\nu) + \cosh^2(\mu) \sin^2(\nu)) d\nu \otimes d\nu$$
$$+ a^2 \cosh^2(\mu) \cos^2(\nu) d\phi \otimes d\phi.$$

Then for the volume form $dx_1 \wedge dx_2 \wedge dx_3$ we find

$$a^3 (\sinh^2(\mu) + \sin^2(\nu)) \cosh(\mu) \cos(\nu) d\mu \wedge d\nu \wedge d\phi.$$

Problem 11. Let $a > 0$ and $r > 0$. Find the *Gaussian curvature* for the *torus* given by the parametrization

$$\mathbf{x}(u_1, u_2) = \begin{pmatrix} x_1(u_1, u_2) \\ x_2(u_1, u_2) \\ x_3(u_1, u_2) \end{pmatrix} = \begin{pmatrix} (a + r \cos(u_1)) \cos(u_2) \\ (a + r \cos(u_1)) \sin(u_2) \\ r \sin(u_1) \end{pmatrix}$$

where $0 < u_1 < 2\pi$ and $0 < u_2 < 2\pi$.

Solution 11. With

$$dx_1 = -a \sin(u_2) du_2 - r \sin(u_1) \cos(u_2) du_1 - r \cos(u_1) \sin(u_2) du_2$$
$$dx_2 = a \cos(u_2) du_2 - r \sin(u_1) \sin(u_2) du_1 + r \cos(u_1) \cos(u_2) du_2$$
$$dx_3 = r \cos(u_1) du_1$$

we find

$$\sum_{j=1}^{3} (dx_j \otimes dx_j) = r^2 du_1 \otimes du_1 + (a + r \cos(u_1))^2 du_2 \otimes du_2.$$

Thus the coefficients of the first fundamental form are given by

$$E = r^2, \quad G = (a + r \cos(u_1))^2, \quad F = 0$$

and $\sqrt{EG - F^2} = r(r\cos(u_1) + a)$. The coefficients of the first fundamental form could also be calculated from

$$
\frac{\partial \mathbf{x}}{\partial u_1} = \begin{pmatrix} -r\sin(u_1)\cos(u_2) \\ -r\sin(u_1)\sin(u_2) \\ r\cos(u_1 \end{pmatrix}, \quad \frac{\partial \mathbf{x}}{\partial u_2} = \begin{pmatrix} -(a + r\cos(u_1))\sin(u_2) \\ (a + r\cos(u_1))\cos(u_2) \\ 0 \end{pmatrix}.
$$

Then

$$
E(\mathbf{u}) = \frac{\partial \mathbf{x}}{\partial u_1} \cdot \frac{\partial \mathbf{x}}{\partial u_1} = r^2, \qquad F(\mathbf{u}) = \frac{\partial \mathbf{x}}{\partial u_1} \cdot \frac{\partial \mathbf{x}}{\partial u_2} = 0
$$

$$
G(\mathbf{u}) = \frac{\partial \mathbf{x}}{\partial u_2} \cdot \frac{\partial \mathbf{x}}{\partial u_2} = (a + r\cos(u_1))^2
$$

where $\cdot$ denotes the scalar product. The normal vector is given by

$$
\mathbf{n}(u_1, u_2) = \frac{(\partial \mathbf{x}/\partial u_1) \times (\partial \mathbf{x}/\partial u_2)}{\|(\partial \mathbf{x}/\partial u_1) \times (\partial \mathbf{x}/\partial u_2)\|}.
$$

To find the coefficients of the second fundamental form we need the second order derivatives

$$
\frac{\partial^2 \mathbf{x}}{\partial u_1^2} = \begin{pmatrix} -r\cos(u_1)\cos(u_2) \\ -r\cos(u_1)\sin(u_2) \\ -r\sin(u_1) \end{pmatrix}, \quad \frac{\partial^2 \mathbf{x}}{\partial u_2^2} = \begin{pmatrix} -(a + r\cos(u_1))\cos(u_2) \\ -(a + r\cos(u_1))\sin(u_2 \\ 0 \end{pmatrix}
$$

$$
\frac{\partial^2 \mathbf{x}}{\partial u_1 u_2} = \begin{pmatrix} r\sin(u_1)\sin(u_2) \\ -r\sin(u_1)\cos(u_2) \\ 0 \end{pmatrix}.
$$

With

$$
L := \mathbf{n} \cdot \frac{\partial \mathbf{x}}{\partial u_1^2}, \quad M := \mathbf{n} \cdot \frac{\partial \mathbf{x}}{\partial u_1 \partial u_2}, \quad N := \mathbf{n} \cdot \frac{\partial \mathbf{x}}{\partial u_2^2}
$$

it follows that $L = r$, $M = 0$, $N = \cos(u_1)(a + r\cos(u_1))$. Thus the Gaussian curvature is given by

$$
K = \frac{LN - M^2}{EG - F^2} = \frac{\cos(u_1)}{r(a + r\cos(u_1))}.
$$

If $u = \pi/2$ and $u = 3\pi/2$ we have $K = 0$.

Problem 12. Let $a > b > 0$ and define $\mathbf{f} : \mathbb{R}^2 \to \mathbb{R}^3$ by

$$
\mathbf{f}(\theta, \phi) = \begin{pmatrix} f_1(\theta, \phi) \\ f_2(\theta, \phi) \\ f_3(\theta, \phi) \end{pmatrix} = \begin{pmatrix} (a + b\cos(\phi))\cos(\theta) \\ (a + b\cos(\phi))\sin(\theta) \\ b\sin(\phi) \end{pmatrix}.
$$

The function **f** is a parametrized *torus* $\mathbb{T}^2$ in $\mathbb{R}^3$. Consider the metric tensor field

$$g = dx_1 \otimes dx_1 + dx_2 \otimes dx_2 + dx_3 \otimes dx_3.$$

(i) Calculate $g|_{T^2}$.
(ii) Calculate the Christoffel symbols Γ^m_{ab} from $g|_{T^2}$.
(iii) Give the differential equations for the geodesics. The differential equations of the *geodesics* are given by

$$\frac{d^2y^\alpha}{ds^2} + \sum_{\beta=1}^{2}\sum_{\gamma=1}^{2} \Gamma^\alpha_{\beta\gamma} \frac{dy^\beta}{ds} \frac{dy^\gamma}{ds} = 0, \quad \alpha = 1, 2$$

with $y^1 = \theta$ and $y^2 = \phi$.

Solution 12.   (i) Since

$$x_1(\theta, \phi) = a\cos(\theta) + b\cos(\theta)\cos(\phi)$$
$$x_2(\theta, \phi) = a\sin(\theta) + b\sin(\theta)\cos(\phi)$$
$$x_3(\theta, \phi) = b\sin(\phi)$$

we find

$$dx_1 = (-a\sin(\theta) - b\cos(\phi)\sin(\theta))d\theta - b\sin(\phi)\cos(\theta)d\phi$$
$$dx_2 = \cos(\theta)(a + b\cos(\phi))d\theta - b\sin(\phi)\sin(\theta)d\phi$$
$$dx_3 = b\cos(\phi)d\phi.$$

Thus

$$dx_1 \otimes dx_1 = -\sin^2(\theta)(a + b\cos(\phi))^2 d\theta \otimes d\theta + b^2\sin^2(\phi)\cos^2(\theta)d\phi \otimes d\phi$$
$$dx_2 \otimes dx_2 = \cos^2(\theta)(a + b\cos(\phi))^2 d\theta \otimes d\theta + b^2\sin^2(\phi)\sin^2(\theta)d\phi \otimes d\phi$$
$$dx_3 \otimes dx_3 = b^2\cos^2(\phi)d\phi \otimes d\phi.$$

Using $\sin^2(\alpha) + \cos^2(\alpha) = 1$ we obtain

$$g_{T^2} = (a + b\cos(\phi))^2 d\theta \otimes d\theta + b^2 d\phi \otimes d\phi.$$

Thus we have

$$(g_{ij}) = \begin{pmatrix} (a + b\cos(\phi))^2 & 0 \\ 0 & b^2 \end{pmatrix}$$

and therefore the inverse matrix is given by

$$(g^{ij}) = \begin{pmatrix} 1/(a + b\cos(\phi))^2 & 0 \\ 0 & 1/b^2 \end{pmatrix}.$$

Let $y^1 = \theta, y^2 = \phi$. The *Christoffel symbols* are defined as $(\alpha, \beta, \gamma, \delta = 1, 2)$

$$\Gamma^\alpha_{\beta\gamma}(y^1, y^2) := \frac{1}{2} \sum_{\delta=1}^{2} g^{\alpha\delta} \left(\frac{\partial g_{\beta\delta}}{\partial y^\gamma} + \frac{\partial g_{\gamma\delta}}{\partial y^\beta} - \frac{\partial g_{\beta\gamma}}{\partial y^\delta} \right).$$

Then we find

$$\Gamma^1_{11} = 0, \quad \Gamma^1_{12} = -\frac{b\sin(\phi)}{a + b\cos(\phi)}, \quad \Gamma^1_{21} = -\frac{b\sin(\phi)}{a + b\cos(\phi)}, \quad \Gamma^1_{22} = 0$$

and

$$\Gamma^2_{11} = \frac{(a + b\cos(\phi))\sin(\phi)}{b}, \quad \Gamma^2_{12} = 0, \quad \Gamma^2_{21} = 0, \quad \Gamma^2_{22} = 0.$$

(iii) We obtain

$$\frac{d^2\theta}{ds^2} = \frac{2b\sin(\phi)}{a + b\cos(\phi)} \frac{d\theta}{ds} \frac{d\phi}{ds}$$

$$\frac{d^2\phi}{ds^2} = -\frac{(a + b\cos(\phi))\sin(\phi)}{b} \left(\frac{d\theta}{ds} \right)^2.$$

Problem 13.   Consider the smooth surface

$$\mathbf{x}(\mathbf{u}) = \begin{pmatrix} x_1(u_1, u_2) \\ x_2(u_1, u_2) \\ x_3(u_1, u_2) \end{pmatrix} = \begin{pmatrix} u_1 \\ u_2 \\ u_1 u_2 \end{pmatrix}$$

embedded in the three-dimensional Euclidean space $\mathbb{E}^3$.
(i) Find the Gaussian curvature and mean curvature of the surface.
(ii) Find the curvature scalar. Owing to

$$dx_1 = du_1, \quad dx_2 = du_2, \quad dx_3 = u_1 du_2 + u_2 du_1$$

we have the metric tensor field for the surface

$$g_S = (1 + u_2^2)du_1 \otimes du_1 + (1 + u_1^2)du_2 \otimes du_2 + u_1 u_2(du_1 \otimes du_2 + du_2 \otimes du_1)$$

with the corresponding symmetric matrix

$$\begin{pmatrix} 1 + u_2^2 & u_1 u_2 \\ u_1 u_2 & 1 + u_1^2 \end{pmatrix}$$

and the inverse matrix

$$\frac{1}{1 + u_1^2 + u_2^2} \begin{pmatrix} 1 + u_1^2 & -u_1 u_2 \\ -u_1 u_2 & 1 + u_2^2 \end{pmatrix}.$$

Solution 13. (i) The rank of the 2×3 matrix

$$\begin{pmatrix} \partial x_1/\partial u_1 & \partial x_2/\partial u_1 & \partial x_3/\partial u_1 \\ \partial x_1/\partial u_2 & \partial x_2/\partial u_2 & \partial x_3/\partial u_2 \end{pmatrix} = \begin{pmatrix} 1 & 0 & u_2 \\ 0 & 1 & u_1 \end{pmatrix}$$

is equal to 2. Now

$$\frac{\partial \mathbf{x}}{\partial u_1} = \begin{pmatrix} 1 \\ 0 \\ u_2 \end{pmatrix}, \qquad \frac{\partial \mathbf{x}}{\partial u_2} = \begin{pmatrix} 0 \\ 1 \\ u_1 \end{pmatrix}$$

and therefore $E(\mathbf{u}) = 1 + u_2^2$, $F(\mathbf{u}) = u_1 u_2$, $G(\mathbf{u}) = 1 + u_1^2$. The vector product is given by

$$\frac{\partial \mathbf{x}}{\partial u_1} \times \frac{\partial \mathbf{x}}{\partial u_2} = \begin{pmatrix} -u_2 \\ -u_1 \\ 1 \end{pmatrix}$$

and thus the normal vector follows as

$$\mathbf{n}(\mathbf{u}) = \frac{1}{\sqrt{1 + u_1^2 + u_2^2}} \begin{pmatrix} -u_2 \\ -u_1 \\ 1 \end{pmatrix}.$$

Now

$$\frac{\partial^2 \mathbf{x}}{\partial u_1^2} = \begin{pmatrix} 0 \\ 0 \\ 0 \end{pmatrix}, \qquad \frac{\partial^2 \mathbf{x}}{\partial u_1 \partial u_2} = \begin{pmatrix} 0 \\ 0 \\ 1 \end{pmatrix}, \qquad \frac{\partial^2 \mathbf{x}}{\partial u_2^2} = \begin{pmatrix} 0 \\ 0 \\ 0 \end{pmatrix}.$$

Then $L(\mathbf{u}) = 0$, $N(\mathbf{u}) = 0$, $M(\mathbf{u}) = 1/\sqrt{1 + u_1^2 + u_2^2}$. The Gaussian curvature follows as

$$\frac{LN - M^2}{EG - F^2} = -\frac{1}{(1 + u_1^2 + u_2^2)^2}$$

and the mean curvature is

$$\frac{NE - 2MF + LG}{EG - F^2} = -\frac{u_1 u_2}{(1 + u_1^2 + u_2^2)^{3/2}}.$$

(ii) The Christoffel symbols are defined as

$$\Gamma^a_{mn} = \frac{1}{2} \sum_{b=1}^{2} g^{ab} \left(\frac{\partial g_{bm}}{\partial x_n} + \frac{\partial g_{bn}}{\partial x_m} - \frac{\partial g_{mn}}{\partial x_b} \right)$$

where $a, m, n = 1, 2$. The nonzero coefficients are

$$\Gamma^1_{12} = \Gamma^1_{21} = \frac{u_2}{1 + u_1^2 + u_2^2}, \qquad \Gamma^2_{12} = \Gamma^2_{21} = \frac{u_1}{1 + u_1^2 + u_2^2}.$$

The Ricci tensor is given by

$$R_{mq} = \sum_{s=1}^{2} \Gamma_{mq,s}^{s} - \sum_{s=1}^{2} \Gamma_{ms,q}^{s} + \sum_{s=1}^{2}\sum_{n=1}^{2} \Gamma_{ns}^{s}\Gamma_{mq}^{n} - \sum_{s=1}^{2}\sum_{n=1}^{2} \Gamma_{nq}^{s}\Gamma_{ms}^{n}.$$

The nonzero coefficients are

$$R_{11} = -\frac{1+u_2^2}{(1+u_1^2+u_2^2)^2}, \quad R_{22} = -\frac{1+u_1^2}{(1+u_1^2+u_2^2)^2}$$

$$R_{12} = R_{21} = -\frac{u_1 u_2}{(1+u_1^2+u_2^2)^2}.$$

Now

$$R_k^j = \sum_{m=1}^{2} g^{jm} R_{km}.$$

We obtain

$$R_1^1 = R_2^2 = -\frac{1}{(1+u_1^2+u_2^2)^2}$$

and $R_1^2 = R_2^1 = 0$. Hence the curvature scalar R is given by

$$R = R_1^1 + R_2^2 = -\frac{2}{(1+u_1^2+u_2^2)^2}$$

which is twice the Gaussian curvature.

Problem 14.   Let $f : \mathbb{R}^2 \to \mathbb{R}$ be a smooth function. Consider the surface

$$x_1(u_1,u_2) = u_1, \quad x_2(u_1,u_2) = u_2, \quad x_3(u_1,u_2) = f(u_1,u_2)$$

in the three-dimensional Euclidean space. Find the *Gaussian curvature* and the *mean curvature*. For $f(u_1,u_2) = u_1 u_2$ the surface is the *Monge patch*.

Solution 14.   With

$$dx_1 = du_1, \quad dx_2 = du_2, \quad dx_3 = \frac{\partial f}{\partial u_1}du_1 + \frac{\partial f}{\partial u_2}du_2$$

we obtain

$$g_S = \left(1 + \left(\frac{\partial f}{\partial u_1}\right)^2\right)du_1 \otimes du_1 + \left(1 + \left(\frac{\partial f}{\partial u_2}\right)^2\right)du_2 \otimes du_2$$

$$+ \left(\frac{\partial f}{\partial u_1}\frac{\partial f}{\partial u_2}\right)du_1 \otimes du_2 + \left(\frac{\partial f}{\partial u_2}\frac{\partial f}{\partial u_1}\right)du_2 \otimes du_1.$$

With

$$\frac{\partial^2 \mathbf{x}}{\partial u_1^2} = \begin{pmatrix} 0 \\ 0 \\ \partial^2 f/\partial u_1^2 \end{pmatrix}, \quad \frac{\partial^2 \mathbf{x}}{\partial u_2^2} = \begin{pmatrix} 0 \\ 0 \\ \partial^2 f/\partial u_2^2 \end{pmatrix}, \quad \frac{\partial^2 \mathbf{x}}{\partial u_1 \partial u_2} = \begin{pmatrix} 0 \\ 0 \\ \partial^2 f/\partial u_1 \partial u_2 \end{pmatrix}$$

it follows that the Gausssian curvature is given by

$$K = \frac{(\partial^2 f/\partial u_1^2)(\partial^2 f/\partial u_2^2) - (\partial^2 f/\partial u_1 \partial u_2)^2}{(1 + (\partial f/\partial u_1)^2 + (\partial f/\partial u_2)^2)^2}$$

and the mean curvature is

$$H = \frac{(1 + (\partial f/\partial u_2)^2)\frac{\partial^2 f}{\partial u_1^2} - 2\frac{\partial f}{\partial u_1}\frac{\partial f}{\partial u_2}\frac{\partial^2 f}{\partial u_1 \partial u_2} + (1 + (\partial f/\partial u_1)^2)\frac{\partial^2 f}{\partial u_2^2}}{2(1 + (\partial f/\partial u_1)^2 + (\partial f/\partial u_2)^2)^{3/2}}.$$

Problem 15. Consider the *Enneper surface* in $\mathbb{R}^3$

$$\mathbf{x}(u_1, u_2) = \begin{pmatrix} x_1(u_1, u_2) \\ x_2(u_1, u_2) \\ x_3(u_1, u_2) \end{pmatrix} = \begin{pmatrix} u_1 - u_1^3/3 + u_1 u_2^2 \\ u_2 - u_2^3/3 + u_1^2 u_2 \\ u_1^2 - u_2^2 \end{pmatrix}.$$

Find dx_1, dx_2, dx_3 and $dx_1 \otimes dx_1 + dx_2 \otimes dx_2 + dx_3 \otimes dx_3$.

Solution 15. We have

$$dx_1 = (1 - u_1^2 + u_2^2)du_1 + 2u_1 u_2 du_2$$
$$dx_2 = 2u_1 u_2 du_1 + (1 + u_1^2 - u_2^2)du_2$$
$$dx_3 = 2u_1 du_1 - 2u_2 du_2.$$

It follows that $dx_1 \otimes dx_1 + dx_2 \otimes dx_2 + dx_3 \otimes dx_3$ is given by

$$g_E = (1 + u_1^2 + u_2^2)^2 du_1 \otimes du_1 + (1 + u_1^2 + u_2^2)^2 du_2 \otimes du_2.$$

The coefficients of the first fundamental form are $E = G = (1 + u_1^2 + u_2^2)^2$ and $F = 0$. The principal curvatures are

$$k_1 = \frac{2}{(1 + u_1^2 + u_2^2)^2}, \quad k_2 = -\frac{2}{(1 + u_1^2 + u_2^2)^2}.$$

Problem 16. Let $\mathbb{E}^3$ be the three-dimensional Euclidean space and $a > 0$. Show that the *helicoid*

$$\mathbf{x}(u_1, u_2) = \begin{pmatrix} x_1(u_1, u_2) \\ x_2(u_1, u_2) \\ x_3(u_1, u_2) \end{pmatrix} = \begin{pmatrix} a \sinh(u_2) \cos(u_1) \\ a \sinh(u_2) \sin(u_1) \\ a u_1 \end{pmatrix}$$

is a *minimal surface* in $\mathbb{E}^3$. The condition for a minimal surface is

$$\frac{\partial \mathbf{x}}{\partial u_1} \cdot \frac{\partial \mathbf{x}}{\partial u_1} - \frac{\partial \mathbf{x}}{\partial u_2} \cdot \frac{\partial \mathbf{x}}{\partial u_2} = 0, \quad \frac{\partial \mathbf{x}}{\partial u_1} \cdot \frac{\partial \mathbf{x}}{\partial u_2} = 0.$$

With the notation given above we have $E - G = 0$, $F = 0$.

Solution 16.   We have

$$dx_1 = -a\sin(u_1)\sinh(u_2)du_1 + a\cos(u_1)\cosh(u_2)du_2$$
$$dx_2 = a\cos(u_1)\sinh(u_2)du_1 + a\sin(u_1)\cosh(u_2)du_2$$
$$dx_3 = adu_1.$$

Then

$$dx_1 \otimes dx_1 + dx_2 \otimes dx_2 = a^2\sinh^2(u_2)du_1 \otimes du_1 + a^2\cosh^2(u_2)du_2 \otimes du_2$$

and $dx_3 \otimes dx_3 = a^2 du_1 \otimes du_1$. Hence

$$g|_H = a^2\cosh^2(u_2)du_1 \otimes du_1 + a^2\cosh^2(u_2)du_2 \otimes du_2.$$

Therefore $E = G = a^2\cosh^2(u_2)$, $F = 0$ and we have a minimal surface.

Problem 17.   Consider the three-dimensional Euclidean space $\mathbb{E}^3$.
(i) Consider the map $\mathbf{f} : \mathbb{R}^2 \to \mathbb{R}^3$

$$\mathbf{f}(u_1, u_2) = \begin{pmatrix} f_1(u_1, u_2) \\ f_2(u_1, u_2) \\ f_3(u_1, u_2) \end{pmatrix} = \begin{pmatrix} u_1 + u_2 \\ u_1 - u_2 \\ u_1 u_2 \end{pmatrix}.$$

Find $\mathbf{f}^*(g)$.
(ii) Consider the *surface of revolution*

$$\mathbf{x}(u_1, u_2) = \begin{pmatrix} x_1(u_1, u_2) \\ x_2(u_1, u_2) \\ x_3(u_1, u_2) \end{pmatrix} = \begin{pmatrix} \cos(u_1)\cosh(u_2) \\ \sin(u_1)\cosh(u_2) \\ u_2 \end{pmatrix}$$

$(0 < u_1 < 2\pi, \; -\infty < u_2 < \infty)$. Find the metric tensor field for the surface of revolution.

Solution 17.   (i) We have

$$dx_1 = du_1 + du_2, \quad dx_2 = du_1 - du_2, \quad dx_3 = u_1 du_2 + u_2 du_1.$$

It follows that

$$dx_1 \otimes dx_1 = du_1 \otimes du_1 + du_1 \otimes du_2 + du_2 \otimes du_1 + du_2 \otimes du_2$$
$$dx_2 \otimes dx_2 = du_1 \otimes du_1 - du_1 \otimes du_2 - du_2 \otimes du_1 + du_2 \otimes du_2$$
$$dx_3 \otimes dx_3 = u_2^2 du_1 \otimes du_1 + u_1 u_2 (du_1 \otimes du_2 + du_2 \otimes du_1) + u_1^2 du_2 \otimes du_2.$$

Hence

$$\mathbf{f}(g) = (2 + u_2^2) du_1 \otimes du_1 + (2 + u_1^2) du_2 \otimes du_2 + u_1 u_2 (du_1 \otimes du_2 + du_2 \otimes du_1).$$

(ii) Since

$$dx_1 = -\sin(u_1)\cosh(u_2) du_1 + \cos(u_1)\sinh(u_2) du_2$$
$$dx_2 = \cos(u_1)\cosh(u_2) du_1 + \sin(u_1)\sinh(u_2) du_2$$
$$dx_3 = du_2$$

and utilizing $\cosh^2(\alpha) - \sinh^2(\alpha) = 1$ we obtain

$$\widetilde{g} = \cosh^2(u_2) du_1 \otimes du_1 + \sinh^2(u_2) du_2 \otimes du_2 + du_2 \otimes du_2$$
$$= \cosh^2(u_2) du_1 \otimes du_1 + \cosh^2(u_2) du_2 \otimes du_2.$$

Problem 18. Consider the three-dimensional Euclidean space $\mathbb{E}^3$ and the invertible *chaotic map* $\mathbf{f} : \mathbb{R}^3 \to \mathbb{R}^3$

$$\mathbf{f}(x_1, x_2, x_3) = \begin{pmatrix} f_1(x_1, x_2, x_3) \\ f_2(x_1, x_2, x_3) \\ f_3(x_1, x_2, x_3) \end{pmatrix} = \begin{pmatrix} x_1 x_2 - x_3 \\ x_1 \\ x_2 \end{pmatrix}.$$

Find the inverse of the map. Find $\mathbf{f}^*(g)$.

Solution 18. With $x_1 = f_2$, $x_2 = f_3$, $x_3 = x_1 x_2 - f_1$ we find

$$f_1^{-1}(x_1, x_2, x_3) = x_2, \quad f_2^{-1}(x_1, x_2, x_3) = x_3, \quad f_3^{-1}(x_1, x_2, x_3) = x_2 x_3 - x_1.$$

With

$$df_1 = x_2 dx_1 + x_1 dx_2 - dx_3, \quad df_2 = dx_1, \quad df_3 = dx_2$$

we obtain

$$\mathbf{f}(g) = (1 + x_2^2) dx_1 \otimes dx_1 + (1 + x_1^2) dx_2 \otimes dx_2 + dx_3 \otimes dx_3$$
$$+ x_1 x_2 dx_1 \otimes dx_2 + x_1 x_2 dx_2 \otimes dx_1 - x_1 dx_2 \otimes dx_3 - x_1 dx_3 \otimes dx_2$$
$$- x_2 dx_1 \otimes dx_3 - x_2 dx_3 \otimes dx_1$$

with the corresponding 3×3 matrix

$$G(x_1, x_2, x_3) = \begin{pmatrix} 1 + x_2^2 & x_1 x_2 & -x_2 \\ x_1 x_2 & 1 + x_1^2 & -x_1 \\ -x_2 & -x_1 & 1 \end{pmatrix}.$$

The determinant of the matrix is given by $\det(G(x_1, x_2, x_3)) = 1$ and the trace is given by $\operatorname{tr}(G(x_1, x_2, x_3)) = 3 + x_1^2 + x_2^2$. The inverse of the matrix is

$$G^{-1}(x_1, x_2, x_3) = \begin{pmatrix} 1 & 0 & x_2 \\ 0 & 1 & x_1 \\ x_2 & x_1 & 1 + x_1 + x_2 \end{pmatrix}.$$

Find the curvature for the metric tensor field $\mathbf{f}^*(g)$.

Problem 19.   Consider the three-dimensional Euclidean space $\mathbb{E}^3$ and the surface

$$\begin{aligned} x_1(u_1, u_2) &= f_1(u_1) \\ x_2(u_1, u_2) &= f_2(u_1) \cos(u_2) \\ x_3(u_1, u_2) &= f_2(u_1) \sin(u_2) \end{aligned}$$

embedded in $\mathbb{E}^3$, where f_1, f_2 are smooth functions. Find the metric tensor field.

Solution 19.   We have

$$dx_1 = \frac{df_1}{du_1} du_1$$

$$dx_2 = \frac{df_2}{du_1} \cos(u_2) du_1 - f_2(u_1) \sin(u_2) du_2$$

$$dx_3 = \frac{df_2}{du_1} \sin(u_2) du_1 + f_2(u_1) \cos(u_2) du_2.$$

Then

$$dx_1 \otimes dx_1 = \left(\frac{df_1}{du_1} \right)^2 du_1 \otimes du_1$$

$$dx_2 \otimes dx_2 = \left(\frac{df_2}{du_1} \right)^2 du_1 \otimes du_1 + f_2^2(u_1) \sin^2(u_2) du_2 \otimes du_2$$

$$\qquad - f_2 \frac{df_2}{du_1} \cos(u_2) \sin(u_2)(du_1 \otimes du_2 + du_2 \otimes du_1)$$

$$dx_3 \otimes dx_3 = \left(\frac{df_2}{du_1} \right)^2 du_1 \otimes du_1 + f_1^2(u_1) \cos^2(u_2) du_2 \otimes du_2$$

$$+ f_2 \frac{df_2}{du_1} \cos(u_2) \sin(u_2)(du_1 \otimes du_2 + du_2 \otimes du_1).$$

Hence

$$\widetilde{g} = \left(\left(\frac{df_1}{du_1} \right)^2 + \left(\frac{df_2}{du_1} \right)^2 \right) du_1 \otimes du_1 + f_2^2(u_1) du_2 \otimes du_2.$$

The condition that we have a *minimal surface* is

$$\left(\frac{df_1}{du_1} \right)^2 + \left(\frac{df_2}{du_1} \right)^2 - f_2^2(u_1) = 0.$$

Problem 20.   For the *Poincaré upper half plane*

$$\mathbb{H}_+^2 := \{ \, (x_1, x_2) \, : \, x_2 > 0, x_1 \in \mathbb{R} \, \}$$

the *Poincaré metric tensor field* is given by

$$g = \frac{1}{x_2^2}(dx_1 \otimes dx_1 + dx_2 \otimes dx_2).$$

It defines a two-dimensional Riemann manifold with Gaussian curvature given by -1.
(i) Show that the metric tensor field admits the symmetry $(x_1, x_2) \to (-x_1, x_2)$ and the transformation $(z = x_1 + ix_2)$

$$z \to z' = \frac{az + b}{cz + d}, \qquad a, b, c, d \in \mathbb{R}, \qquad ad - bc = 1$$

preserve the metric tensor field.
(ii) For the Poincaré metric tensor field the corresponding matrix is

$$g = (g_{mn}) = \begin{pmatrix} 1/x_2^2 & 0 \\ 0 & 1/x_2^2 \end{pmatrix}$$

with the inverse matrix

$$g^{-1} = (g^{mn}) = \begin{pmatrix} x_2^2 & 0 \\ 0 & x_2^2 \end{pmatrix}.$$

Find the Christoffel symbols given by

$$\Gamma_{mn}^a := \frac{1}{2} \sum_{b=1}^{2} g^{ab} \left(\frac{\partial g_{bm}}{\partial x_n} + \frac{\partial g_{bn}}{\partial x_m} - \frac{\partial g_{mn}}{\partial x_b} \right).$$

Find the *geodesic equation* given by

$$\frac{d^2x_1}{ds^2} + \sum_{\beta=1,\gamma=1}^{2} \Gamma^1_{\beta,\gamma} \frac{dx_\beta}{ds} \frac{dx_\gamma}{ds} = 0, \qquad \frac{d^2x_2}{ds^2} + \sum_{\beta=1,\gamma=1}^{2} \Gamma^2_{\beta,\gamma} \frac{dx_\beta}{ds} \frac{dx_\gamma}{ds} = 0.$$

(iii) Find the surface element dS and the Laplace operator Δ.
(iv) Consider the conformal mapping from the upper half plane $\{z = x_1 + ix_2 : x_2 > 0\}$ to the unit disk $\{w = re^{i\theta} : r \leq 1\}$

$$w(z) = \frac{iz+1}{z+i}.$$

Express g in r and θ.

Solution 20.   (i) We have

$$g = \frac{-4dz \otimes d\bar{z}}{(z - \bar{z})^2}, \quad \bar{z} = x_1 - ix_2.$$

(ii) For the eight Christoffel symbols we find

$$\Gamma^1_{11} = 0, \quad \Gamma^1_{12} = -\frac{1}{x_2}, \quad \Gamma^1_{21} = -\frac{1}{x_2}, \quad \Gamma^1_{22} = 0$$

$$\Gamma^2_{11} = \frac{1}{x_2}, \quad \Gamma^2_{12} = 0, \quad \Gamma^2_{21} = 0, \quad \Gamma^2_{22} = -\frac{1}{x_2}.$$

(ii) Then the geodesic equations are given by

$$\frac{d^2x_1}{ds^2} + \Gamma^1_{12}\frac{dx_1}{ds}\frac{dx_2}{ds} + \Gamma^1_{21}\frac{dx_2}{ds}\frac{dx_1}{ds} = 0$$

$$\frac{d^2x_2}{ds^2} + \Gamma^2_{11}\frac{dx_1}{ds}\frac{dx_1}{ds} + \Gamma^2_{22}\frac{dx_2}{ds}\frac{dx_2}{ds} = 0.$$

Hence

$$x_2\frac{d^2x_1}{ds^2} - 2\frac{dx_1}{ds}\frac{dx_2}{ds} = 0, \qquad x_2\frac{d^2x_2}{ds^2} + \frac{dx_1}{ds}\frac{dx_1}{ds} - \frac{dx_2}{ds}\frac{dx_2}{ds} = 0.$$

The first equation can be integrated and we arrive at

$$\frac{1}{x_2^2}\frac{dx_1}{ds} = \frac{1}{R}$$

where R is the constant of integration. Further integration provides

$$ds = \frac{dx_2}{x_2\sqrt{1 - x_2^2/R^2}} = \frac{dt}{\sin(t)}.$$

where we set $x_2 = R\sin(t)$. Thus

$$x_1 = -R\cos(t), \quad x_2 = \sin(t) \quad x_2 > 0.$$

(iii) We find

$$dS = \frac{1}{x_2^2}(dx_1 \wedge dx_2), \qquad \Delta = x_2^2\left(\frac{\partial^2}{\partial x_1^2} + \frac{\partial^2}{\partial x_2^2}\right).$$

(iv) We obtain

$$g = \frac{4}{(1-r^2)^2}(dr \otimes dr + r^2 d\theta \otimes d\theta).$$

Problem 21. Let $a > 0$ and $0 \leq r \leq a$, $0 \leq \phi \leq 2\pi$. Consider the three-dimensional Euclidean space $\mathbb{E}^3$ and the transformation

$$x_1(r, \phi) = r\cos(\phi)$$
$$x_2(r, \phi) = r\sin(\phi)$$
$$x_3(r, \phi) = \frac{1}{a}x_1(r, \phi)x_2(r, \phi) = \frac{r^2}{a}\cos(\phi)\sin(\phi).$$

(i) Express the metric tensor field applying this transformation and find g_{rr}, $g_{\phi\phi}$, $g_{r\phi} = g_{\phi r}$ and $\sqrt{g_{rr}g_{\phi\phi} - g_{r\phi}}$.
(ii) Calculate the integral

$$S = \int_{\phi=0}^{2\pi}\int_{r=0}^{a}\sqrt{g_{rr}g_{\phi\phi} - g_{r\phi}}\,drd\phi.$$

Discuss.

Solution 21. (i) We obtain

$$dx_1 = \cos(\phi)dr - r\sin(\phi)d\phi, \qquad dx_2 = \sin(\phi)dr + r\cos(\phi)d\phi$$

$$dx_3 = \frac{r}{a}\sin(2\phi)dr + \frac{r^2}{a}\cos(2\phi)d\phi$$

and

$$g = \left(1 + \frac{r^2}{a^2}\sin^2(2\phi)\right)dr \otimes dr + \left(r^2 + \frac{r^4}{a^2}\cos^2(2\phi)\right)d\phi \otimes d\phi$$

$$+ \frac{r^3}{a^2}\cos(2\phi)\sin(2\phi)(dr \otimes d\phi + d\phi \otimes dr)$$

and hence

$$g_{rr} = \left(1 + \frac{r^2}{a^2}\sin^2(2\phi)\right), \quad g_{\phi\phi} = \left(r^2 + \frac{r^4}{a^2}\cos^2(2\phi)\right)$$

$$g_{r\phi} = g_{\phi r} = \frac{r^3}{a^2}\cos(2\phi)\sin(2\phi).$$

Thus

$$\sqrt{g_{rr}g_{\phi\phi} - g_{r\phi}} = r\sqrt{1 + \frac{r^2}{a^2}}.$$

The integral provides

$$S = \int_{\phi=0}^{2\pi}\int_{r=0}^{a} r\sqrt{1 + \frac{r^2}{a^2}} = \frac{2}{3}\pi a^2(2\sqrt{2} - 1).$$

Thus we calculated the surface given by the intersection of a cylinder $x_1^2 + x_2^2 = a^2$ and a hyperbolic paraboloid $x_3 = x_1 x_2/a$.

Problem 22. Consider the metric tensor field of $\mathbb{E}^4$ and the compact manifold

$$\mathbb{S}^3 = \{(x_1, x_2, x_3, x_4) \in \mathbb{R}^4 : x_1^2 + x_2^2 + x_3^2 + x_4^2 = 1\}.$$

We set

$$x_1 + ix_2 = \cos(\theta/2)\exp(i(\psi+\phi)/2), \quad x_3 + ix_4 = \sin(\theta/2)\exp(i(\psi-\phi)/2)$$

to satisfy the constraint $x_1^2 + x_2^2 + x_3^2 + x_4^2 = 1$ with $0 \le \theta < \pi$, $0 \le \psi < 4\pi$, $0 \le \phi < 2\pi$. Find the metric tensor field with the coordinates (θ, ψ, ϕ).

Solution 22. With $e^{i\alpha} \equiv \cos(\alpha) + i\sin(\alpha)$ we have

$$\begin{aligned}
x_1 &= \cos(\theta/2)\cos((\psi+\phi)/2) \\
x_2 &= \cos(\theta/2)\sin((\psi+\phi)/2) \\
x_3 &= \sin(\theta/2)\cos((\psi-\phi)/2) \\
x_4 &= \sin(\theta/2)\sin((\psi-\phi)/2).
\end{aligned}$$

Then the metric tensor field $g = \sum_{j=1}^{4}(dx_j \otimes dx_j)$ with the coordinates θ, ψ, ϕ takes the form

$$\tilde{g} = \frac{1}{4}\left(d\theta \otimes d\theta + d\psi \otimes d\psi + d\phi \otimes d\phi + \cos(\theta)d\psi \otimes d\phi + \cos(\theta)d\phi \otimes d\psi\right).$$

Problem 23.   Consider the compact differentiable manifold

$$\mathbb{S}^3 := \{\, (x_1, x_2, x_3, x_4) \, : \, x_1^2 + x_2^2 + x_3^2 + x_4^2 = 1 \,\}$$

and the metric tensor field

$$g = dx_1 \otimes dx_1 + dx_2 \otimes dx_2 + dx_3 \otimes dx_3 + dx_4 \otimes dx_4.$$

(i) Express g using the following parametrization

$$x_1(\alpha, \beta, \theta) = \cos(\alpha) \cos(\theta), \quad x_2(\alpha, \beta, \theta) = \sin(\alpha) \cos(\theta)$$
$$x_3(\alpha, \beta, \theta) = \cos(\beta) \sin(\theta), \quad x_4(\alpha, \beta, \theta) = \sin(\beta) \sin(\theta)$$

where $0 \le \theta \le \pi/2$, $0 \le \alpha, \beta \le 2\pi$.
(ii) Now $\mathbb{S}^3$ is the manifold of the compact Lie group $SU(2)$. Thus we can define the vector fields (angular momentum operators)

$$L_1 = \frac{1}{2} \cos(\alpha + \beta)(\tan(\theta)\frac{\partial}{\partial \alpha} - \cot(\theta)\frac{\partial}{\partial \beta}) - \sin(\alpha + \beta)\frac{\partial}{\partial \theta}$$

$$L_2 = \frac{1}{2} \sin(\alpha + \beta)(\tan(\theta)\frac{\partial}{\partial \alpha} - \cot(\theta)\frac{\partial}{\partial \beta}) + \cos(\alpha + \beta)\frac{\partial}{\partial \theta}$$

$$L_3 = -\left(\frac{\partial}{\partial \alpha} + \frac{\partial}{\partial \beta}\right).$$

Find the commutation relation $[L_j, L_k]$ for $j, k = 1, 2, 3$. Find the dual basis of L_1, L_2, L_3.

Solution 23.   (i) We obtain

$$g = \cos^2(\theta)d\alpha \otimes d\alpha + \sin^2(\theta)d\beta \otimes d\beta + d\theta \otimes d\theta.$$

(ii) We obtain $[L_1, L_2] = L_3$, $[L_2, L_3] = L_1$, $[L_3, L_1] = L_2$.
(iii) We obtain

$$\omega_1 = \cos(\alpha + \beta) \sin(2\theta)(d\alpha - d\beta) - \sin(\alpha + \beta)d\theta$$
$$\omega_2 = \sin(\alpha + \beta) \sin(2\theta)(d\alpha - d\beta) + \cos(\alpha + \beta)d\theta$$
$$\omega_3 = -(\cos^2(\theta)d\alpha + \sin^2(\theta)d\beta)$$

with $L_1 \rfloor \omega_1 = 1$, $L_2 \rfloor \omega_2 = 1$, $L_3 \rfloor \omega_3 = 1$.

Problem 24.   Let $x_0 = ct$. Consider the metric tensor field

$$g = dx_0 \otimes dx_0 - dx_1 \otimes dx_1 - dx_2 \otimes dx_2 - dx_3 \otimes dx_3$$

and the transformation

$$x_0 = \tilde{x}_0, \quad x_1 = r\cos(\phi + \omega\tilde{x}_0/c), \quad x_2 = r\sin(\phi + \omega\tilde{x}_0/c), \quad x_3 = \tilde{x}_3.$$

Express g in the new coordinates $\tilde{x}_0$, r, ϕ, $\tilde{x}_3$.

Solution 24.   We obtain

$$\tilde{g} = (1 - \omega^2 r^2/c^2)d\tilde{x}_0 \otimes d\tilde{x}_0 - \frac{\omega r^2}{c}d\phi \otimes d\tilde{x}_0 - \frac{\omega r^2}{c}d\tilde{x}_0 \otimes d\phi$$
$$- d\tilde{x}_3 \otimes d\tilde{x}_3 - dr \otimes dr - r^2 d\phi \otimes d\phi.$$

This metric tensor field plays a role for *rotating platforms*.

Problem 25.   Let $x_0 = ct$. Consider the Minkowski metric tensor field in cylindrical form in the laboratory system $S(t, r, \theta, x_3)$

$$g = dx_0 \otimes dx_0 - dr \otimes dr - r^2 d\theta \otimes d\theta - dx_3 \otimes dx_3.$$

Find the metric tensor field in the rotating frame of reference $S'(t', r', \theta', x_3')$ given by

$$t'(t) = (1 - R^2\omega^2/c^2)^{1/2}t$$
$$r'(r) = (1 - R^2\omega^2/c^2)^{1/2}r$$
$$\theta'(\theta, t) = \frac{\theta - \omega t}{(1 - R^2\omega^2/c^2)^{1/2}}$$
$$x_3'(x_3) = x_3$$

with the inverse transformation given by

$$t(t') = (1 + R'^2\omega'^2/c^2)^{1/2}t'$$
$$r(r') = (1 - R'^2\omega'^2/c^2)^{1/2}r'$$
$$\theta(\theta', t') = \frac{\theta' + \omega't'}{(1 + R'^2\omega'^2/c^2)^{1/2}}$$
$$x_3(x_3') = x_3'$$

where $(1 - R^2\omega^2/c^2) = (1 + R'^2\omega'^2/c^2)^{-1}$ and R, ω, R', ω' are positive constants (ω, ω' frequencies).

Solution 25.   The metric tensor field takes the form

$$\tilde{g} = (1 + \omega'^2(R'^2 - r'^2)/c^2)dx_0' \otimes dx_0' - (1 + R'^2\omega'^2/c^2)dr' \otimes dr'$$
$$- r'^2 d\theta' \otimes d\theta' - \omega'r'^2 d\theta' \otimes dt' - \omega'r'^2 dt' \otimes d\theta' - dx_3' \otimes dx_3'.$$

Problem 26. Consider the metric tensor field

$$g = dX_0 \otimes dX_0 - dX_1 \otimes dX_1$$

where $0 < X_1 < \infty$ and $-\infty < X_0 < \infty$. Show that under the transformation

$$X_0(r, \eta) = r \sinh(\eta), \qquad X_1(r, \eta) = r \cosh(\eta)$$

$(0 < r < \infty, -\infty < \eta < \infty)$ the metric tensor field takes the form (*Rindler chart*) $g = r^2 d\eta \otimes d\eta - dr \otimes dr$.

Solution 26. Since

$$dX_0 = \sinh(\eta)dr + r\cosh(\eta)d\eta, \qquad dX_1 = \cosh(\eta)dr + r\sinh(\eta)d\eta$$

we obtain

$$dX_0 \otimes dX_0 = \sinh^2(\eta)dr \otimes dr + r^2\cosh^2(\eta)d\eta \otimes d\eta$$
$$+ r\sinh(\eta)\cosh(\eta)(dr \otimes d\eta + d\eta \otimes dr)$$
$$dX_1 \otimes dX_1 = \cosh^2(\eta)dr \otimes dr + r^2\sinh^2(\eta)d\eta \otimes d\eta$$
$$+ r\sinh(\eta)\cosh(\eta)(dr \otimes d\eta + d\eta \otimes dr).$$

Utilizing $\cosh^2(\eta) - \sinh^2(\eta) = 1$ we arrive at $r^2 d\eta \otimes d\eta - dr \otimes dr$.

Problem 27. Consider the metric tensor field

$$g = dx_0 \otimes dx_0 - dx_1 \otimes dx_1 - dx_2 \otimes dx_2 - dx_3 \otimes dx_3$$
$$- \frac{(x_1 dx_1 + x_2 dx_2 + x_3 dx_3) \otimes (x_1 dx_1 + x_2 dx_2 + x_3 dx_3)}{R^2 - (x_1^2 + x_2^2 + x_3^2)}$$

where R is a positive constant and $x_0 = ct$. Apply the transformation

$$x_1(r, \alpha, \beta, u) = R\sin(r/R)\sin(\alpha)\cos(\beta)$$
$$x_2(r, \alpha, \beta, u) = R\sin(r/R)\sin(\alpha)\sin(\beta)$$
$$x_3(r, \alpha, \beta, u) = R\sin(r/R)\cos(\alpha)$$
$$x_0(r, \alpha, \beta, u) = u + r.$$

Solution 27. We have

$$x_1^2 + x_2^2 + x_3^2 = R^2\sin^2(r/R)$$
$$x_1 dx_1 + x_2 dx_2 + x_3 dx_3 = \sin(r/R)\cos(r/R)dr$$
$$dx_0 = du + dr$$
$$dx_1 \otimes dx_1 + dx_2 \otimes dx_2 + dx_3 \otimes dx_3 = R^2\cos^2(r/R)dr \otimes dr.$$

Hence the metric tensor field takes the form

$$\widetilde{g} = du \otimes dr + dr \otimes du + du \otimes du - R^2 \sin^2(r/R)(d\alpha \otimes d\alpha + \sin^2(\alpha)d\beta \otimes d\beta).$$

Problem 28. Let $x_0 = ct$. Consider the metric tensor field

$$g = -e^{f(r,x_0)}dx_0 \otimes dx_0 + e^{h(r,x_0)}dr \otimes dr + r^2(d\theta \otimes d\theta + \sin^2(\theta)d\phi \otimes d\phi)$$

i.e. we have the signature $(-,+,+,+)$ and $f(r,x_0)$, $h(r,x_0)$ are differentiable functions. The corresponding 4×4 matrix is given by

$$\begin{pmatrix} -e^f & 0 & 0 & 0 \\ 0 & e^h & 0 & 0 \\ 0 & 0 & r^2 & 0 \\ 0 & 0 & 0 & r^2 \sin^2(\theta) \end{pmatrix}$$

with the inverse matrix given by

$$\begin{pmatrix} -e^{-f} & 0 & 0 & 0 \\ 0 & e^{-h} & 0 & 0 \\ 0 & 0 & 1/r^2 & 0 \\ 0 & 0 & 0 & 1/(r^2 \sin^2(\theta)) \end{pmatrix}.$$

Find the Christoffel symbols, Ricci tensor and curvature scalar. Solve

$$R_{\mu\nu} - \frac{1}{2}g_{\mu\nu}R = 0, \quad \mu,\nu = 0,1,2,3.$$

Solution 28. The Christoffel symbols $\Gamma^a_{mn} = \Gamma^a_{nm}$ are given by

$$\Gamma^a_{mn} = \frac{1}{2}\sum_{b=0}^{3} g^{ab}(g_{bm,n} + g_{bn,m} - g_{mn,b})$$

with $a,m,n = 0,1,2,3$ and $(r = 1, \theta = 2, \phi = 3)$. We obtain for the nonzero Christoffel symbols

$$\Gamma^0_{00} = \frac{1}{2}\frac{\partial f}{\partial x_0}, \quad \Gamma^r_{00} = \frac{1}{2}e^{f-h}\frac{\partial f}{\partial r}, \quad \Gamma^0_{r0} = \frac{1}{2}\frac{\partial f}{\partial r}, \quad \Gamma^r_{r0} = \frac{1}{2}\frac{\partial h}{\partial x_0}$$

$$\Gamma^0_{rr} = \frac{1}{2}e^{h-f}\frac{\partial h}{\partial x_0}, \quad \Gamma^r_{rr} = \frac{1}{2}\frac{\partial h}{\partial r}, \quad \Gamma^\theta_{r\theta} = \frac{1}{r}$$

$$\Gamma^r_{\theta\theta} = -re^{-h}, \quad \Gamma^\theta_{\phi\phi} = -\sin(\theta)\cos(\theta), \quad \Gamma^r_{\phi\phi} = -re^{-h}\sin^2(\theta)$$

$$\Gamma^{\phi}_{r\phi} = \frac{1}{r}, \quad \Gamma^{\phi}_{\theta\phi} = \cot(\theta).$$

The Ricci tensor follows as

$$R_{mq} = \sum_{s=0}^{3} \Gamma^{s}_{mq,s} - \sum_{s=0}^{3} \Gamma^{s}_{ms,q} + \sum_{s=0}^{3}\sum_{n=0}^{3} \Gamma^{s}_{ns}\Gamma^{n}_{mq} - \sum_{s=0}^{3}\sum_{n=0}^{3} \Gamma^{s}_{nq}\Gamma^{n}_{ms}.$$

The nonzero components of the Ricci tensor are

$$R_{00} = \frac{1}{4}e^{f-h}\left(-\frac{\partial h}{\partial r}\frac{\partial f}{\partial r} + \frac{4}{r}\frac{\partial f}{\partial r} + 2\frac{\partial^2 f}{\partial r^2} + \left(\frac{\partial f}{\partial r}\right)^2\right)$$

$$+ \frac{1}{4}\left(\frac{\partial h}{\partial x_0}\frac{\partial f}{\partial x_0} - \left(\frac{\partial h}{\partial x_0}\right)^2 - 2\frac{\partial^2 h}{\partial x_0^2}\right)$$

$$R_{11} = \frac{1}{4}e^{h-f}\left(2\frac{\partial^2 h}{\partial x_0^2} + \left(\frac{\partial h}{\partial x_0}\right)^2 - \frac{\partial f}{\partial x_0}\frac{\partial h}{\partial x_0}\right)$$

$$+ \frac{1}{4}\left(\frac{\partial f}{\partial r}\frac{\partial h}{\partial r} + \frac{4}{r}\frac{\partial h}{\partial r} - 2\frac{\partial^2 f}{\partial r^2} - \left(\frac{\partial f}{\partial r}\right)^2\right)$$

$$R_{22} = 1 + \frac{1}{2}e^{-h}\left(r\frac{\partial h}{\partial r} - 2 - r\frac{\partial f}{\partial r}\right)$$

$$R_{33} = \sin^2(\theta)\left(1 + \frac{1}{2}e^{-h}\left(r\frac{\partial h}{\partial r} - 2 - r\frac{\partial f}{\partial r}\right)\right).$$

With

$$R^{j}_{k} = \sum_{m=0}^{3} g^{jm}R_{km}$$

and $R = R^0_0 + R^1_1 + R^2_2 + R^3_3$ we obtain

$$R = \frac{2}{r^2} + \frac{1}{2}e^{-f}\left(2\frac{\partial^2 h}{\partial x_0^2} + \left(\frac{\partial h}{\partial x_0}\right)^2 - \frac{\partial h}{\partial x_0}\frac{\partial f}{\partial x_0}\right)$$

$$+ e^{-h}\left(\frac{2}{r}\left(\frac{\partial h}{\partial r} - \frac{\partial f}{\partial r}\right) - \frac{2}{r^2} + \frac{1}{2}\frac{\partial h}{\partial r}\frac{\partial f}{\partial r} - \frac{\partial^2 f}{\partial r^2} - \frac{1}{2}\left(\frac{\partial f}{\partial r}\right)^2\right).$$

From $R_{\mu\nu} - \frac{1}{2}g_{\mu\nu}R = 0$ we obtain the system of partial differential equations

$$e^{-h}\left(r\frac{\partial f}{\partial r} + 1\right) = 1, \quad e^{-h}\left(1 - r\frac{\partial h}{\partial r}\right) = 1, \quad \frac{\partial h}{\partial x_0} = 0.$$

Integration yields

$$e^{-h} = e^f = 1 - \frac{r_s}{r}$$

with $r_s = 2Gm/c^2$ the *Schwarzschild radius*. Then the metric tensor field takes the form

$$g = -\left(1 - \frac{r_s}{r}\right) dx_0 \otimes dx_0 + \left(1 - \frac{r_s}{r}\right)^{-1} dr \otimes dr + r^2 (d\theta \otimes d\theta + \sin^2(\theta) d\phi \otimes d\phi)$$

which is the Schwarzschild metric tensor field.

Problem 29.   Let $x_0 = ct$. The *Gödel metric tensor field* is given by

$$g = -dx_0 \otimes dx_0 + dx_1 \otimes dx_1 - e^{x_1/a} dx_0 \otimes dx_2 - e^{x_1/a} dx_2 \otimes dx_0$$
$$- \frac{1}{2} e^{2x_1/a} dx_2 \otimes dx_2 + dx_3 \otimes dx_3$$

with a a positive constant of dimension length. The corresponding 4×4 matrix is

$$\begin{pmatrix} -1 & 0 & -e^{x_1/a} & 0 \\ 0 & 1 & 0 & 0 \\ -e^{x_1/a} & 0 & -\frac{1}{2}e^{2x_1/a} & 0 \\ 0 & 0 & 0 & 1 \end{pmatrix}.$$

Find the inverse of this matrix. Find the Christoffel symbols, the Riemann tensor, the Ricci tensor and the scalar curvature.

Solution 29.   The inverse of this matrix is given by

$$\begin{pmatrix} 1 & 0 & -2e^{-x_1/a} & 0 \\ 0 & 1 & 0 & 0 \\ -2e^{-x_1/a} & 0 & 2e^{-2x_1/a} & 0 \\ 0 & 0 & 0 & 1 \end{pmatrix}.$$

The Christoffel symbols are $\Gamma^a_{mn} = \Gamma^a_{nm}$ are defined as

$$\Gamma^a_{mn} = \frac{1}{2} \sum_{b=0}^{3} g^{ab} \left(\frac{\partial g_{bm}}{\partial x_n} + \frac{\partial g_{bn}}{\partial x_m} - \frac{\partial g_{mn}}{\partial x_b} \right)$$

where $a, m, n = 0, 1, 2, 3$. The nonzero coefficients are

$$\Gamma^0_{01} = \Gamma^0_{10} = \frac{1}{a}, \quad \Gamma^0_{12} = \Gamma^0_{21} = \frac{1}{2a} e^{x_1/a},$$

$$\Gamma^1_{02} = \Gamma^1_{20} = \frac{1}{2a} e^{x_1/a}, \quad \Gamma^1_{22} = \frac{1}{2a} e^{2x_1/a},$$

$$\Gamma^2_{10} = \Gamma^2_{01} = -\frac{1}{a}e^{-x_1/a}.$$

The Ricci tensor is defined as

$$R_{mq} = \sum_{s=0}^{3} \Gamma^s_{mq,s} - \sum_{s=0}^{3} \Gamma^s_{ms,q} + \sum_{s=0}^{3}\sum_{n=0}^{3} \Gamma^s_{ns}\Gamma^n_{mq} - \sum_{s=0}^{3}\sum_{n=0}^{3} \Gamma^s_{nq}\Gamma^n_{ms}.$$

The nonzero terms are

$$R_{00} = \frac{1}{a^2}, \quad R_{22} = \frac{1}{a^2}e^{2x_1/a}, \quad R_{20} = R_{02} = \frac{1}{a^2}e^{x_1/a}.$$

Now

$$R^j_k := \sum_{m=0}^{3} g^{jm}R_{km}$$

with

$$R^0_0 = -\frac{1}{a^2}, \quad R^1_1 = R^2_2 = R^3_3 = 0.$$

Hence $R = R^0_0 + R^1_1 + R^2_2 + R^3_3 = -1/a^2$. Therefore $R_{\mu\nu} - \frac{1}{2}g_{\mu\nu}R$ can be written as

$$\frac{1}{a^2}\begin{pmatrix} 1 & 0 & e^{x_1/a} & 0 \\ 0 & 0 & 0 & 0 \\ e^{x_1/a} & 0 & e^{2x_1/a} & 0 \\ 0 & 0 & 0 & 0 \end{pmatrix} + \frac{1}{2a^2}\begin{pmatrix} -1 & 0 & -e^{x_1/a} & 0 \\ 0 & 1 & 0 & 0 \\ -e^{x_1/a} & 0 & -\frac{1}{2}e^{2x_1/a} & 0 \\ 0 & 0 & 0 & 1 \end{pmatrix}$$

$$= \frac{1}{2a^2}\begin{pmatrix} 1 & 0 & e^{x_1/a} & 0 \\ 0 & 1 & 0 & 0 \\ e^{x_1/a} & 0 & 3e^{2x_1/a}/2 & 0 \\ 0 & 0 & 0 & 1 \end{pmatrix}.$$

An energy momentum tensor can be found to satisfy the Einstein field equations.

Problem 30. The two-dimensional *de Sitter space* $\mathbb{V}$ with the topology $\mathbb{R} \times \mathbb{S}$ may be visualized as a one-sheet *hyperboloid* $\mathbb{H}_{r_0}$ embedded in three-dimensional Minkowski space $\mathbb{M}$, i.e.

$$\mathbb{H}_{r_0} = \{(y_0, y_1, y_2) \in \mathbb{M} : (y_2)^2 + (y_1)^2 - (y_0)^2 = r_0^2, \ r_0 > 0\}$$

where r_0 is the parameter of the one-sheet hyperboloid $\mathbb{H}_{r_0}$. The induced metric, $g_{\mu\nu}$ ($\mu, \nu = 0, 1$), on $\mathbb{H}_{r_0}$ is the de Sitter metric.
(i) Show that we can parametrize (parameters ρ and θ) the hyperboloid as follows

$$y_0(\rho, \theta) = -\frac{r_0\cos(\rho/r_0)}{\sin(\rho/r_0)}, \quad y_1(\rho, \theta) = \frac{r_0\cos(\theta/r_0)}{\sin(\rho/r_0)}, \quad y_2(\rho, \theta) = \frac{r_0\sin(\theta/r_0))}{\sin(\rho/r_0)}$$

where $0 < \rho < \pi r_0$ and $0 \leq \theta < 2\pi r_0$.

(ii) Using this parametrization find the metric tensor field induced on $\mathbb{H}_{r_0}$.

Solution 30. (i) Inserting y_0, y_1, y_2 into $(y_2)^2 + (y_1)^2 - (y_0)^2$ and using $\sin^2(\rho/r_0) + \cos^2(\rho/r_0) = 1$ we obtain r_0^2.

(ii) We start from the metric tensor field

$$g = dy_0 \otimes dy_0 - dy_1 \otimes dy_1 - dy_2 \otimes dy_2.$$

Since

$$\frac{\partial y_0}{\partial \rho} = \csc^2(\rho/r_0), \quad \frac{\partial y_0}{\partial \theta} = 0$$

$$\frac{\partial y_1}{\partial \rho} = -\frac{\cos(\theta/r_0)\cos(\rho/r_0)}{\sin^2(\rho/r_0)}, \quad \frac{\partial y_1}{\partial \theta} = -\frac{\sin(\theta/r_0)}{\sin(\rho/r_0)}$$

$$\frac{\partial y_2}{\partial \rho} = -\frac{\sin(\theta/r_0)\cos(\rho/r_0)}{\sin^2(\rho/r_0)}, \quad \frac{\partial y_2}{\partial \theta} = \frac{\cos(\theta/r_0)}{\sin(\rho/r_0)}$$

we find

$$g = \left(\frac{1}{\sin^4(\rho/r_0)} - \frac{\cos^2(\theta/r_0)\cos^2(\rho/r_0)}{\sin^4(\rho/r_0)} - \frac{\sin^2(\theta/r_0)\cos^2(\rho/r_0)}{\sin^4(\rho/r_0)} \right) d\rho \otimes d\rho$$

$$+ \left(-\frac{\sin^2(\theta/r_0)}{\sin^2(\rho/r_0)} - \frac{\cos^2(\theta/r_0)}{\sin^2(\rho/r_0)} \right) d\theta \otimes d\theta.$$

Using $\sin^2(\theta/r_0) + \cos^2(\theta/r_0) = 1$ we arrive at

$$g = \frac{1}{\sin^2(\rho/r_0)} d\rho \otimes d\rho - \frac{1}{\sin^2(\rho/r_0)} d\theta \otimes d\theta.$$

Problem 31. Let $R > 0$. The equation of the two-dimensional single-sheeted hyperboloid $\mathbb{H}_{-1}^{(3)}$ is given by $x_0^2 - x_1^2 - x_2^2 = -R^2$. Consider *pseudo-spherical polar coordinates*

$$x_0(\tau, \phi) = R\sinh(\tau),$$

$$x_1(\tau, \phi) = R\cosh(\tau)\sin(\phi), \quad x_2(\tau, \phi) = R\cosh(\tau)\cos(\phi)$$

where $-\infty < \tau < \infty$, $\phi \in [0, 2\pi)$. Express the metric tensor field

$$g = dx_0 \otimes dx_0 - dx_1 \otimes dx_1 - dx_2 \otimes dx_2$$

in pseudo-spherical polar coordinates.

Solution 31.   We have

$$dx_0 = R\cosh(\tau)d\tau$$
$$dx_1 = R\sinh(\tau)\sin(\phi)d\tau + R\cosh(\tau)\cos(\phi)d\phi$$
$$dx_2 = R\sinh(\tau)\cos(\phi)d\tau - R\cosh(\tau)\sin(\phi)d\phi.$$

Thus

$$dx_0 \otimes dx_0 = R^2\cosh^2(\tau)d\tau \otimes d\tau$$
$$dx_1 \otimes dx_1 + dx_2 \otimes dx_2 = R^2(\sinh^2(\tau)d\tau \otimes d\tau + \cosh^2(\tau)d\phi \otimes d\phi).$$

Then

$$g_{\mathbb{H}^{(3)}_{-1}} = R^2 d\tau \otimes d\tau - R^2\cosh^2(\tau)d\phi \otimes d\phi.$$

Problem 32.   The *anti-de Sitter space* is defined as the surface

$$X_1^2 + X_2^2 + X_3^2 - U_1^2 - U_2^2 = -1$$

embedded in a five-dimensional flat space with the metric tensor field

$$g = dX_1 \otimes dX_1 + dX_2 \otimes dX_2 + dX_3 \otimes dX_3 - dU_1 \otimes dU_1 - dU_2 \otimes dU_2.$$

This is a solution of Einstein's field equations with the cosmological constant $\Lambda = -3$. Its intrinsic curvature is constant and negative. Find the metric tensor field in terms of the intrinsic coordinates $(\rho, \theta, \phi, \alpha)$ where

$$X_1(\rho, \theta, \phi, \alpha) = \frac{2\rho}{1 - \rho^2}\sin(\theta)\cos(\phi)$$

$$X_2(\rho, \theta, \phi, \alpha) = \frac{2\rho}{1 - \rho^2}\sin(\theta)\sin(\phi)$$

$$X_3(\rho, \theta, \phi, \alpha) = \frac{2\rho}{1 - \rho^2}\cos(\theta)$$

$$U_1(\rho, \theta, \phi, \alpha) = \frac{1 + \rho^2}{1 - \rho^2}\cos(\alpha)$$

$$U_2(\rho, \theta, \phi, \alpha) = \frac{1 + \rho^2}{1 - \rho^2}\sin(\alpha)$$

where $0 \le \rho < 1$, $0 \le \phi < 2\pi$, $0 \le \theta < \pi$, $-\pi \le \alpha < \pi$.

Solution 32.   We find

$$g = -\left(\frac{1 + \rho^2}{1 - \rho^2}\right)^2 d\alpha \otimes d\alpha + \frac{4}{(1 - \rho^2)^2}(d\rho \otimes d\rho + \rho^2 d\theta \otimes d\theta + \rho^2\sin^2(\theta)d\phi \otimes d\phi).$$

Programming Problems

Problem 33. Let $\mathbb{E}^3$ be the three-dimensional Euclidean space. The two-dimensional unit sphere

$$\mathbb{S}^2 := \{\, (x_1, x_2, x_3) \ : \ x_1^2 + x_2^2 + x_3^2 = 1 \,\}$$

is embedded into $\mathbb{E}^3$. We evaluate the metric tensor field for the unit sphere. A parameter representation is given by

$$x_1(u, v) = \cos(u)\sin(v), \quad x_2(u, v) = \sin(u)\sin(v), \quad x_3(u, v) = \cos(v)$$

where $0 < v < \pi$. Since

$$dx_1 = -\sin(u)\sin(v)du + \cos(u)\cos(v)dv$$
$$dx_2 = \cos(u)\sin(v)du + \sin(u)\cos(v)dv$$
$$dx_3 = -\sin(v)dv$$

we obtain for the metric tensor field of the unit sphere

$$\widetilde{g} = \sin^2(v)du \otimes du + dv \otimes dv.$$

Provide a SymbolicC++ program that finds $\widetilde{g}$. The tensor product $\otimes$ is denoted by $*$.

Solution 33. The program is

```
// metrics.cpp
#include <iostream>
#include "symbolicc++.h"
using namespace std;

int main(void)
{
 Symbolic u("u"), v("v");
 Symbolic du("du"), dv("dv");
 du = ~du; dv = ~dv; // du and dv are not commutative
 Symbolic x1 = cos(u)*sin(v);
 Symbolic x2 = sin(u)*sin(v);
 Symbolic x3 = cos(v);
 Symbolic dx1 = df(x1,u)*du + df(x1,v)*dv;
 Symbolic dx2 = df(x2,u)*du + df(x2,v)*dv;
 Symbolic dx3 = df(x3,u)*du + df(x3,v)*dv;
 Symbolic GT = dx1*dx1 + dx2*dx2 + dx3*dx3;
 GT = GT[(cos(u)^2)==1-(sin(u)^2),(cos(v)^2)==1-(sin(v)^2)];
 cout << GT << endl;
 return 0;
}
```

The output is `dv^(2)+sin(v)^(2)*du^(2)` or $dv \otimes dv + \sin^2(v)du \otimes du$.

Problem 34. The equation of the *monkey saddle* surface in $\mathbb{E}^3$ is given by

$$x_3 = x_1(x_1^2 - 3x_2^2)$$

with the parameter representation

$$x_1(u_1, u_2) = u_1, \quad x_2(u_1, u_2) = u_2, \quad x_3(u_1, u_2) = u_1^3 - 3u_1u_2^2.$$

Find the mean and Gaussian curvature. Apply a Maxima program. Find g restricted to the monkey saddle surface.

Solution 34. We have

$$\frac{\partial \mathbf{x}}{\partial u_1} = \begin{pmatrix} 1 \\ 0 \\ 3u_1^2 - 3u_2^2 \end{pmatrix}, \quad \frac{\partial \mathbf{x}}{\partial u_2} = \begin{pmatrix} 0 \\ 1 \\ -6u_1u_2 \end{pmatrix}.$$

The Maxima program is

```
/* Monkey.mac */
x1: u1; x2: u2; x3: u1*u1*u1-3*u1*u2*u2;
x1u1: diff(x1,u1); x2u1: diff(x2,u1); x3u1: diff(x3,u1);
x1u2: diff(x1,u2); x2u2: diff(x2,u2); x3u2: diff(x3,u2);
x1u1u1: diff(x1u1,u1); x2u1u1: diff(x2u1,u1); x3u1u1: diff(x3u1,u1);
x1u2u2: diff(x1u2,u2); x2u2u2: diff(x2u2,u2); x3u2u2: diff(x3u2,u2);
x1u1u2: diff(x1u1,u2); x2u1u2: diff(x2u1,u2); x3u1u2: diff(x3u1,u2);
n1: x2u1*x3u2-x3u1*x2u2; n2: x3u1*x1u2-x1u1*x3u2;
n3: x1u1*x2u2-x2u1*x1u2;
temp: n1*n1 + n2*n2 + n3*n3; temp: ratsimp(temp);
norm: sqrt(temp);
E: x1u1*x1u1+x2u1*x2u1+x3u1*x3u1;
F: x1u1*x1u2+x2u1*x2u2+x3u1*x3u2;
G: x1u2*x1u2+x2u2*x2u2+x3u2*x3u2;
n1: n1/norm; n2: n2/norm; n3: n3/norm;
L: n1*x1u1u1 + n2*x2u1u1 + n3*x3u1u1;
M: n1*x1u1u2 + n2*x2u1u2 + n3*x3u1u2;
N: n1*x1u2u2 + n2*x2u2u2 + n3*x3u2u2;
```

The output is

$$E = 1 + 9(u_1^2 - u_2^2)^2, \quad F = -18u_1u_2(u_1^2 - u_2^2), \quad G = 1 + 36u_1^2u_2^2$$

and

$$L = \frac{6u_1}{D}, \quad M = -\frac{6u_2}{D}, \quad N = -\frac{6u_1}{D}$$

with

$$D = \sqrt{1 + 9(u_1^2 + u_2^2)^2}.$$

Problem 35.   Consider the *sine-Gordon equation*

$$\frac{\partial^2 u}{\partial x_1 \partial x_2} = \sin(u).$$

The metric tensor field for the sine-Gordon equation is given by

$$g = dx_1 \otimes dx_1 + \cos(u(x_1, x_2)) dx_1 \otimes dx_2 + \cos(u(x_1, x_2)) dx_2 \otimes dx_1 + dx_2 \otimes dx_2$$

i.e. the line element is

$$\left(\frac{ds}{d\lambda}\right)^2 = \left(\frac{dx_1}{d\lambda}\right)^2 + 2\cos(u(x_1, x_2))\frac{dx_1}{d\lambda}\frac{dx_2}{d\lambda} + \left(\frac{dx_2}{d\lambda}\right)^2.$$

Here u is a smooth function of x_1 and x_2. First we have to calculate the Riemann curvature scalar R from g. Then the sine-Gordon equation follows when we impose the condition $R = -2$. We have

$$g_{11} = g_{22} = 1, \qquad g_{12} = g_{21} = \cos(u(x_1, x_2)).$$

The quantity g can be written in matrix form

$$g = \begin{pmatrix} g_{11} & g_{12} \\ g_{21} & g_{22} \end{pmatrix}.$$

Then the inverse of g is given by

$$g^{-1} = \begin{pmatrix} g^{11} & g^{12} \\ g^{21} & g^{22} \end{pmatrix}$$

where

$$g^{11} = g^{22} = \frac{1}{\sin^2(u)}, \qquad g^{12} = g^{21} = -\frac{\cos(u)}{\sin(u)}.$$

Next we have to calculate the *Christoffel symbols*. They are defined as

$$\Gamma^a_{mn} := \frac{1}{2}\sum_{b=1}^{2} g^{ab}(g_{bm,n} + g_{bn,m} - g_{mn,b})$$

with

$$g_{bm,1} := \frac{\partial g_{bm}}{\partial x_1} \qquad g_{bm,2} := \frac{\partial g_{bm}}{\partial x_2}.$$

Next we have to calculate the *Riemann curvature tensor* which is given by

$$R^r_{msq} := \Gamma^r_{mq,s} - \Gamma^r_{ms,q} + \Gamma^r_{ns}\Gamma^n_{mq} - \Gamma^r_{nq}\Gamma^n_{ms}.$$

The *Ricci tensor* follows as

$$R_{mq} := \sum_{a=1}^{2} R^a_{maq} = -\sum_{a=1}^{2} R^a_{mqa}$$

i.e. the Ricci tensor is constructed by contraction. From R_{nq} we obtain R^m_q via

$$R^m_q = \sum_{n=1}^{2} g^{mn} R_{nq}.$$

Finally the *curvature scalar* R is given by

$$R := \sum_{m=1}^{2} R^m_m.$$

With the metric tensor field given above we find that

$$R = -\frac{2}{\sin(u)} \frac{\partial^2 u}{\partial x_1 \partial x_2}.$$

If $R = -2$, then we obtain the sine-Gordon equation. Give a Maxima implementation to find the curvature from the metric tensor field.

Solution 35.   The Maxima implementation is

```
/* tensor.mac */
g[1,1]:1; g[2,2]:1;
g[1,2]:cos(u(x[1],x[2])); g[2,1]:cos(u(x[1],x[2]));
G: genmatrix(g,2,2);
G1: trigsimp(invert(G));
for a:1 thru 2 do
 for m:1 thru 2 do
  for n:1 thru 2 do
   gmma[a,m,n]: (1/2)*
               trigsimp(sum(G1[a,b]*(diff(G[b,m],x[n])+
               diff(G[b,n],x[m])-diff(G[m,n],x[b])),b,1,2));
for a:1 thru 2 do
 for m:1 thru 2 do
  for n:1 thru 2 do ldisplay(gmma[a,m,n]);
for b:1 thru 2 do
 for m:1 thru 2 do
```

```
   for s:1 thru 2 do
    for q:1 thru 2 do
     R[b,m,s,q]: trigsimp(diff(gmma[b,m,q],x[s])
                          -diff(gmma[b,m,s],x[q])
                          +sum(gmma[b,n,s]*gmma[n,m,q],n,1,2)
                          -sum(gmma[b,n,q]*gmma[n,m,s],n,1,2));
 for m:1 thru 2 do
  for q:1 thru 2 do
   Ricci[m,q]: trigsimp(sum(R[s,m,s,q],s,1,2));
 for m:1 thru 2 do
  for q:1 thru 2 do ldisplay(Ricci[m,q]);
 for m:1 thru 2 do
  for q:1 thru 2 do
   Ricci1[m,q]: trigsimp(sum(G1[m,b]*Ricci[q,b],b,1,2));
 CS: trigsimp(sum(Ricci1[m,m],m,1,2));
```

The output is

$$-\frac{\frac{\partial^2}{\partial x_1 \partial x_2}(u(x_1, x_2))}{\sin(u(x_1, x_2))}.$$

3.3 Supplementary Problems

Problem 1. Consider the metric tensor field $g = dx_0 \otimes dx_0 - dx_1 \otimes dx_1$. Express the metric in the coordinates u, v with

$$x_0 = a\sinh(u)\cosh(v), \quad x_1 = a\cosh(u)\sinh(v)$$

and $a > 0$ with dimension *meter*. Note that

$$dx_0 = a\cosh(u)\cosh(v)du + a\sinh(u)\sinh(v)dv$$

$$dx_1 = a\sinh(u)\sinh(v)du + a\cosh(u)\cosh(v)dv.$$

Problem 2. Consider the metric tensor field

$$g = -dZ \otimes dZ - dT \otimes dT + dW \otimes dW.$$

Consider the parametrization

$$Z(z,t) = \cosh(\epsilon z)\cos(\epsilon t), \ T(z,t) = \cosh(\epsilon z)\sin(\epsilon t), \ W(z,t) = \sinh(\epsilon z).$$

Show that $Z^2 + T^2 - W^2 = 1$. Express g using this parametrization.

Problem 3.   Let $a > 0$, $b > 0$, $c > 0$. Consider the *ellipsoid S*

$$\frac{x_1^2}{a^2} + \frac{x_2^2}{b^2} + \frac{x_3^2}{c^2} = 1.$$

Show that the Gaussian curvature is given by

$$K(p) = \frac{1}{a^2b^2c^2} \left(\frac{x_1^2}{a^4} + \frac{x_2^2}{b^4} + \frac{x_3^2}{c^4} \right)$$

where $p \in S$. If $a = b = c$, then $K(p) = 1/a^2$.

Problem 4.   Let $z = x + iy$, $\bar{z} = x - iy$ with $x, y \in \mathbb{R}$. Then

$$dz \otimes d\bar{z} = dx \otimes dx - idx \otimes dy + idy \otimes dx + dy \otimes dy$$

with the corresponding 2×2 matrix

$$\begin{pmatrix} 1 & -i \\ i & 1 \end{pmatrix}.$$

Show that the eigenvalues are 2 and 0 and thus the matrix is not invertible.

Problem 5.   Find solutions of

$$dx_1(\tau) \otimes dx_1(\tau) + dx_2(\tau) \otimes dx_2(\tau) + dx_3(\tau) \otimes dx_3(\tau) = dx_4(\tau) \otimes dx_4(\tau).$$

For example a solution is $x_1(\tau) = x_2(\tau) = x_3(\tau) = \tau$, $x_4(\tau) = \sqrt{3}\tau$.

Problem 6.   Consider the *right conoid*

$$\mathbf{x}(u_1, u_2) = \begin{pmatrix} x_1(u_1, u_2) \\ x_2(u_1, u_2) \\ x_3(u_1, u_2) \end{pmatrix} = \begin{pmatrix} u_1 \cos(u_2) \\ u_1 \sin(u_2) \\ u_2 \end{pmatrix}$$

$(0 < u_2 < 2\pi, \ -\infty < u_1 < \infty)$ in the three-dimensional Euclidean space. Find the metric tensor field for the right conoid.

Problem 7.   Consider the Euclidean space $\mathbb{E}^3$.
(i) Express the metric tensor field in *spheroidal coordinates*

$$x_1(\xi, \eta, \phi) = \frac{1}{2} R((\xi^2 - 1)(1 - \eta^2))^{1/2} \cos(\phi)$$

$$x_2(\xi, \eta, \phi) = \frac{1}{2} R((\xi^2 - 1)(1 - \eta^2))^{1/2} \sin(\phi)$$

$$x_3(\xi, \eta, \phi) = \frac{1}{2} R(\xi\eta + 1)$$

where $1 \leq \xi < \infty$, $-1 \leq \eta \leq 1$, $0 \leq \phi \leq 2\pi$. As $R \to 0$, we have $\xi \to 2r/R$, $\eta \to \cos(\theta)$ and as $R \to \infty$ we have $\xi \to 1+\mu/R$, $\eta \to -1+\nu/R$ with r and θ spherical coordinates and μ and ν are parabolic coordinates. (ii) Let $R > 0$ and fixed. The *oblate spheroidal coordinates* are given by

$$x_1(\eta, \xi, \phi) = R\sqrt{(1-\eta^2)(\xi^2+1)}\cos(\phi)$$
$$x_2(\eta, \xi, \phi) = R\sqrt{(1-\eta^2)(\xi^2+1)}\sin(\phi)$$
$$x_3(\eta, \xi, \phi) = R\eta\xi$$

where $-1 \leq \eta \leq 1$, $0 \leq \xi < \infty$, $0 \leq \phi \leq 2\pi$ with the x_3 axis as the axis of revolution. Express the metric tensor field in oblate spheroidal coordinates. Show that

$$g_O = h_\eta^2 d\eta \otimes d\eta + h_\chi^2 d\xi \otimes d\xi + h_\phi^2 d\phi \otimes d\phi$$

where

$$h_\eta = R\sqrt{\frac{\xi^2+\eta^2}{1-\eta^2}}, \quad h_\chi = R\sqrt{\frac{\xi^2+\eta^2}{\xi^2+1}}, \quad h_\phi = R\sqrt{(1-\eta^2)(\xi^2+1)}.$$

(iii) Express the metric tensor field in *prolate spheroidal coordinates* given by

$$x_1(\mu, \nu, \phi) = a\sinh(\mu)\sin(\nu)\cos(\phi)$$
$$x_2(\mu, \nu, \phi) = a\sinh(\mu)\sin(\nu)\sin(\phi)$$
$$x_3(\mu, \nu, \phi) = a\cosh(\mu)\cos(\nu)$$

where $\nu \in [-\pi/2, \pi/2]$, $\pi \in [-\pi, \pi]$, $\mu \geq 0$ and $a > 0$ with dimension length. Express the volume form $dx_1 \wedge dx_2 \wedge dx_3$ in prolate spheroidal coordinates.

Problem 8. Consider the subset of $\mathbb{R}^3$ (called *Roman surface* or *Steiner surface*)

$$S = \{(x_2x_3, x_3x_1, x_1x_2) : x_1, x_2, x_3 \in \mathbb{R}, x_1^2 + x_2^2 + x_3^2 = 1\}.$$

(i) Show that S is not a submanifold of $\mathbb{R}^3$.
(ii) Show that the Roman surface can also be represented as the set of solutions $(x_1, x_2, x_3) \in \mathbb{R}^3$ of the equation $x_2^2 x_3^2 + x_3^2 x_1^2 + x_1^2 x_2^2 = x_1 x_2 x_3$.
(iii) Show that

$$d(x_2x_3) \wedge d(x_3x_1) \wedge d(x_1x_2) = 2x_1x_2x_3 dx_1 \wedge dx_2 \wedge dx_3.$$

Problem 9.   Consider the Euclidean space $\mathbb{E}^3$. Let $c_1, c_2, c_3 > 0$. The *hyperboloid*

$$\frac{x_1^2}{c_1^2} + \frac{x_2^2}{c_2^2} - \frac{x_3^2}{c_3^2} = 1$$

can be written in parameter form as

$$x_1(\theta, \phi) = c_1 \cos(\theta) \sec(\phi) \quad x_2(\theta, \phi) = c_2 \sin(\theta) \sec(\phi) \quad x_3(\theta, \phi) = c_3 \tan(\phi)$$

where $\sec(\phi) \equiv 1/\cos(\phi)$. Find the metric tensor field for the hyperboloid.

Problem 10.   Consider the three-dimensional Euclidean space and the *surface of revolution* in $\mathbb{E}^3$ parametrized by

$$\mathbf{x}(u_1, u_2) = \begin{pmatrix} x_1(u_1, u_2) \\ x_2(u_1, u_2) \\ x_3(u_1, u_2) \end{pmatrix} = \begin{pmatrix} f(u_2) \cos(u_1) \\ f(u_2) \sin(u_1) \\ h(u_2) \end{pmatrix}$$

with the smooth functions $f : \mathbb{R} \to \mathbb{R}$, $h : \mathbb{R} \to \mathbb{R}$. Assume that $f \neq 0$. Note that

$$dx_1 = -\sin(u_1) f(u_2) du_1 + \frac{df}{du_2} \cos(u_1) du_2$$

$$dx_2 = \cos(u_1) f(u_2) du_1 + \frac{df}{du_2} \sin(u_1) du_2$$

$$dx_3 = \frac{dh}{du_2} du_2.$$

Show that the metric tensor field is given by

$$f^2(u_2) du_1 \otimes du_1 + \left(\left(\frac{df}{du_2} \right)^2 + \left(\frac{dh}{du_2} \right)^2 \right) du_2 \otimes du_2$$

with the corresponding 2×2 matrix

$$\begin{pmatrix} f^2 & 0 \\ 0 & (df/du_2)^2 + (dh/du_2)^2 \end{pmatrix}$$

and its inverse

$$\begin{pmatrix} 1/f^2 & 0 \\ 0 & 1/((df/du_2)^2 + (dh/du_2)^2) \end{pmatrix}.$$

Show that the Christoffel symbols

$$\Gamma^a_{mn} := \frac{1}{2} \sum_{b=1}^{2} g^{ab} \left(\frac{\partial g_{bm}}{\partial u_n} + \frac{\partial g_{bn}}{\partial u_m} - \frac{\partial g_{mn}}{\partial u_b} \right)$$

are given by $(a, m, n \in \{1, 2\})$ $\Gamma_{11}^1 = \Gamma_{22}^1 = 0$,

$$\Gamma_{12}^1 = \Gamma_{21}^1 = \frac{1}{f(u_2)} \frac{df}{du_2},$$

$\Gamma_{12}^2 = \Gamma_{21}^2 = 0$,

$$\Gamma_{11}^2 = \frac{-f \, df/du_2}{(df/du_2)^2 + (dh/du_2)^2},$$

$$\Gamma_{22}^2 = \frac{(df/du_2)(d^2f/du_2^2) + (dh/du_2)(d^2h/du_2^2)}{(df/du_2)^2 + (dh/du_2)^2}$$

utilizing that f and h only depend on u_2 and using

$$\frac{d}{du_2} f^2 = 2f \frac{df}{du_2}$$

$$\frac{d}{du_2} \left(\left(\frac{df}{du_2} \right)^2 + \left(\frac{dh}{du_2} \right)^2 \right) = 2 \left(\frac{df}{du_2} \frac{d^2f}{du_2^2} + \frac{dh}{du_2} \frac{d^2h}{du_2^2} \right).$$

Find the Ricci tensor and the curvature scalar.

Problem 11. Let $M = \mathbb{R}^n \setminus \{(0, \ldots, 0)\}$. Show that the metric tensor field

$$g = \frac{1}{\|\mathbf{x}\|^2} \sum_{j=1}^{n} (dx_j \otimes dx_j)$$

is invariant under $x_j \mapsto -x_j$ $j = 1, \ldots, n$.

Problem 12. Consider the three-dimensional Euclidean space. Let $a > 0$. The parametrization (u_1, u_2) of the *catenoid* is given by

$$\mathbf{x}(u_1, u_2) = \begin{pmatrix} x_1(u_1, u_2) \\ x_2(u_1, u_2) \\ x_3(u_1, u_2) \end{pmatrix} = \begin{pmatrix} a \cosh(u_2) \cos(u_1) \\ a \cosh(u_2) \sin(u_1) \\ a u_2 \end{pmatrix}$$

where $0 \leq u_1 \leq 2\pi$ and $-\infty < u_2 < \infty$. With

$$dx_1 = -a \cosh(u_2) \sin(u_1) du_1 + a \sinh(u_2) \cos(u_1) du_2$$
$$dx_2 = a \cosh(u_2) \cos(u_1) du_1 + a \sinh(u_2) \sin(u_1) du_2$$
$$dx_3 = a \, du_2$$

the metric tensor field takes the form utilizing $\cosh^2(u_2) = 1 + \sinh^2(u_2)$

$$g_C = a^2 \cosh^2(u_2) du_1 \otimes du_1 + a^2 \cosh^2(u_2) du_2 \otimes du_2.$$

The parametrization (v_1, v_2) of the *helicoid* is given by

$$\mathbf{x}(v_1, v_2) = \begin{pmatrix} x_1(v_1, v_2) \\ x_2(v_1, v_2) \\ x_3(v_1, v_2) \end{pmatrix} = \begin{pmatrix} v_2 \cos(v_1) \\ v_2 \sin(v_1) \\ av_1 \end{pmatrix}$$

where $0 \leq v_1 < 2\pi$ and $-\infty < v_2 < \infty$. With

$$dx_1 = -v_2 \sin(v_1)dv_1 + \cos(v_1)dv_2$$
$$dx_2 = v_2 \cos(v_1)dv_1 + \sin(v_1)dv_2$$
$$dx_3 = adv_1$$

the metric tensor field takes the form utilizing $\cosh^2(u_2) = 1 + \sinh^2(u_2)$

$$g_H = (v_2^2 + a^2)dv_1 \otimes dv_1 + dv_2 \otimes dv_2.$$

Show that a helocoid and catenoid are locally isometric. Apply the map

$$(u_1, u_2) \mapsto (u_1, a \sinh(u_2)).$$

Problem 13. Consider the metric tensor field

$$g = e^{2f_1(x_0)}dx_0 \otimes dx_0 - e^{2f_2(x_0)}(dx_1 \otimes dx_1 + dx_2 \otimes dx_2) - e^{2f_3(x_0)}dx_3 \otimes dx_3$$

where f_1, f_2, f_3 are smooth functions of x_0. Find the Christoffel symbols, the Ricci tensor and the curvature.

Problem 14. Any static spherically symmetric metric tensor field can be written in the form $(x_0 = ct)$

$$g = -f(r)dx_0 \otimes dx_0 + h(r)dr \otimes dr + r^2(\sin^2(\theta)d\phi \otimes d\phi + d\theta \otimes d\theta).$$

Show that spherically symmetric tensor fields can be characterized by its invariance property with respect to arbitrary rotation of the unit sphere given by the coordinates (θ, ϕ) and the map $(\theta \to -\theta, \phi \to -\phi)$.

Problem 15. Consider the metric tensor field

$$g = -e^{2\Phi(x_3)}dx_0 \otimes dx_0 + dx_3 \otimes dx_3.$$

The proper acceleration of a test particle at rest with respect to this metric tensor field is given by $\partial\Phi/\partial x_3$. Hence if the gravitational potential has the form $\Phi(x_3) = ax_3$ $(a > 0)$ then all the test particles at rest have the same acceleration of magnitude a in the positive x_3-direction. Show that the metric tensor takes the form

$$g = -(a\rho)^2 dx_0 \otimes dx_0 + \frac{1}{a\rho^2} d\rho \otimes d\rho$$

under the transformation

$$\rho(x_3) = \frac{1}{a} e^{ax_3}.$$

Problem 16.   The *Heisenberg group* is defined as $\mathbb{R}^3$ with the group operation ·

$$(x_1, x_2, x_3) \cdot (\bar{x}_1, \bar{x}_2, \bar{x}_3) := (x_1 + \bar{x}_1, x_2 + \bar{x}_2, x_3 + \bar{x}_3 + \frac{1}{2}x_1\bar{x}_2 - \frac{1}{2}\bar{x}_1x_2).$$

The identity element of the group is obviously $(0, 0, 0)$. The inverse element of (x_1, x_2, x_3) is $(-x_1, -x_2, -x_3)$. Consider the metric tensor field

$$g = (1 + x_2^2/4)dx_1 \otimes dx_1 + (1 + x_1^2/4)dx_2 \otimes dx_2 + dx_3 \otimes dx_3$$
$$- \frac{1}{4}x_1x_2(dx_1 \otimes dx_2 + dx_2 \otimes dx_1) + \frac{1}{2}x_2(dx_1 \otimes dx_3 + dx_3 \otimes dx_1)$$
$$- \frac{1}{2}x_1(dx_2 \otimes dx_3 + dx_3 \otimes dx_2)$$

with the corresponding symmetric 3×3 matrix

$$\begin{pmatrix} 1 + x_2^2/4 & -x_1x_2/4 & x_2/2 \\ -x_1x_2/4 & 1 + x_1^2/4 & -x_1/2 \\ x_2/2 & -x_1/2 & 1 \end{pmatrix}.$$

The inverse matrix is given by

$$\begin{pmatrix} 1 & 0 & -x_2/2 \\ 0 & 1 & x_1/2 \\ -x_2/2 & x_1/2 & (x_1^2 + x_2^2 + 4)/4 \end{pmatrix}.$$

Show that the metric tensor field is left-invariant. Find the Christoffel symbols. Show that the three vector fields

$$V_1 = \frac{\partial}{\partial x_1} - \frac{1}{2}x_2\frac{\partial}{\partial x_3}, \quad V_2 = \frac{\partial}{\partial x_2} + \frac{1}{2}x_1\frac{\partial}{\partial x_3}, \quad V_3 = \frac{\partial}{\partial x_3}$$

satisfy the commutation relations

$$[V_1, V_2] = V_3, \quad [V_2, V_3] = 0, \quad [V_3, V_1] = 0.$$

Problem 17. Let $r > 0$. The *Klein bagel* (figure 8 immersion) is a specific immersion of the Klein bottle manifold into three dimensions. The figure 8 immersion has the parametrization

$$x_1(u, v) = (r + \cos(u/2)\sin(v) - \sin(u/2)\sin(2v))\cos(u)$$
$$x_2(u, v) = (r + \cos(u/2)\sin(v) - \sin(u/2)\sin(2v))\sin(u)$$
$$x_3(u, v) = \sin(u/2)\sin(v) + \cos(u/2)\sin(2v)$$

where r is a positive constant and $0 \le u < 2\pi$, $0 \le v < 2\pi$. Find the Riemann curvature of the Klein bagel. Starting from the metric tensor field in $\mathbb{E}^3$

$$g = dx_1 \otimes dx_1 + dx_2 \otimes dx_2 + dx_3 \otimes dx_3$$

and $dx_1(u, v)$, $dx_2(u, v)$, $dx_3(u, v)$.

Problem 18. The embedding of the *Klein bottle* in the Euclidean space $\mathbb{E}^4$ with metric tensor field

$$g = dx_1 \otimes dx_1 + dx_2 \otimes dx_2 + dx_3 \otimes dx_3 + dx_4 \otimes dx_4$$

is given by

$$x_1(\alpha, \beta) = \cos(\alpha)\cos(\beta)$$
$$x_2(\alpha, \beta) = \sin(\alpha)\cos(\beta)$$
$$x_3(\alpha, \beta) = 2\cos(\alpha/2)\sin(\beta)$$
$$x_4(\alpha, \beta) = 2\sin(\alpha/2)\sin(\beta).$$

Show that

$$g_K = d\alpha \otimes d\alpha + (1 + 3\cos^2(\beta))d\beta \otimes d\beta.$$

Show that

$$dx_1 \wedge dx_2 = \sin(\beta)\cos(\beta)d\alpha \wedge d\beta, \qquad dx_3 \wedge dx_4 = -2\sin(\beta)\cos(\beta)d\alpha \wedge d\beta.$$

Find $dx_1 \wedge dx_3$ and $dx_2 \wedge dx_4$.

Problem 19. Let n_1, n_2 be positive integers and assume that n_1 and n_2 have the common divisor 1. Consider the map $\mathbf{f}_{n_1,n_2} : \mathbb{R}^2 \mapsto \mathbb{S}^3$ given by

$$\mathbf{f}_{n_1,n_2}(u_1, u_2) = \begin{pmatrix} f_{1,n_1,n_2}(u_1, u_2) \\ f_{2,n_1,n_2}(u_1, u_2) \\ f_{3,n_1,n_2}(u_1, u_2) \\ f_{4,n_1,n_2}(u_1, u_2) \end{pmatrix} = \begin{pmatrix} \cos(n_1 u_1)\cos(u_2) \\ \sin(n_1 u_1)\cos(u_2) \\ \cos(n_2 u_1)\sin(u_2) \\ \sin(n_2 u_1)\sin(u_2) \end{pmatrix}.$$

Thus we have a normalized vector in $\mathbb{R}^4$

$$\mathbf{v}_{n_1,n_2}(u_1,u_2) = \begin{pmatrix} \cos(n_1 u_1)\cos(u_2) \\ \sin(n_1 u_1)\cos(u_2) \\ \cos(n_2 u_1)\sin(u_2) \\ \sin(n_2 u_1)\sin(u_2) \end{pmatrix}.$$

Let

$$g = dx_1 \otimes dx_1 + dx_2 \otimes dx_2 + dx_3 \otimes dx_3 + dx_4 \otimes dx_4.$$

Find $\mathbf{f}^*_{n_1,n_2}(g)$. Then calculate the curvature. Discuss.

Problem 20. (i) The *Enneper surface* is given by

$$x_1(u_1,u_2) = 3u_2 - 3u_1^2 u_2 + u_2^3$$
$$x_2(u_1,u_2) = 3u_1 - 3u_1 u_2^2 + u_1^3$$
$$x_3(u_1,u_2) = -6u_1 u_2.$$

Show that the affine invariants are given by

$$F(u_1,u_2) = k(1 + u_1^2 + u_2^2), \quad A(u_1,u_2) = 2ku_2, \quad B(u_1,u_2) = 2ku_1$$

where $k = 3\sqrt{6}$.
(ii) Let $\tau \in (0,1)$. *Minimal Thomson surfaces* are given by

$$x_1(u_1,u_2) = -(1 - \tau^2)^{-1/2}(\tau u_2 + \cos(u_1)\sinh(u_2))$$
$$x_2(u_1,u_2) = (1 - \tau^2)^{-1/2}(u_1 + \tau \sin(u_1)\cosh(u_2))$$
$$x_3(u_1,u_2) = \sin(u_1)\sinh(u_2).$$

Show that the corresponding affine invariants are

$$F(u_1,u_2) = (1 - \tau^2)^{-1/2}(\cosh(u_2) + \tau \cos(u_1))$$
$$A(u_1,u_2) = (1 - \tau^2)^{-1/2}\sinh(u_2)$$
$$B(u_1,u_2) = -\tau(1 - \tau^2)^{-1/2}\sin(u_1).$$

Problem 21. Consider the manifold $\mathbb{R}^3$. Let $0 \le u_1, u_2 < \pi$. Find the curvature for the metric tensor field

$$g = \frac{9 + (1 + 8\cos^2(u_2))^2}{1 + 8\cos^2(u_2)}(du_1 \otimes du_1 + \frac{1}{1 + 8\cos^2(u_2)}du_2 \otimes du_2).$$

Problem 22. The unit three sphere $\mathbb{S}^3$ can be thought of unitary 2×2 matrices U with $\det(U) = 1$, i.e. the Lie group $SU(2)$. Show that the tangent bundle is

$$T\mathbb{S}^3 = \mathbb{S}^3 \times \mathbb{R}^3.$$

Problem 23.   Consider the manifold $\mathbb{R}^3$. Let $a, b, c > 0$ and $a \neq b$, $a \neq c$, $b \neq c$. The *sphero-conical coordinates* s_2, s_3 are defined to be the roots of the quadratic equation

$$\frac{x_1^2}{a+s} + \frac{x_2^2}{b+s} + \frac{x_3^2}{c+s} = 0.$$

The first sphero-conical coordinate s_1 is given as the sum of the squares

$$s_1 = x_1^2 + x_2^2 + x_3^2.$$

The formula that expresses the Cartesian coordinates x_1, x_2, x_3 through s_1, s_2, s_3 are

$$x_1^2 = \frac{s_1(a+s_2)(a+s_3)}{(a-b)(a-c)}, \quad x_2^2 = \frac{s_1(b+s_2)(b+s_3)}{(b-a)(b-c)}$$

$$x_3^2 = \frac{s_1(c+s_2)(c+s_3)}{(c-a)(c-b)}.$$

Given the metric tensor field $g = dx_1 \otimes dx_1 + dx_2 \otimes dx_2 + dx_3 \otimes dx_3$. Show that this metric tensor field using sphero-conical coordinates is given by

$$\frac{1}{4}\left(\frac{1}{s_1}T_1 - \frac{s_1(s_2-s_3)}{(a+s_2)(b+s_2)(c+s_2)}T_2 - \frac{s_1(s_3-s_2)}{(a+s_3)(b+s_3)(c+s_3)}T_3\right)$$

where

$$T_1 = ds_1 \otimes ds_1, \quad T_2 = ds_2 \otimes ds_2, \quad T_3 = ds_3 \otimes ds_3.$$

Problem 24.   Consider the manifold M of the upper space $x_n > 0$ of $\mathbb{R}^n$ endowed with the metric tensor field

$$g = \frac{dx_1 \otimes dx_1 + \cdots + dx_n \otimes dx_n}{x_n^2}.$$

Show that the Gaussian curvature is given by -1.

Problem 25.   The *Poincaré upper half plane* is defined as

$$\mathbb{H}_+^2 := \{\, z = x_1 + ix_2 \, : \, x_1 \in \mathbb{R}, \, x_2 > 0 \,\}$$

together with the metric tensor field

$$g = \frac{1}{x_2^2}(dx_1 \otimes dx_1 + dx_2 \otimes dx_2).$$

Show that under the *Cayley transform*

$$\zeta = \frac{-iz + i}{z + 1}, \qquad z = x_1 + ix_2 = \frac{-\zeta + i}{\zeta + i}$$

the Poincaré upper half-plane is mapped onto the Poincaré disc with metric

$$g_{jk} = \frac{2}{1 - r^2} \text{diag}(1, r^2), \qquad r^2 = x_1^2 + x_2^2.$$

Problem 26.   (i) Consider the metric tensor field

$$g(u_1, u_2) = du_1 \otimes du_1 + e^{2u_1} du_2 \otimes du_2$$

where $-\infty < u_1, u_2 < +\infty$. Show that Gaussian curvature $K(u_1, u_2)$ has the value -1.

(ii) Consider the transformation

$$x_1(u_1, u_2) = u_2, \qquad x_2(u_1, u_2) = e^{-u_1}.$$

Show that

$$g(x_1, x_2) = \frac{1}{x_2^2}(dx_1 \otimes dx_1 + dx_2 \otimes dx_2)$$

where $x_2 > 0$ and $-\infty < x_1 < +\infty$.

(iii) Consider the transformation

$$x_1(\rho, \phi) = x_{10} + \rho\cos(\phi), \qquad x_2(\rho, \phi) = \rho\sin(\phi)$$

where x_{10} is a constant. Show that

$$g = \frac{1}{\rho^2 \sin^2(\phi)} d\rho \otimes d\rho + \frac{1}{\sin^2(\phi)} d\phi \otimes d\phi.$$

Problem 27.   Let $N \geq 2$ and $a > 0$. An N-dimensional Riemann manifold of constant *negative Gaussian curvature* $K = -1/a^2$ is described by the metric tensor field

$$g = dr \otimes dr + a^2 \sinh\left(\frac{r}{a}\right) d\sigma_{N-1} \otimes d\sigma_{N-1}$$

where $r \in [0, \infty)$ measures the distance to the origin and $d\sigma_{N-1} \otimes d\sigma_{N-1}$ denotes the metric tensor field of the unit sphere $\mathbb{S}^{N-1}$.

(i) Show that the volume element dV is covariantly defined as

$$dV_N = \left(a \sinh\left(\frac{r}{a}\right)\right)^{N-1} dr d\Omega_{N-1}$$

where $d\Omega_{N-1}$ is the surface element of the unit-sphere S_{N-1}.
(ii) Show that the radial part Δ_r of the Laplace operator for the metric tensor field given above is

$$\Delta_r = \frac{1}{(\sinh(r/a))^{N-1}} \frac{\partial}{\partial r} \left((\sinh(r/a))^{N-1} \frac{\partial}{\partial r} \right).$$

Problem 28. The cosmological constant Λ is a parameter with unit of $1/(length)^2$. Show that the metric tensor field

$$g = -dx_0 \otimes dx_0 + e^{2c\tau/a} dx \otimes dx + a^2 (d\theta \otimes d\theta + \sin^2(\theta) d\phi \otimes d\phi)$$

where $a > 0$ has the dimension of a length. Show that this metric tensor field satisfies the vacuum *Einstein field equation* with a positive cosmological constant Λ

$$R_{\mu\nu} - \frac{1}{2} R g_{\mu\nu} + \Lambda g_{\mu\nu} = 0$$

where $a = 1/\sqrt{(\Lambda)}$.

Problem 29. Let $x_0 = ct$. Show that the metric tensor field

$$g = (1 - 2a/r) dx_0 \otimes dx_0 - dr \otimes dr - r^2 d\phi \otimes d\phi - dx_3 \otimes dx_3$$

is not a solution of Einstein's equation.

Problem 30. Let $x_0 = ct$. The metric tensor field g of a weak, plane, elliptically polarized gravitational wave propagating in the x_1-direction can be written as

$$\begin{aligned}
g = {} & dx_0 \otimes dx_0 - dx_1 \otimes dx_1 \\
& - (1 - h_{22}(x_1, x_0)) dx_2 \otimes dx_2 - (1 + h_{22}(x_1, x_0)) dx_3 \otimes dx_3 \\
& + h_{23}(x_1, x_0) dx_2 \otimes dx_3 + h_{23}(x_1, x_0) dx_3 \otimes dx_2
\end{aligned}$$

where

$$h_{22}(x_1, x_0) = h \sin(k(x_0 - x_1) + \phi), \quad h_{23}(x_1, x_0) = \widetilde{h} \sin(k(x_0 - x_1) + \widetilde{\phi})$$

with k the wave vector, h, $\widetilde{h}$ the amplitudes and ϕ, $\widetilde{\phi}$ the initial phase. They determine the state of the polarization of the gravitational wave. Show that in terms of the retarded and advanced coordinates

$$u(x_1, x_0) = \frac{1}{2}(x_0 - x_1), \qquad v(x_1, x_0) = \frac{1}{2}(x_0 + x_1)$$

the coordinates x_2, x_3 and v can be omitted.

Problem 31.   Consider the *Kähler potential*

$$K = \ln(1 + \bar{z}z)$$

and

$$g_{z\bar{z}} = \frac{\partial^2 K}{\partial z \partial \bar{z}}.$$

Show that the metric tensor field is given by

$$g_{z\bar{z}}d\bar{z} \otimes dz = \frac{d\bar{z} \otimes dz}{(1 + \bar{z}z)^2}.$$

Problem 32.   Show that the *Burgers equation*

$$\frac{\partial u}{\partial t} = (1 + u)\frac{\partial u}{\partial x} + \frac{\partial^2 u}{\partial x^2}$$

can be derived from the metric tensor field

$$g = \left(\frac{u^2}{4} + \eta^2\right) dx \otimes dx + \left(\frac{\eta^2 u}{2} + \frac{u}{4}\left(\frac{u^2}{2} + \frac{\partial u}{\partial x}\right)\right) dx \otimes dt$$

$$+ \left(\frac{\eta^2 u}{2} + \frac{u}{4}\left(\frac{u^2}{2} + \frac{\partial u}{\partial x}\right)\right) dt \otimes dx + \left(\left(\frac{u^2}{4} + \frac{1}{2}\frac{\partial u}{\partial x}\right)^2 + \frac{\eta^2}{4}u\right) dt \otimes dt$$

by setting the curvature R of g equal to 1. Here η is a real parameter.

Problem 33.   Consider the Robertson-Walker metric tensor field

$$g = -dx_0 \otimes dx_0 + a^2(x_0)\left(\frac{dr \otimes dr}{1 - kr^2} + r^2(d\theta \otimes d\theta + \sin^2(\theta)d\phi \otimes d\phi)\right).$$

Find the Christoffel symbols, the Riemann tensor, the Ricci tensor and the scalar curvature.

Problem 34.   Show that the de Sitter space is an exact solution of the vacuum *Einstein field equation* with a positive cosmological constant Λ

$$R_{\mu\nu} - \frac{1}{2}Rg_{\mu\nu} + \Lambda g_{\mu\nu} = 0.$$

Problem 35. Let $x_0 = ct$ and thus $dx_0 = cd$. The Reissner-Nordstrom metric tensor field is given by

$$g = \left(1 - \frac{2Gm}{c^2r^2} + \frac{\alpha}{r^2}\right) dx_0 \otimes dx_0$$
$$- \left(1 - \frac{2Gm}{c^2r^2} + \frac{\alpha}{r^2}\right)^{-1} dr \otimes dr - r^2(d\theta \otimes d\theta + \sin^2(\theta)d\phi \otimes d\phi)$$

where (r, θ, ϕ) are the spherical coordinates and t the time. The corresponding 4×4 matrix is

$$\begin{pmatrix} 1-2Gm/(c^2r)+\alpha/r^2 & 0 & 0 & 0 \\ 0 & -(1-2Gm/(c^2r)+\alpha/r^2)^{-1} & 0 & 0 \\ 0 & 0 & -r^2 & 0 \\ 0 & 0 & 0 & -r^2\sin(\theta) \end{pmatrix}.$$

Find the inverse of this matrix. Find the Christoffel symbols, the Riemann tensor, the Ricci tensor and the scalar curvature.

Problem 36. Consider the metric tensor field

$$g = e^{u(x_0,x_1)}(dx_0 \otimes dx_0 - dx_1 \otimes dx_1).$$

Find the Ricci tensor and the curvature.

Problem 37. Consider the metric tensor field

$$g = \sum_{j=1}^{4}(dz_j \otimes dz_j).$$

Let $z_j = (2R)u_j e^{i\phi_j}$ for $j = 1, \dots, 4$ with $u_j \in \mathbb{R}$. Show that g takes the form

$$g = \sum_{j=1}^{4}(2R)^2(du_j \otimes du_j + u_j^2 d\phi_j \otimes d\phi_j).$$

Problem 38. Consider the metric tensor field with coordinates (x, y, u, v)

$$g = du \otimes du + du \otimes dv + dv \otimes du - kxdy \otimes du - kxdu \otimes dy$$
$$- kxdy \otimes dv - kxdv \otimes dy + (k^2x^2 - e^{kv})dy \otimes dy - e^{-kv}dx \otimes dx.$$

Find the curvature.

Problem 39.   Consider the metric tensor field

$$g = g_{11}(\mathbf{x})dx_1 \otimes dx_1 + g_{12}(\mathbf{x})dx_1 \otimes dx_2 + g_{21}(\mathbf{x})dx_2 \otimes dx_1 + g_{22}(\mathbf{x})dx_2 \otimes dx_2$$

with g_{jk} be smooth functions. Find the system of differential equations from

$$R_{jk} = \frac{1}{4}Rg_{jk} = 0, \quad j,k = 1,2.$$

Chapter 4

Differential Forms and Applications

4.1 Notations and Definitions

We denote by $\wedge$ the *exterior product* (also called the wedge product or Grassmann product). The exterior product is associative. Let α be a differential r-form, β, β_1, β_2 be a differential s-form and γ be a differential t-form. We have the properties

$$\alpha \wedge (\beta \wedge \gamma) = (\alpha \wedge \beta) \wedge \gamma$$
$$\alpha \wedge (\beta_1 + \beta_2) = \alpha \wedge \beta_1 + \alpha \wedge \beta_2$$
$$\alpha \wedge \beta = (-1)^{rs}\beta \wedge \alpha.$$

Hence for a differential one-form α we have $\alpha \wedge \alpha = 0$. We denote by d the *exterior derivative*. The exterior derivative d is linear, i.e.

$$d(c_1\beta_1 + c_2\beta_2) = c_1 d\beta_1 + c_2 d\beta_2.$$

Let α be a differential r-form and β be a s-form then

$$d(\alpha \wedge \beta) = (d\alpha) \wedge \beta + (-1)^r \alpha \wedge d\beta$$
$$dd\alpha = 0$$

$$dx_j \wedge dx_k = -dx_k \wedge dx_j$$
$$dx_j \wedge (f(x))dx_k = f(x)(dx_j \wedge dx_k).$$

Let $M = \mathbb{R}^n$, $\mathbf{p} \in \mathbb{R}^n$ and $T_{\mathbf{p}}(\mathbb{R}^n)$ be the tangent space at $\mathbf{p}$. A differential one-form at $\mathbf{p}$ is a linear map h from $T_{\mathbf{p}}(\mathbb{R}^n)$ into $\mathbb{R}$. This map satisfies the properties

$$h(V_{\mathbf{p}}) \in \mathbb{R} \quad \text{for all } V_{\mathbf{p}} \in \mathbb{R}^n$$
$$h(aV_{\mathbf{p}} + bW_{\mathbf{p}}) = ah(V_{\mathbf{p}}) + bh(W_{\mathbf{p}}) \quad \text{for all } a, b \in \mathbb{R}, \ V_{\mathbf{p}}, W_{\mathbf{p}} \in T_{\mathbf{p}}(\mathbb{R}^n).$$

A differential one-form is a smooth choice of a linear map ϕ given above for each point $\mathbf{p}$ in the manifold $\mathbb{R}^n$. Let $f : \mathbb{R}^n \to \mathbb{R}$ be a real-valued smooth function. Let $\rfloor$ be the *interior product* (also called *contraction*). One defines $V \rfloor f = 0$ and df of the function f as the differential one-form such that

$$df(V) \equiv V \rfloor df = V(f)$$

for every smooth vector field V in $\mathbb{R}^n$. Thus at any point $\mathbf{p}$, the differential df of a smooth function f is an operator that assigns to a tangent vector $V_{\mathbf{p}}$ the directional derivative of the function f in the direction of this vector, i.e. $df(V)(\mathbf{p}) = V_{\mathbf{p}}(f) = \nabla f(\mathbf{p}) \cdot V(\mathbf{p})$. Applying the differential of the coordinate functions x_j $(j = 1, \ldots, n)$ provides

$$dx_j \left(\frac{\partial}{\partial x_k} \right) \equiv \frac{\partial}{\partial x_j} \rfloor dx_k = \frac{\partial x_j}{\partial x_k} = \delta_{jk}.$$

Let γ be a differential form and V a vector field. Then $V \rfloor \gamma$ denotes the interior product (contraction). Moreover

$$V \rfloor (\gamma + \delta) = V \rfloor \gamma + V \rfloor \delta$$
$$V \rfloor (\gamma_1 \wedge \gamma_2) = (V \rfloor \gamma_1) \wedge \gamma_2 + (-1)^p \gamma_1 \wedge (V \rfloor \gamma_2)$$
$$V_1 \rfloor (V_2 \rfloor \alpha) = -V_2 \rfloor (V_1 \rfloor \alpha)$$

where γ and δ are p-forms and γ_1 is a differential p-form. Let Ω be the volume form in $\mathbb{R}^n$. Then

$$\left(\sum_{j=1}^{n} f_j(x) \frac{\partial}{\partial x_j} \right) \rfloor (dx_1 \wedge \cdots \wedge dx_n) = \sum_{j=1}^{n} (-1)^j f_j(x) dx_1 \wedge \cdots \wedge \widehat{dx_j} \wedge \cdots \wedge dx_n$$

where $\widehat{dx_j}$ indicates omission.

For Riemannian or pseudo-Riemannian manifolds the *Hodge duality operator* $\star$ is defined as a linear transformation between the spaces $\bigwedge_k(T^*M)$ and $\bigwedge_{m-k}(T^*M)$, i.e.

$$\star : \bigwedge_k (T^*M) \to \bigwedge_{m-k} (T^*M)$$

where $k = 0, 1, \ldots, m$. The operator has the $\star$-linearity which can be expressed as

$$\star(f\omega + g\sigma) = f(\star\omega) + g(\star\sigma)$$

for all $f, g \in \bigwedge_0(T^*M)$ and all $\omega, \sigma \in \bigwedge_k(T^*M)$. The $\star$ operator applied to a p-form $(p \leq m)$ defined on an arbitrary Riemannian (or pseudo-Riemannian) manifold with metric tensor field g is given by

$$\star(dx_{i_1} \wedge dx_{i_2} \wedge \cdots \wedge dx_{i_p})$$

$$:= \sum_{j_1 \cdots j_m = 1}^{m} g^{i_1 j_1} \cdots g^{i_p j_p} \frac{1}{(m-p)!} \frac{g}{\sqrt{|g|}} \epsilon_{j_1 \cdots j_m} dx_{j_{p+1}} \wedge \cdots \wedge dx_{j_m}$$

where $\epsilon_{j_1 \cdots j_m}$ is the totally antisymmetric tensor. $\epsilon_{12 \ldots m} = +1$, $g \equiv \det(g_{ij})$ and

$$\sum_{j=1}^{m} g^{ij} g_{jk} = \delta_{ik} \quad \text{(Kronecker symbol)}.$$

The double duality (composition of $\star$ with itself) for $\omega \in \bigwedge_k T^*M$ is given by

$$\star\star(\omega) = (-1)^{k(m-k)} \omega.$$

The inverse of $\star$ is $(\star)^{-1} = (-1)^{k(m-k)} \star$. Moreover, $\omega \wedge (\star\sigma) = \sigma \wedge (\star\omega)$ for $\omega, \sigma \in \bigwedge_k T^*M$. A differential form γ is called *harmonic* if

$$(\star d \star d + d \star d\star)\gamma = 0.$$

Let M^m and N^m be an m-dimensional differentiable manifolds and $f : M^n \to N^n$ be a diffeomorphism. Then

$$\int_{N^n} \alpha = \int_{M^n} f^*(\alpha)$$

where α is a differential form in N^m.

Let S be a $(p+1)$-surface in M with boundary ∂S, then (Stokes theorem)

$$\int_S d\alpha = \int_{\partial S} \alpha.$$

4.2   Solved Problems

Problem 1.   (i) Consider the differential one-form

$$\alpha_1 = x_1 dx_1 + x_2 dx_2$$

in $\mathbb{R}^2$. Find $d\alpha_1$. Can one find a smooth function $f : \mathbb{R}^2 \to \mathbb{R}$ such that $df = \alpha_1$?
(ii) Consider the differential one-form

$$\alpha_2 = x_1 dx_2 - x_2 dx_1$$

in $\mathbb{R}^2$. Find $d\alpha_2$. Let $c \in \mathbb{R}$. Show that $x_2 - cx_1 = 0$ satisfies $\alpha_2 = 0$. Can one find a smooth function $f : \mathbb{R}^2 \to \mathbb{R}$ such that $df = \alpha_2$?
(iii) Consider the differential one-forms

$$\alpha_1 = x_1 dx_1 + x_2 dx_2, \qquad \alpha_2 = x_1 dx_2 - x_2 dx_1.$$

Find $\alpha_1 \wedge \alpha_2$.

Solution 1.   (i) We have $d\alpha_1 = dx_1 \wedge dx_1 + dx_2 \wedge dx_2 = 0$ and find $f(x_1, x_2)) = \frac{1}{2}(x_1^2 + x_2^2)$.
(ii) We have $d\alpha_2 = dx_1 \wedge dx_2 - dx_2 \wedge dx_1 = 2dx_1 \wedge dx_2$. Since $dx_2 = cdx_1$ we have

$$\alpha_2 = x_1 dx_2 - x_2 dx_1 = x_1 d(cx_1) - cx_1 dx_1 = cx_1 dx_1 - cx_1 dx_1.$$

From $df = (\partial f / \partial x_1) dx_1 + (\partial f / \partial x_2) dx_2$ and $df = \alpha_2$ we obtain the two equations

$$\frac{\partial f}{\partial x_1} = -x_2, \qquad \frac{\partial f}{\partial x_2} = x_1.$$

It follows that

$$\frac{\partial^2 f}{\partial x_2 \partial x_1} = -1, \qquad \frac{\partial^2 f}{\partial x_1 \partial x_2} = 1.$$

Hence there is no such f.
(iii) With $dx_1 \wedge dx_2 = -dx_2 \wedge dx_1$ we obtain

$$\alpha_1 \wedge \alpha_2 = x_1^2 dx_1 \wedge dx_2 - x_2^2 dx_2 \wedge dx_1 = (x_1^2 + x_2^2) dx_1 \wedge dx_2.$$

Problem 2.   (i) Let f, g be two smooth functions defined on $\mathbb{R}^2$. Find the differential two-form $df \wedge dg$.
(ii) Consider the analytic functions $f_1 : \mathbb{R}^2 \to \mathbb{R}$, $f_2 : \mathbb{R}^2 \to \mathbb{R}$

$$f_1(x_1, x_2) = x_1 + x_2, \qquad f_2(x_1, x_2) = x_1^2 + x_2^2 - 1.$$

Find df_1 and df_2. Then calculate $df_1 \wedge df_2$. Solve the system of equations $df_1 \wedge df_2 = 0$, $x_1^2 + x_2^2 - 1 = 0$.

Solution 2.   (i) We have

$$df = \frac{\partial f}{\partial x_1}dx_1 + \frac{\partial f}{\partial x_2}dx_2, \qquad dg = \frac{\partial g}{\partial x_1}dx_1 + \frac{\partial g}{\partial x_2}dx_2.$$

Since $dx_1 \wedge dx_1 = 0$, $dx_2 \wedge dx_2 = 0$, $dx_1 \wedge dx_2 = -dx_2 \wedge dx_1$ we obtain

$$df \wedge dg = \left(\frac{\partial f}{\partial x_1}\frac{\partial g}{\partial x_2} - \frac{\partial f}{\partial x_2}\frac{\partial g}{\partial x_1} \right) dx_1 \wedge dx_2.$$

(ii) We have $df_1 = dx_1 + dx_2$, $df_2 = 2x_1 dx_1 + 2x_2 dx_2$. With $dx_1 \wedge dx_1 = 0$, $dx_2 \wedge dx_2 = 0$, $dx_1 \wedge dx_2 = -dx_2 \wedge dx_1$ we arrive at

$$df_1 \wedge df_2 = 2(x_2 - x_1)dx_1 \wedge dx_2.$$

From $df_1 \wedge df_2 = 0$ we obtain $x_2 = x_1$. Inserting this into $x_1^2 + x_2^2 = 1$ provides $x^2 = 1/2$ with $x = x_1 = x_2$. Thus we have the two solutions $(x_1, x_2) = (1/\sqrt{2}, 1/\sqrt{2})$ and $(x_1, x_2) = (-1/\sqrt{2}, -1/\sqrt{2})$.

Problem 3.   Let $M = \mathbb{R}^2$. Consider the differential one-form

$$\alpha = (2x_1^3 + 3x_2)dx_1 + (3x_1 + x_2 - 1)dx_2.$$

(i) Find $d\alpha$.
(ii) Can one find a function $f : \mathbb{R}^2 \to \mathbb{R}$ such that $df = \alpha$?

Solution 3.   (i) Since $dx_1 \wedge dx_2 = -dx_2 \wedge dx_1$ we have

$$d\alpha = 3dx_2 \wedge dx_1 + 3dx_1 \wedge dx_2 = 0.$$

(ii) Owing to the result of (i) we find the function

$$f(x_1, x_2) = \frac{1}{2}x_1^4 + 3x_1 x_2 + \frac{1}{2}x_2^2 - x_2.$$

Problem 4.   Consider the differential one-form in $\mathbb{R}^2$

$$\alpha = x_1 x_2 dx_1 + (2x_1^3 + 3x_2^2 - 10)dx_2.$$

(i) Show that $d\alpha \neq 0$.

(ii) Can one find a smooth function $f : \mathbb{R}^2 \to \mathbb{R}$ (*integrating factor*) such that $d(f(x_1, x_2)\alpha) = 0$?

Solution 4.   (i) We have

$$d\alpha = -x_1 dx_1 \wedge dx_2 + 4x_1 dx_1 \wedge dx_2 = 3x_1 dx_1 \wedge dx_2 \neq 0.$$

(ii) Since

$$df = \frac{\partial f}{\partial x_1} dx_1 + \frac{\partial f}{\partial x_2} dx_2$$

and $d(f(x_1, x_2)\alpha) = df \wedge \alpha + f d\alpha$ we obtain

$$d(f(x_1, x_2)\alpha) = \left(-\frac{\partial f}{\partial x_2} x_1 x_2 + \frac{\partial f}{\partial x_1}(2x_1^3 + 3x_2^2 + 10) + 3x_1 f \right) dx_1 \wedge dx_2.$$

Then $f(x_1, x_2) = x_2^3$ with $\partial f/\partial x_1 = 0$, $\partial f/\partial x_2 = 3x_2^2$ satisfies

$$d(f(x_1, x_2)\alpha) = 0.$$

Problem 5.   (i) Consider the three differential one-forms in $\mathbb{R}^2$

$$\alpha_1 = (\, x_1 \quad x_2 \,) \begin{pmatrix} 1 & 0 \\ 0 & 1 \end{pmatrix} \begin{pmatrix} dx_1 \\ dx_2 \end{pmatrix} = x_1 dx_1 + x_2 dx_2$$

$$\alpha_2 = (\, x_1 \quad x_2 \,) \begin{pmatrix} 0 & 1 \\ 1 & 0 \end{pmatrix} \begin{pmatrix} dx_1 \\ dx_2 \end{pmatrix} = x_1 dx_2 + x_2 dx_1$$

$$\alpha_3 = (\, x_1 \quad x_2 \,) \begin{pmatrix} 1 & 1 \\ 1 & 1 \end{pmatrix} \begin{pmatrix} dx_1 \\ dx_2 \end{pmatrix} = (x_1 + x_2)dx_1 + (x_1 + x_2)dx_2.$$

Find $\alpha_1 \wedge \alpha_2$, $\alpha_2 \wedge \alpha_3$, $\alpha_3 \wedge \alpha_1$.
(ii) Consider the differential one-forms ($a_{jk}, b_{jk} \in \mathbb{R}$)

$$\alpha_1 = (\, x_1 \quad x_2 \,) \begin{pmatrix} a_{11} & a_{12} \\ a_{21} & a_{22} \end{pmatrix} \begin{pmatrix} dx_1 \\ dx_2 \end{pmatrix},$$

$$\alpha_2 = (\, x_1 \quad x_2 \,) \begin{pmatrix} b_{11} & b_{12} \\ b_{21} & b_{22} \end{pmatrix} \begin{pmatrix} dx_1 \\ dx_2 \end{pmatrix}.$$

Find $\alpha_1 \wedge \alpha_2$.

Solution 5.   (i) We obtain

$$\alpha_1 \wedge \alpha_2 = (x_1^2 - x_2^2)dx_1 \wedge dx_2$$
$$\alpha_2 \wedge \alpha_3 = (x_2^2 - x_1^2)dx_1 \wedge dx_2$$
$$\alpha_3 \wedge \alpha_1 = (x_2^2 - x_1^2)dx_1 \wedge dx_2.$$

(ii) With

$$\alpha_1 = (a_{11}x_1 + a_{21}x_2)dx_1 + (a_{12}x_1 + a_{22}x_2)dx_2$$
$$\alpha_2 = (b_{11}x_1 + b_{21}x_2)dx_1 + (b_{12}x_1 + b_{22}x_2)dx_2$$

we obtain

$$\alpha_1 \wedge \alpha_2 = (a_{11}b_{12} - a_{12}b_{11})x_1^2 dx_1 \wedge dx_2 + (a_{21}b_{22} - a_{22}b_{21})x_2^2 dx_1 \wedge dx_2$$
$$+ (a_{11}b_{22} - a_{22}b_{11} + a_{21}b_{12} - a_{12}b_{21})x_1 x_2 dx_1 \wedge dx_2.$$

Problem 6. Consider the differential one-form $\alpha = x_1 dx_2 - x_2 dx_1$ on $\mathbb{R}^2$. Show that α is *invariant* under the transformation

$$\begin{pmatrix} x_1' \\ x_2' \end{pmatrix} = \begin{pmatrix} \cos(\alpha) & -\sin(\alpha) \\ \sin(\alpha) & \cos(\alpha) \end{pmatrix} \begin{pmatrix} x_1 \\ x_2 \end{pmatrix}.$$

Show that $\Omega = dx_1 \wedge dx_2$ is invariant under this transformation.

Solution 6. We have

$$dx_1' = \cos(\alpha)dx_1 - \sin(\alpha)dx_2, \qquad dx_2' = \sin(\alpha)dx_1 + \cos(\alpha)dx_2.$$

Thus

$$x_1' dx_2' - x_2' dx_1' = x_1 dx_2 - x_2 dx_1$$

and with $dx_1 \wedge dx_1 = dx_2 \wedge dx_2 = 0$, $dx_1 \wedge dx_2 = -dx_2 \wedge dx_1$ we have

$$dx_1' \wedge dx_2' = dx_1 \wedge dx_2.$$

Problem 7. Consider the differential two-form $dx_1 \wedge dx_2$ in $\mathbb{R}^2$ and the transformation

$$\begin{pmatrix} x_1' \\ x_2' \end{pmatrix} = \begin{pmatrix} \cosh(\alpha) & \sinh(\alpha) \\ \sinh(\alpha) & \cosh(\alpha) \end{pmatrix} \begin{pmatrix} x_1 \\ x_2 \end{pmatrix}.$$

The determinant of the matrix is $+1$. Find $dx_1' \wedge dx_2'$.

Solution 7. Since $\cosh^2(\alpha) - \sinh^2(\alpha) = 1$ we obtain

$$dx_1' \wedge dx_2' = dx_1 \wedge dx_2$$

i.e. $dx_1 \wedge dx_2$ is invariant under the linear transformation.

Problem 8.   (i) Consider the differential one-form

$$\alpha = f_1(\mathbf{x})dx_1 + f_2(\mathbf{x})dx_2$$

in $\mathbb{R}^2$. Find a differential one-form $\gamma = g_1(\mathbf{x})dx_1 + g_2(\mathbf{x})dx_2$ such that $d\alpha = \alpha \wedge \gamma$.
(ii) Let α be a smooth differential one-form

$$\alpha = f_1(\mathbf{x})dx_1 + f_2(\mathbf{x})dx_2$$

and $g : \mathbb{R}^2 \to \mathbb{R}$ be a smooth function. Find the conditions on f_1, f_2, g such that $d(g(\mathbf{x})\alpha) = 0$.

Solution 8.   (i) We have

$$d\alpha = \left(-\frac{\partial f_1}{\partial x_2} + \frac{\partial f_2}{\partial x_1} \right) dx_1 \wedge dx_2$$

and $\alpha \wedge \gamma = (f_1 g_2 - f_2 g_1)dx_1 \wedge dx_2$. It follows that

$$-\frac{\partial f_1}{\partial x_2} + \frac{\partial f_2}{\partial x_1} = f_1 g_2 - f_2 g_1.$$

For example if $f_1 = c_1$, $f_2 = c_2$ are constants, then $c_1 g_2 - c_2 g_1 = 0$.
(ii) We have $d(g(\mathbf{x})\alpha) = (dg) \wedge \alpha + g(\mathbf{x})d\alpha$ and

$$d\alpha = \left(-\frac{\partial f_1}{\partial x_2} + \frac{\partial f_2}{\partial x_1} \right) dx_1 \wedge dx_2, \quad dg(\mathbf{x}) = \frac{\partial g}{\partial x_1}dx_1 + \frac{\partial g}{\partial x_2}dx_2.$$

Hence

$$(dg) \wedge \alpha = \left(\frac{\partial g}{\partial x_1}f_2 - \frac{\partial g}{\partial x_2}f_1 \right) dx_1 \wedge dx_2$$

$$g(\mathbf{x})d\alpha = \left(-g\frac{\partial f_1}{\partial x_2} + g\frac{\partial f_2}{\partial x_1} \right) dx_1 \wedge dx_2$$

and the condition follows as

$$\frac{\partial g}{\partial x_1}f_2 - \frac{\partial g}{\partial x_2}f_1 - g\frac{\partial f_1}{\partial x_2} + g\frac{\partial f_2}{\partial x_1} = 0.$$

Problem 9.   Let $M = \mathbb{R}^2$ and $a > b > 0$. Consider the transformation

$$x_1(\theta, \phi) = (a + b\cos(\phi))\cos(\theta), \quad x_2(\theta, \phi) = (a + b\cos(\phi))\sin(\theta).$$

Find $dx_1 \wedge dx_2$ and $dx_1 \otimes dx_1 + dx_2 \otimes dx_2$.

Solution 9. We have

$$dx_1 = -a\sin(\theta)d\theta - b\sin(\phi)\cos(\theta)d\phi - b\cos(\phi)\sin(\theta)d\theta$$
$$dx_2 = a\cos(\theta)d\theta - b\sin(\phi)\sin(\theta)d\phi + b\cos(\phi)\cos(\theta)d\theta.$$

Thus

$$dx_1 \wedge dx_2 = ab\sin(\phi)d\theta \wedge d\phi + b^2\sin(\phi)\cos(\phi)d\theta \wedge d\phi$$
$$= b\sin(\phi)(a + b\cos(\phi))d\theta \wedge d\phi$$

and

$$dx_1 \otimes dx_1 + dx_2 \otimes dx_2 = (a + b\cos(\phi))^2 d\theta \otimes d\theta + b^2\sin^2(\phi)d\phi \otimes d\phi.$$

The corresponding matrix is

$$\begin{pmatrix} (a + b\cos(\phi))^2 & 0 \\ 0 & b^2\sin^2(\phi) \end{pmatrix}.$$

Problem 10. (i) Let $z \in \mathbb{C}$ and $z = x + iy$ with $x, y \in \mathbb{R}$. Find

$$\alpha = \overline{z}dz - zd\overline{z}.$$

(ii) Let $z = x + iy$, $x, y \in \mathbb{R}$. Calculate $-idz \wedge d\overline{z}$.
(iii) Let $z \in \mathbb{C}$. Find $dz \otimes d\overline{z}$ and $dz \wedge d\overline{z}$.
(iv) Let $z = re^{i\phi}$. Find $dz \wedge d\overline{z}$.
(v) Consider the complex number $z = re^{i\phi}$ with $r \geq 0$. Then $\overline{z} = re^{-i\phi}$.
Calculate

$$\frac{dz \wedge d\overline{z}}{z}.$$

Solution 10. (i) With $dz = dx + idy$ and $d\overline{z} = dx - idy$ we obtain

$$\alpha = 2i(xdy - ydx).$$

(ii) We have $dz = dx + idy$, $d\overline{z} = dx - idy$. Since $dx \wedge dx = 0$, $dy \wedge dy = 0$, $dy \wedge dx = -dx \wedge dy$ we obtain

$$-idz \wedge d\overline{z} = 2dx \wedge dy.$$

(iii) With $z = x + iy$, $\overline{z} = x - iy$ ($x, y \in \mathbb{R}$) we have

$$dz \otimes d\overline{z} = d(x + iy) \otimes d(x - iy) = dx \otimes dx - idx \otimes dy + idy \otimes dx + dy \otimes dy$$

with the corresponding 2×2 hermitian matrix

$$\begin{pmatrix} 1 & -i \\ i & 1 \end{pmatrix}.$$

The eigenvalues of this matrix are 0 and 2 and thus the matrix is not invertible. For $dz \wedge d\bar{z}$ we find

$$dz \wedge d\bar{z} = (dx + idy) \wedge (dx - idy) = -idx \wedge dy + idy \wedge dx = -2idx \wedge dy.$$

(iv) Since $dz = e^{i\phi}dr + rie^{i\phi}d\phi$, $d\bar{z} = e^{-i\phi}dr - rie^{-i\phi}d\phi$ we obtain

$$dz \wedge d\bar{z} = -irdr \wedge d\phi + rid\phi \wedge dr = -2irdr \wedge d\phi.$$

(v) We have

$$dz = e^{i\phi}dr + ie^{i\phi}rd\phi, \qquad d\bar{z} = e^{-i\phi}dr - ie^{-i\phi}rd\phi.$$

Thus

$$dz \wedge d\bar{z} = (e^{i\phi}dr + ie^{i\phi}rd\phi) \wedge (e^{-i\phi}dr - ie^{-i\phi}rd\phi)$$
$$= -irdr \wedge d\phi + ird\phi \wedge dr = -2irdr \wedge d\phi.$$

It follows that

$$\frac{dz \wedge d\bar{z}}{z} = -2ie^{-i\phi}dr \wedge d\phi.$$

Problem 11.   Consider the differential 2-form

$$\beta = \frac{4dz \wedge d\bar{z}}{(1 + |z|^2)^2}$$

and the *linear fractional transformations*

$$z = \frac{aw + b}{cw + d}, \qquad ad - bc = 1.$$

What is the conditions on a, b, c, d such that β is invariant under the transformation?

Solution 11.   Straightforward calculation yields

$$\frac{4dz \wedge d\bar{z}}{(1 + |z|^2)^2} = \frac{4dw \wedge d\bar{w}}{(|aw + b|^2 + |cw + d|^2)^2}$$
$$= \frac{4dw \wedge d\bar{w}}{(|b|^2 + |d|^2 + w(a\bar{b} + c\bar{d}) + \bar{w}(\bar{a}b + \bar{c}d) + (|a|^2 + |c|^2)|w|^2)^2}.$$

Thus the conditions are $|b|^2 + |d|^2 = 1$, $a\bar{b} + c\bar{d} = 0$, $|a|^2 + |c|^2 = 1$.

Problem 12. Find the solution of the *Pfaffian equation*

$$\alpha = (x_1^2 x_3 - x_2^3)dx_1 + 3x_1 x_2^2 dx_2 + x_1^3 dx_3 = 0.$$

First show that $\alpha \wedge d\alpha = 0$.

Solution 12. We have

$$d\alpha = -2x_1^2 dx_3 \wedge dx_1 + 6x_2^2 dx_1 \wedge dx_2.$$

Then

$$\alpha \wedge d\alpha = -6x_1^3 x_2^2 dx_2 \wedge dx_3 \wedge dx_1 + 6x_1^3 x_2^2 dx_3 \wedge dx_1 \wedge dx_2 = 0.$$

The solution of $\alpha = 0$ is $x_1^2 x_3 + x_2^3 = cx_1$, where c is a constant. This can be seen by rewritting the equation as

$$d(x_1 x_3) + d\left(\frac{x_2^3}{x_1}\right) = 0.$$

Note that with the vector field

$$V(\mathbf{x}) = \begin{pmatrix} x_1^2 x_3 - x_2^3 \\ 3x_1 x_2^2 \\ x_1^3 \end{pmatrix}$$

we have

$$\text{curl}(V) = \begin{pmatrix} 0 \\ -2x_1^2 \\ 6x_2^2 \end{pmatrix}$$

and $V \cdot \text{curl}(V) = 0$, where $\cdot$ denotes the scalar product.

Problem 13. (i) Consider the differential one-form in $\mathbb{R}^3$

$$\alpha = (x_2^2 + x_2 x_3)dx_1 + (x_1 x_3 + x_3^2)dx_2 + (x_2^2 - x_1 x_2)dx_3.$$

Find $d\alpha$ and $\alpha \wedge d\alpha$.
(ii) Consider the vector field

$$V = \begin{pmatrix} x_2^2 + x_2 x_3 \\ x_1 x_3 + x_3^2 \\ x_2^2 - x_1 x_2 \end{pmatrix}.$$

Find curl(V) and $V \cdot$ curl(V), where $\cdot$ denotes the scalar product.

Solution 13.   (i) We find

$$d\alpha = -2x_2 dx_1 \wedge dx_2 + (-2x_1 + 2x_2 - 2x_3)dx_2 \wedge dx_3 + 2x_2 dx_3 \wedge dx_1.$$

With $dx_1 \wedge dx_1 = 0$, $dx_2 \wedge dx_2 = 0$, $dx_3 \wedge dx_3 = 0$ we obtain $\alpha \wedge d\alpha = 0$.
(ii) We have

$$\text{curl}(V) = \begin{pmatrix} -2x_1 + 2x_2 - 2x_3 \\ 2x_2 \\ -2x_2 \end{pmatrix} \quad \text{and} \quad V \cdot \text{curl}(V) = 0.$$

Problem 14.   (i) Consider the differential one-form in $\mathbb{R}^3$

$$\alpha_1 = dx_1 + x_1 dx_2 + x_1 x_2 dx_3.$$

Find $d\alpha_1$, $\alpha_1 \wedge d\alpha_1$ and solve $\alpha_1 \wedge d\alpha_1 = 0$.
(ii) Consider the differential one-form in $\mathbb{R}^3$

$$\alpha_2 = x_1 dx_2 + x_2 dx_3 + x_3 dx_1.$$

Find $\alpha_2 \wedge d\alpha_2$. Find the solutions of the equation $\alpha_2 \wedge d\alpha_2 = 0$.
(iii) Consider the differential one-form in $\mathbb{R}^3$

$$\alpha_3 = dx_3 - x_2 dx_1 - dx_2.$$

Show that $\alpha_3 \wedge d\alpha_3 \neq 0$.

Solution 14.   (i) Since

$$d\alpha_1 = dx_1 \wedge dx_2 + x_2 dx_1 \wedge dx_3 + x_1 dx_2 \wedge dx_3$$

we obtain $\alpha_1 \wedge d\alpha_1 = x_1 dx_1 \wedge dx_2 \wedge dx_3$. Thus the solution of $\alpha_1 \wedge d\alpha_1 = 0$
is $x_1 = 0$.
(ii) We have

$$d\alpha_2 = dx_1 \wedge dx_2 + dx_2 \wedge dx_3 + dx_3 \wedge dx_1$$

and thus

$$\alpha_2 \wedge d\alpha_2 = (x_1 + x_2 + x_3)dx_1 \wedge dx_2 \wedge dx_3.$$

From $\alpha_2 \wedge d\alpha_2 = 0$ it follows that $x_1 + x_2 + x_3 = 0$ which describes a
plane in $\mathbb{R}^3$.

(iii) We have $d\alpha_3 = dx_1 \wedge dx_2 \Rightarrow \alpha_3 \wedge d\alpha_3 = dx_1 \wedge dx_2 \wedge dx_3$.

Problem 15. Consider the analytic functions $f, g : \mathbb{R}^3 \to \mathbb{R}$ given by

$$f(x_1, x_2, x_3) = x_1^2 + x_2^2 + x_3^2, \quad g(x_1, x_2, x_3) = x_1 x_2 x_3.$$

Find df, dg and then $df \wedge dg$. Solve $df \wedge dg = 0$.

Solution 15. We have

$$df = 2x_1 dx_1 + 2x_2 dx_2 + 2x_3 dx_3, \quad dg = x_2 x_3 dx_1 + x_1 x_3 dx_2 + x_1 x_2 dx_3.$$

Then

$$df \wedge dg = 2x_3(x_1^2 - x_2^2)dx_1 \wedge dx_2 + 2x_2(x_3^2 - x_1^2)dx_3 \wedge dx_1 + 2x_1(x_2^2 - x_3^2)dx_2 \wedge dx_3.$$

With $x_1, x_2, x_3 \neq 0$ we obtain $x_1^2 = x_2^2$, $x_2^2 = x_3^2$, $x_3^2 = x_1^2$ and $x_1^2 = x_2^2 = x_3^2$.

Problem 16. Consider the manifold $\mathbb{R}^3$.
(i) Given the smooth differential one-form

$$\alpha = f_1(\mathbf{x})dx_1 + f_2(\mathbf{x})dx_2 + f_3(\mathbf{x})dx_3.$$

Find $d\alpha$ and the differential equations $d\alpha = 0$.
(ii) Given the smooth differential two-form

$$\beta = f_{12}(\mathbf{x})dx_1 \wedge dx_2 + f_{23}(\mathbf{x})dx_2 \wedge dx_3 + f_{31}(\mathbf{x})dx_3 \wedge dx_1.$$

Find $d\beta$ and the differential equations $d\beta = 0$.
(iii) Let β be the differential two-form in $\mathbb{R}^3$

$$\beta = x_1 dx_2 \wedge dx_3 + x_2 dx_3 \wedge dx_1 + x_3 dx_1 \wedge dx_2.$$

Find $d\beta$.

Solution 16. (i) We have

$$d\alpha = \left(\frac{\partial f_2}{\partial x_1} - \frac{\partial f_1}{\partial x_2}\right) dx_1 \wedge dx_2 + \left(\frac{\partial f_1}{\partial x_3} - \frac{\partial f_3}{\partial x_1}\right) dx_3 \wedge dx_1$$

$$+ \left(\frac{\partial f_3}{\partial x_2} - \frac{\partial f_2}{\partial x_3}\right) dx_2 \wedge dx_3.$$

Thus we obtain three equations

$$\left(\frac{\partial f_2}{\partial x_1} - \frac{\partial f_1}{\partial x_2}\right) = 0, \quad \left(\frac{\partial f_1}{\partial x_3} - \frac{\partial f_3}{\partial x_1}\right) = 0, \quad \left(\frac{\partial f_3}{\partial x_2} - \frac{\partial f_2}{\partial x_3}\right) = 0.$$

A solution is $f_1(\mathbf{x}) = x_2 x_3$, $f_2(\mathbf{x}) = x_1 x_3$, $f_3(\mathbf{x}) = x_1 x_2$.
(ii) We have

$$d\beta = \left(\frac{\partial f_{12}}{\partial x_3} + \frac{\partial f_{23}}{\partial x_1} + \frac{\partial f_{31}}{\partial x_2} \right) dx_1 \wedge dx_2 \wedge dx_3.$$

Hence we find the equation

$$\left(\frac{\partial f_{12}}{\partial x_3} + \frac{\partial f_{23}}{\partial x_1} + \frac{\partial f_{31}}{\partial x_2} \right) = 0.$$

(iii) We have $d\beta = 3 dx_1 \wedge dx_2 \wedge dx_3$.

Problem 17. Consider the smooth differential one-form in $\mathbb{R}^3$

$$\alpha = f_1(\mathbf{x}) dx_1 + f_2(\mathbf{x}) dx_2 + f_3(\mathbf{x}) dx_3$$

with

$$d\alpha = \left(\frac{\partial f_2}{\partial x_1} - \frac{\partial f_1}{\partial x_2} \right) dx_1 \wedge dx_2 + \left(\frac{\partial f_3}{\partial x_2} - \frac{\partial f_2}{\partial x_3} \right) dx_2 \wedge dx_3$$
$$+ \left(\frac{\partial f_1}{\partial x_3} - \frac{\partial f_3}{\partial x_1} \right) dx_3 \wedge dx_1.$$

(i) Find the conditions of the functions f_j such that $\alpha \wedge d\alpha = 0$.
(ii) Consider the smooth differential one-form α and smooth differential two-form β, respectively

$$\alpha = f_1(\mathbf{x}) dx_1 + f_2(\mathbf{x}) dx_2 + f_3(\mathbf{x}) dx_3$$
$$\beta = g_1(\mathbf{x}) dx_2 \wedge dx_3 + g_2(\mathbf{x}) dx_3 \wedge dx_1 + g_3(\mathbf{x}) dx_1 \wedge dx_2.$$

Find the conditions on f_1, f_2, f_3 and g such that $d\alpha = \beta$.

Solution 17.   (i) Let $\Omega = dx_1 \wedge dx_2 \wedge dx_3$. We have

$$\left(\left(\frac{\partial f_2}{\partial x_1} - \frac{\partial f_1}{\partial x_2} \right) f_3 + \left(\frac{\partial f_3}{\partial x_2} - \frac{\partial f_2}{\partial x_3} \right) f_1 + \left(\frac{\partial f_1}{\partial x_3} - \frac{\partial f_3}{\partial x_1} \right) f_2 \right) \Omega.$$

Hence we find the partial differential equation

$$\left(\frac{\partial f_2}{\partial x_1} - \frac{\partial f_1}{\partial x_2} \right) f_3 + \left(\frac{\partial f_3}{\partial x_2} - \frac{\partial f_2}{\partial x_3} \right) f_1 + \left(\frac{\partial f_1}{\partial x_3} - \frac{\partial f_3}{\partial x_1} \right) f_2 = 0.$$

Besides the trivial solution $f_1(\mathbf{x} = f_2(\mathbf{x}) = f_3(\mathbf{x}) = 0$ we also find other solutions, for example $f_1(\mathbf{x}) = x_2 x_3$, $f_2(\mathbf{x}) = x_1 x_3$, $f_3(\mathbf{x}) = x_1 x_2$.

(ii) We obtain the three conditions

$$\left(\frac{\partial f_2}{\partial x_1} - \frac{\partial f_1}{\partial x_2}\right) = g_3, \quad \left(\frac{\partial f_3}{\partial x_2} - \frac{\partial f_2}{\partial x_3}\right) = g_1, \quad \left(\frac{\partial f_1}{\partial x_3} - \frac{\partial f_3}{\partial x_1}\right) = g_2.$$

Since $dd\alpha = 0$ we also have $d\beta = 0$.

Problem 18. Consider the differential two-form

$$\beta = x_1 dx_2 \wedge dx_3 + x_2 dx_3 \wedge dx_1$$

in $\mathbb{R}^3$ and the map $\mathbf{f} : \mathbb{R}^2 \to \mathbb{R}^3$

$$\mathbf{f}(u_1, u_2) = \begin{pmatrix} f_1(u_1, u_2) \\ f_2(u_1, u_2) \\ f_3(u_1, u_2) \end{pmatrix} = \begin{pmatrix} (3 + \cos(u_2)) \cos(u_1) \\ (3 + \cos(u_2)) \sin(u_1) \\ \sin(u_2) \end{pmatrix}.$$

Find $d\beta$ and $\mathbf{f}^*(\beta)$.

Solution 18. We have

$$d\beta = dx_1 \wedge dx_2 \wedge dx_3 + dx_2 \wedge dx_3 \wedge dx_1 = 2dx_1 \wedge dx_2 \wedge dx_3.$$

With

$$
\begin{aligned}
df_1 &= -(3 + \cos(u_2)) \sin(u_1) du_1 - \sin(u_2) \cos(u_1) du_2 \\
df_2 &= (3 + \cos(u_2)) \cos(u_1) du_1 - \sin(u_2) \sin(u_1) du_2 \\
df_3 &= \cos(u_2) du_2
\end{aligned}
$$

we obtain

$$\mathbf{f}^*(\beta) = f_1 df_2 \wedge df_3 + f_2 df_3 \wedge df_1 = (3 + \cos(u_2))^2 \cos(u_2) du_1 \wedge du_2.$$

Problem 19. Let β be a smooth differential two-form in $\mathbb{R}^3$

$$\beta = g_{12}(\mathbf{x})dx_1 \wedge dx_2 + g_{23}(\mathbf{x})dx_2 \wedge dx_3 + g_{31}(\mathbf{x})dx_3 \wedge dx_1$$

and α_1, α_2, α_3 be smooth differential one-forms in $\mathbb{R}^3$

$$
\begin{aligned}
\alpha_1 &= f_{11}(\mathbf{x})dx_1 + f_{12}(\mathbf{x})dx_2 + f_{13}(\mathbf{x})dx_3 \\
\alpha_2 &= f_{21}(\mathbf{x})dx_1 + f_{22}(\mathbf{x})dx_2 + f_{23}(\mathbf{x})dx_3 \\
\alpha_3 &= f_{31}(\mathbf{x})dx_1 + f_{32}(\mathbf{x})dx_2 + f_{33}(\mathbf{x})dx_3.
\end{aligned}
$$

Find the condition on the functions g_{12}, g_{23}, g_{31}, f_{jk} $(j, k = 1, 2, 3)$ such that

$$d\beta = \alpha_1 \wedge \alpha_2 \wedge \alpha_3.$$

Solution 19.   We have

$$d\beta = \left(\frac{\partial g_{12}}{\partial x_3} + \frac{\partial g_{23}}{\partial x_1} + \frac{\partial g_{31}}{\partial x_2} \right) dx_1 \wedge dx_2 \wedge dx_3$$

and $\alpha_1 \wedge \alpha_2 \wedge \alpha_3 = \det(F) dx_1 \wedge dx_2 \wedge dx_3$, where $\det(F)$ is the determinant of the 3×3 matrix $F = (f_{jk})$. It follows that

$$\left(\frac{\partial g_{12}}{\partial x_3} + \frac{\partial g_{23}}{\partial x_1} + \frac{\partial g_{31}}{\partial x_2} \right) = \det(F).$$

Problem 20.   (i) Consider the three differential one-forms in $\mathbb{R}^3$

$$\alpha_1 = \sin(x_3) dx_1 - \cos(x_3) dx_2$$
$$\alpha_2 = \cos(x_3) dx_1 + \sin(x_3) dx_2$$
$$\alpha_3 = dx_3.$$

Show that these differential one-forms satisfy

$$d\alpha_1 = -\alpha_2 \wedge \alpha_3, \quad d\alpha_2 = -\alpha_3 \wedge \alpha_1, \quad d\alpha_3 = 0.$$

(ii) Consider the three differential one-forms

$$\alpha_1 = dx_1 + e^{x_2} dx_3$$
$$\alpha_2 = \cos(x_1) dx_2 + e^{x_2} \sin(x_1) dx_3$$
$$\alpha_3 = -\sin(x_1) dx_2 + e^{x_2} \cos(x_1) dx_3.$$

Show that these differential forms satisfy

$$d\alpha_1 = \alpha_2 \wedge \alpha_3, \quad d\alpha_2 = -\alpha_3 \wedge \alpha_1, \quad d\alpha_3 = -\alpha_1 \wedge \alpha_2.$$

(iii) Consider the three differential one-forms in $\mathbb{R}^3$

$$\alpha_1 = \frac{1}{2} e^{-x_3} dx_1 - \frac{1}{2} e^{x_3} dx_2$$
$$\alpha_2 = \frac{1}{2} e^{-x_3} dx_1 + \frac{1}{2} e^{x_3} dx_2$$
$$\alpha_3 = dx_3.$$

Show that these differential forms satisfy

$$d\alpha_1 = \alpha_2 \wedge \alpha_3, \quad d\alpha_2 = -\alpha_3 \wedge \alpha_1, \quad d\alpha_3 = 0.$$

Solution 20. (i) We have

$$d\alpha_1 = \cos(x_3)dx_3 \wedge dx_1 + \sin(x_3)dx_3 \wedge dx_2$$
$$d\alpha_2 = -\sin(x_3)dx_3 \wedge dx_1 + \cos(x_3)dx_3 \wedge dx_2$$
$$d\alpha_3 = 0$$

and

$$\alpha_2 \wedge \alpha_3 = -\cos(x_3)dx_3 \wedge dx_1 - \sin(x_3)dx_3 \wedge dx_2$$
$$\alpha_3 \wedge \alpha_1 = \sin(x_3)dx_3 \wedge dx_1 - \cos(x_3)dx_3 \wedge dx_2.$$

(ii) We have

$$d\alpha_1 = e^{x_2}dx_2 \wedge dx_3$$
$$d\alpha_2 = -\sin(x_1)dx_1 \wedge dx_2 + e^{x_2}\sin(x_1)dx_2 \wedge dx_3 + e^{x_2}\cos(x_1)dx_1 \wedge dx_3$$
$$d\alpha_3 = -\cos(x_1)dx_1 \wedge dx_2 + e^{x_2}\cos(x_1)dx_2 \wedge dx_3 - e^{x_2}\sin(x_1)dx_1 \wedge dx_3$$

and utilizing $\sin^2(x_1) + \cos^2(x_1) = 1$ provides

$$\alpha_2 \wedge \alpha_3 = e^{x_2}dx_2 \wedge dx_3$$
$$\alpha_3 \wedge \alpha_1 = -\sin(x_1)dx_2 \wedge dx_1 - e^{x_2}\sin(x_1)dx_2 \wedge dx_3 + e^{x_2}\cos(x_1)dx_3 \wedge dx_1$$
$$\alpha_1 \wedge \alpha_2 = \cos(x_1)dx_1 \wedge dx_2 + e^{x_2}\sin(x_1)dx_1 \wedge dx_3 + e^{x_2}\cos(x_1)dx_3 \wedge dx_2.$$

(iii) We have

$$d\alpha_1 = -\frac{1}{2}e^{-x_3}dx_1 \wedge dx_3 - \frac{1}{2}e^{x_3}dx_3 \wedge dx_2$$
$$d\alpha_2 = -\frac{1}{2}e^{-x_3}dx_3 \wedge dx_1 + \frac{1}{2}e^{x_3}dx_3 \wedge dx_2$$
$$d\alpha_3 = 0$$

and

$$\alpha_2 \wedge \alpha_3 = \frac{1}{2}e^{-x_3}dx_1 \wedge dx_3 + \frac{1}{2}e^{x_3}dx_2 \wedge dx_3$$
$$\alpha_3 \wedge \alpha_1 = \frac{1}{2}e^{-x_3}dx_3 \wedge dx_1 + \frac{1}{2}e^{x_3}dx_3 \wedge dx_2.$$

Problem 21.   Consider the three differential one-forms in $\mathbb{R}^3$

$$\alpha_1 = \cos(\psi)d\theta + \sin(\psi)\sin(\theta)d\phi$$
$$\alpha_2 = -\sin(\psi)d\theta + \cos(\psi)\sin(\theta)d\phi$$
$$\alpha_3 = d\psi + \cos(\theta)d\phi$$

where (ϕ, θ, ψ) are the *Euler angles*. Find $d\alpha_1, d\alpha_2, d\alpha_3$ and

$$\alpha_2 \wedge \alpha_3, \quad \alpha_1 \wedge \alpha_2,, \quad \alpha_3 \wedge \alpha_1.$$

Solution 21.   We have

$$d\alpha_1 = -\sin(\psi)d\psi \wedge d\theta + \cos(\psi)\sin(\theta)d\psi \wedge d\phi + \sin(\psi)\cos(\theta)d\theta \wedge d\phi$$
$$d\alpha_2 = -\cos(\psi)d \wedge d\theta - \sin(\psi)\sin(\theta)d\psi \wedge d\phi + \cos(\psi)\cos(\theta)d\theta \wedge d\phi$$
$$d\alpha_3 = -\sin(\theta)d\theta \wedge d\phi$$

and

$$\alpha_1 \wedge \alpha_2 = \sin(\theta)d\theta \wedge d\phi$$
$$\alpha_1 \wedge \alpha_3 = \cos(\psi)d\theta \wedge d\psi + \cos(\psi)\cos(\theta)d\theta \wedge d\phi$$
$$\qquad + \sin(\psi)\sin(\theta)d\phi \wedge d\psi$$
$$\alpha_2 \wedge \alpha_3 = -\sin(\psi)d\theta \wedge d\psi - \sin(\psi)\cos(\theta)d\theta \wedge d\phi + \cos(\psi)\sin(\theta)d\phi \wedge d\psi.$$

Thus it follows that $d\alpha_3 = \alpha_2 \wedge \alpha_1$, $d\alpha_2 = \alpha_1 \wedge \alpha_3$, $d\alpha_1 = \alpha_3 \wedge \alpha_2$.

Problem 22.   (i) Consider the smooth differential one-form in $\mathbb{R}^3$

$$\alpha = -e^{x_1}x_3 dx_1 + \sin(x_3)dx_2 + (x_2\cos(x_3) - e^{x_1})dx_3.$$

Find $d\alpha$. Can one find a smooth function $f : \mathbb{R}^3 \to \mathbb{R}$ such that $df = \alpha$?
(ii) Consider the smooth differential one-form in $\mathbb{R}^3$

$$\alpha = (3x_1x_3 + 2x_2)dx_1 + x_1 dx_2 + x_1^2 dx_3.$$

Calculate $d\alpha$. Can one find a smooth function $f : \mathbb{R}^3 \to \mathbb{R}$ such that $df = \alpha$? Discuss.
(iii) Consider the smooth differential one-form in $\mathbb{R}^3$

$$\alpha = x_2 dx_1 + dx_2 + dx_3.$$

Find $d\alpha$. Can one find a smooth function $f : \mathbb{R}^3 \to \mathbb{R}$ such that $df = \alpha$? Discuss. Consider the differential one-form $\widetilde{\alpha} = x_1\alpha$.

Solution 22.   (i) With $d(\sin(x_3)) = \cos(x_3)dx_3$, $de^{x_1} = e^{x_1}dx_1$ we obtain $d\alpha = 0$. Thus we can find an f such that $\alpha = df$ which is given by

$$f(x_1, x_2, x_3) = -e^{x_1}x_3 + x_2 \sin(x_3).$$

(ii) We obtain

$$d\alpha = -dx_1 \wedge dx_2 + x_1 dx_3 \wedge dx_1.$$

So there is no f such that $df = \alpha$. For $\tilde{\alpha} = x_1\alpha$ we have $d\tilde{\alpha} = 0$. Thus we can find a function f such that $df = \tilde{\alpha}$. Note that x_1 is called an *integrating factor*.

(iii) We obtain $d\alpha = -dx_1 \wedge dx_2$. Thus no f exists with $df = \alpha$. There is also no integrating factor.

Problem 23.   Consider the differential two-form in $\mathbb{R}^{2n}$

$$\beta = dx_1 \wedge dx_{n+1} + dx_2 \wedge dx_{n+2} + \cdots + dx_n \wedge dx_{2n}.$$

Find $\beta \wedge \beta \wedge \cdots \wedge \beta$ (n-times).

Solution 23.   We obtain

$$\beta \wedge \beta \wedge \cdots \wedge \beta = (-1)^{n(n-1)/2}n!dx_1 \wedge \cdots \wedge dx_{2n}.$$

Problem 24.   (i) Let $f : \mathbb{R}^2 \to \mathbb{R}$, $f(x_1, x_2) = x_1^2 + x_2^2$ and

$$V = x_1\frac{\partial}{\partial x_1} + x_2\frac{\partial}{\partial x_2}.$$

Find the function $df(V) \equiv V\rfloor df$.

(ii) Let $f : \mathbb{R}^2 \to \mathbb{R}$, $f(x_1, x_2) = x_1^2 + x_2^2$ and

$$V = x_1\frac{\partial}{\partial x_2} - x_2\frac{\partial}{\partial x_1}.$$

Find the function $df(V) \equiv V\rfloor df$.

(iii) Consider the manifold $\mathbb{R}^n$. Calculate

$$\frac{\partial}{\partial x_j}\rfloor(dx_k \wedge dx_\ell)$$

where $j, k, \ell = 1, \ldots, n$.

(iv) Consider $\mathbb{R}^2$. Find the differential one-form

$$\frac{\partial}{\partial x_1}\rfloor(dx_1 \wedge dx_2), \qquad \frac{\partial}{\partial x_2}\rfloor(dx_1 \wedge dx_2).$$

(v) Let $n \geq 2$ and $i, j, k, \ell = 1, \ldots, n$. Find

$$\frac{\partial}{\partial x_k} \rfloor \left(\frac{\partial}{\partial x_j} \rfloor (dx_i \wedge dx_\ell) \right).$$

Solution 24.   (i) We obtain the function

$$df(V) = V(f) = \left(x_1 \frac{\partial}{\partial x_1} + x_2 \frac{\partial}{\partial x_2} \right) (x_1^2 + x_2^2) = 2(x_1^2 + x_2^2).$$

We could also calculate $df = 2x_1 dx_1 + 2x_2 dx_2$. Then using

$$dx_j \left(\frac{\partial}{\partial x_k} \right) = \frac{\partial x_j}{\partial x_k} = \delta_{jk}$$

we have

$$V \rfloor df = \left(x_1 \frac{\partial}{\partial x_1} + x_2 \frac{\partial}{\partial x_2} \right) \rfloor (2x_1 dx_1 + 2x_2 dx_2) = 2(x_1^2 + x_2^2).$$

(ii) We have

$$df(V) = V(f) = \left(x_1 \frac{\partial}{\partial x_2} - x_2 \frac{\partial}{\partial x_1} \right) (x_1^2 + x_2^2) = 0.$$

We could also calculate $df = 2x_1 dx_1 + 2x_2 dx_2$. Then using

$$dx_j \left(\frac{\partial}{\partial x_k} \right) = \frac{\partial x_j}{\partial x_k} = \delta_{jk}$$

yields

$$V \rfloor df = \left(x_1 \frac{\partial}{\partial x_2} - x_2 \frac{\partial}{\partial x_1} \right) \rfloor (2x_1 dx_1 + 2x_2 dx_2) = 0.$$

(iii) We find

$$\frac{\partial}{\partial x_j} \rfloor (dx_k \wedge dx_\ell) = \left(\frac{\partial}{\partial x_j} \rfloor dx_k \right) \wedge dx_\ell - dx_k \wedge \left(\frac{\partial}{\partial x_j} \rfloor dx_\ell \right)$$

$$= \delta_{jk} dx_\ell - \delta_{j\ell} dx_k$$

where δ_{jk}, $\delta_{j\ell}$ are the Kronecker delta.

(iv) With $\partial/\partial x_j \rfloor dx_k = \delta_{jk}$ we obtain

$$\frac{\partial}{\partial x_1} \rfloor (dx_1 \wedge dx_2) = dx_2, \quad \frac{\partial}{\partial x_2} \rfloor (dx_1 \wedge dx_2) = -dx_1.$$

(v) We have

$$\frac{\partial}{\partial x_k}\rfloor \left(\frac{\partial}{\partial x_j}\rfloor (dx_i \wedge dx_\ell)\right) = \frac{\partial}{\partial x_k}\rfloor (\delta_{ji}dx_\ell - \delta_{j\ell}dx_i) = \delta_{ji}\delta_{k\ell} - \delta_{ik}\delta_{j\ell}.$$

Problem 25. Consider the manifold $M = \mathbb{R}^2$, the differential two-form $\Omega = dx_1 \wedge dx_2$ and the smooth vector field

$$V = V_1(x_1, x_2)\frac{\partial}{\partial x_1} + V_2(x_1, x_2)\frac{\partial}{\partial x_2}.$$

Find the condition on a smooth function $f : \mathbb{R}^2 \to \mathbb{R}$ such that $V\rfloor\Omega = df$.

Solution 25. We have

$$V\rfloor\Omega = (V\rfloor dx_1) \wedge dx_2 - dx_1 \wedge (V\rfloor dx_2) = V_1 dx_2 - V_2 dx_1$$

and

$$df = \frac{\partial f}{\partial x_1}dx_1 + \frac{\partial f}{\partial x_2}dx_2.$$

Hence

$$V_1 = \frac{\partial f}{\partial x_2}, \qquad -V_2 = \frac{\partial f}{\partial x_1}.$$

Problem 26. Let $A = (a_{jk})$ be a 2×2 matrix over $\mathbb{R}$. Consider the vector field

$$V = (x_1 \quad x_2)\, A \begin{pmatrix} \partial/\partial x_1 \\ \partial/\partial x_2 \end{pmatrix} = (a_{11}x_1 + a_{21}x_2)\frac{\partial}{\partial x_1} + (a_{12}x_1 + a_{22}x_2)\frac{\partial}{\partial x_2}$$

and the differential one-form

$$\alpha = (x_1 \quad x_2)\, A \begin{pmatrix} dx_1 \\ dx_2 \end{pmatrix} = (a_{11}x_1 + a_{21}x_2)dx_1 + (a_{12}x_1 + a_{22}x_2)dx_2.$$

Find the function $V\rfloor\alpha$.

Solution 26. With $\partial/\partial x_j\rfloor dx_k = \delta_{jk}$ we obtain

$$V\rfloor\alpha = (a_{11}x_1 + a_{21}x_2)^2 + (a_{12}x_1 + a_{22}x_2)^2.$$

Problem 27. Let $k \in \mathbb{R}$ and $k \neq 0$. Consider the three differential one-forms

$$\alpha_1 = e^{-kx_1}dx_2, \qquad \alpha_2 = dx_3, \qquad \alpha_3 = dx_1.$$

(i) Find $d\alpha_1$, $d\alpha_2$, $d\alpha_3$, $\alpha_1 \wedge \alpha_2$, $\alpha_2 \wedge \alpha_1$, $\alpha_2 \wedge \alpha_3$, $\alpha_3 \wedge \alpha_2$, $\alpha_3 \wedge \alpha_1$, $\alpha_1 \wedge \alpha_3$.
Then show that $d\alpha_1 = \alpha_1 \wedge \alpha_3$.
(ii) Consider the vector fields

$$V_1 = e^{kx_1} \frac{\partial}{\partial x_2}, \quad V_2 = \frac{\partial}{\partial x_3}, \quad V_3 = \frac{\partial}{\partial x_1}$$

with $V_1 \rfloor \alpha_1 = 1$, $V_2 \rfloor \alpha_2 = 1$, $V_3 \rfloor \alpha_3 = 1$. Find the commutators $[V_1, V_2]$, $[V_2, V_3]$, $[V_3, V_1]$.

Solution 27.   (i) We obtain

$$d\alpha_1 = -ke^{-kx_1} dx_1 \wedge dx_2, \quad d\alpha_2 = 0, \quad d\alpha_3 = 0$$

and

$$\alpha_1 \wedge \alpha_2 = -\alpha_2 \wedge \alpha_1 = e^{-kx_1} dx_2 \wedge dx_3$$
$$\alpha_3 \wedge \alpha_1 = -\alpha_1 \wedge \alpha_3 = e^{-kx_1} dx_1 \wedge dx_2$$
$$\alpha_2 \wedge \alpha_3 = -\alpha_3 \wedge \alpha_2 = dx_3 \wedge dx_1.$$

Then $d\alpha_1 = -k\alpha_3 \wedge \alpha_1 = k\alpha_1 \wedge \alpha_3$.
(ii) The commutators of the other vector fields are

$$[V_1, V_2] = 0, \quad [V_3, V_1] = ke^{kx_1} \frac{\partial}{\partial x_2} = kV_1, \quad [V_2, V_3] = 0.$$

Problem 28.   Consider all 2×2 matrices with $UU^* = I_2$, $\det(U) = 1$
i.e. $U \in SU(2)$. Then U can be written as

$$U = \begin{pmatrix} a & b \\ -\bar{b} & \bar{a} \end{pmatrix}, \quad a, b \in \mathbb{C}$$

with the constraint $a\bar{a} + b\bar{b} = 1$. Let

$$\begin{pmatrix} z_1' \\ z_2' \end{pmatrix} = \begin{pmatrix} a & b \\ -\bar{b} & \bar{a} \end{pmatrix} \begin{pmatrix} z_1 \\ z_2 \end{pmatrix}.$$

Show that $(z_1')(z_1')^* + (z_2')(z_2')^* = z_1 z_1^* + z_2 z_2^*$.
(ii) Consider

$$\begin{pmatrix} z_1' \\ z_2' \end{pmatrix} = \begin{pmatrix} a & b \\ -\bar{b} & \bar{a} \end{pmatrix} \begin{pmatrix} z_1 \\ z_2 \end{pmatrix}.$$

Show that $dz_1' \wedge dz_2' = dz_1 \wedge dz_2$.

Solution 28. (i) We have $z_1' = az_1 + bz_2$, $z_2' = -\bar{b}z_1 + \bar{a}z_2$. Thus

$$(z_1')(z_1')^* + (z_2')(z_2')^* = (a\bar{a} + b\bar{b})z_1 z_1^* + (b\bar{b} + a\bar{a})z_2 z_2^* = z_1 z_1^* + z_2 z_2^*.$$

(ii) We have

$$dz_1' = d(az_1 + bz_2) = adz_1 + bdz_2$$
$$dz_2' = d(-\bar{b}z_1 + \bar{a}z_2) = -\bar{b}dz_1 + \bar{a}dz_2.$$

Since $dz_1 \wedge dz_1 = 0$, $dz_2 \wedge dz_2 = 0$ and $dz_1 \wedge dz_2 = -dz_2 \wedge dz_1$ we obtain

$$dz_1' \wedge dz_2' = (a\bar{a} + b\bar{b})(dz_1 \wedge dz_2) = dz_1 \wedge dz_2.$$

Problem 29. Let $z_j = x_j + iy_j$ with $x_j, y_j \in \mathbb{R}$. Consider the differential one-form

$$\alpha = \frac{i}{4} \sum_{j=0}^{n} (z_j d\bar{z}_j - \bar{z}_j dz_j).$$

Let $z_j = x_j + iy_j$. Find α and $d\alpha$.

Solution 29. We have

$$z_j d\bar{z}_j - \bar{z}_j dz_j = (x_j + iy_j)(dx_j - idy_j) - (x_j - iy_j)(dx_j + idy_j)$$
$$= 2i(y_j dx_j - x_j dy_j).$$

Thus

$$\alpha = \frac{i}{4} \sum_{j=0}^{n} (z_j d\bar{z}_j - \bar{z}_j dz_j) = -\frac{1}{2} \sum_{j=0}^{n} (y_j dx_j - x_j dy_j).$$

Using this result and $dy_j \wedge dx_j = -dx_j \wedge dy_j$ we obtain

$$d\alpha = \sum_{j=0}^{n} (dx_j \wedge dy_j).$$

Problem 30. In *vector analysis* in $\mathbb{R}^3$ we have the identity

$$\nabla(\mathbf{A} \times \mathbf{B}) \equiv \mathbf{B}\mathrm{curl}(\mathbf{A}) - \mathbf{A}\mathrm{curl}(\mathbf{B})$$

where $\times$ denotes the vector product. Express this identity using differential forms, the exterior derivative and the exterior product.

Solution 30.   Consider the differential one-forms

$$\alpha_1 = A_1(\mathbf{x})dx_1 + A_2(\mathbf{x})dx_2 + A_3(\mathbf{x})dx_3,$$
$$\alpha_2 = B_1(\mathbf{x})dx_1 + B_2(\mathbf{x})dx_2 + B_3(\mathbf{x})dx_3.$$

Then the identity is $d(\alpha_1 \wedge \alpha_2) \equiv (d\alpha_1) \wedge \alpha_2 - \alpha_1 \wedge (d\alpha_2)$.

Problem 31.   A necessary and sufficient condition for the Pfaffian system of equations

$$\alpha_j = 0, \quad j = 1, \dots, r$$

to be completely integrable is $d\alpha_j \equiv 0 \mod (\alpha_1, \dots, \alpha_r)$, $j = 1, \dots, r$. Let

$$\alpha \equiv P_1(\mathbf{x})dx_1 + P_2(\mathbf{x})dx_2 + P_3(\mathbf{x})dx_3 = 0 \tag{1}$$

be a *total differential equation* in $\mathbb{R}^3$, where P_1, P_2, P_3 are analytic functions on $\mathbb{R}^3$. Complete integrability of α means that in every sufficiently small neighbourhood there exists a smooth function f such that

$$f(x_1, x_2, x_3) = \text{const}$$

is a first integral of (1). A necessary and sufficient condition for (1) to be completely integrable is $d\alpha \wedge \alpha = 0$.

Solution 31.   We have

$$d\alpha = \left(\frac{\partial P_2}{\partial x_1} - \frac{\partial P_1}{\partial x_2} \right) dx_1 \wedge dx_2 + \left(\frac{\partial P}{\partial x} - \frac{\partial P}{\partial x} \right) dx_2 \wedge dx_3$$
$$+ \left(\frac{\partial P}{\partial x} - \frac{\partial P}{\partial x} \right) dx_3 \wedge dx_1.$$

Then $d\alpha \wedge \alpha = 0$ yields the condition

$$P_1 \left(\frac{\partial P_3}{\partial x_2} - \frac{\partial P_2}{\partial x_3} \right) + P_2 \left(\frac{\partial P_1}{\partial x_3} - \frac{\partial P_3}{\partial x_1} \right) + P_3 \left(\frac{\partial P_2}{\partial x_1} - \frac{\partial P_1}{\partial x_2} \right) = 0.$$

A solution of this partial differential equation is

$$P_1(x_1, x_2, x_3) = x_1, \quad P_2(x_1, x_2, x_3) = x_3, \quad P_3(x_1, x_2, x_3) = x_2.$$

Problem 32.   (i) Consider the differential two-form in $\mathbb{R}^4$

$$\beta = dx_1 \wedge dx_2 + dx_3 \wedge dx_4.$$

Find $d\beta$ and $\beta \wedge \beta$.

(ii) Consider the manifold $M = \mathbb{R}^4$ and the differential two-forms

$$\omega = dq_1 \wedge dp_1 + dq_2 \wedge dp_2.$$

$$\beta = (a^2 + p_1^2)dq_1 \wedge dp_2 - p_1 p_2 (dq_1 \wedge dp_1 - dq_2 \wedge dp_2) - (b^2 + p_2^2)dq_2 \wedge dp_1$$

where a and b are constants. Find $d\beta$. Can $d\beta$ be written in the form

$$d\beta = \alpha \wedge \omega$$

where α is a differential one-form?

Solution 32.   (i) We obtain $d\beta = 0$ and $\beta \wedge \beta = 2dx_1 \wedge dx_2 \wedge dx_3 \wedge dx_4$.
(ii) Since

$$d\beta = 2dp_1 \wedge dq_1 \wedge dp_2 - p_1 dp_2 \wedge dq_1 \wedge dp_1 + p_2 dp_1 \wedge dp_1 \wedge dq_2 \wedge dp_2 - 2dp_2 dq_2 \wedge dp_1$$

we have

$$d\beta = -3(p_1 dp_2 + p_2 dp_1) \wedge \omega.$$

Problem 33.   Consider the differential one-forms in $\mathbb{R}^4$

$$\alpha_1 = \sum_{j=1}^{4} f_j(\mathbf{x})dx_j, \qquad \alpha_2 = \sum_{k=1}^{4} g_k(\mathbf{x})dx_k.$$

(i) Find $\alpha_1 \wedge \alpha_2$.
(ii) Find $\alpha_1 \wedge \alpha_2 \wedge \alpha_1 \wedge \alpha_2$ and the equation from

$$\alpha_1 \wedge \alpha_2 \wedge \alpha_1 \wedge \alpha_2 = 0.$$

Solution 33.   (i) With $dx_j \wedge dx_k = -dx_k \wedge dx_j$ we obtain

$$\begin{aligned}
\alpha_1 \wedge \alpha_2 = {} & (f_1 g_2 - f_2 g_1)dx_1 \wedge dx_2 + (f_1 g_3 - f_3 g_1)dx_1 \wedge dx_3 \\
& + (f_1 g_4 - f_4 g_1)dx_1 \wedge dx_4 + (f_2 g_3 - f_3 g_2)dx_2 \wedge dx_3 \\
& + (f_2 g_4 - f_4 g_2)dx_2 \wedge dx_4 + (f_3 g_4 - f_4 g_3)dx_3 \wedge dx_4.
\end{aligned}$$

(ii) $\delta = \alpha_1 \wedge \alpha_2 \wedge \alpha_1 \wedge \alpha_2$ is a four form and we obtain

$$\begin{aligned}
\delta = {} & 2((f_1 g_2 - f_2 g_1)(f_3 g_4 - f_4 g_3) + (f_4 g_2 - f_2 g_4)(f_1 g_3 - f_3 g_1) \\
& + (f_1 g_4 - f_4 g_1)(f_2 g_3 - f_3 g_2))dx_1 \wedge dx_2 \wedge dx_3 \wedge dx_4.
\end{aligned}$$

Hence the equation is

$$(f_1g_2 - f_2g_1)(f_3g_4 - f_4g_3) + (f_4g_2 - f_2g_4)(f_1g_3 - f_3g_1)$$
$$+ (f_1g_4 - f_4g_1)(f_2g_3 - f_3g_2) = 0.$$

Problem 34.   Let α, β be smooth differential one-forms. The linear operator $d_\alpha(.)$ is defined by

$$d_\alpha(\beta) := d\beta + \alpha \wedge \beta.$$

Let $\alpha = x_1 dx_2 + x_2 dx_3 + x_3 dx_1$, $\beta = x_1 x_2 dx_3$. Find $d_\alpha(\beta)$. Solve

$$d_\alpha(\beta) = 0.$$

Solution 34.   We have

$$d\beta = x_2 dx_1 \wedge dx_3 + x_1 dx_2 \wedge dx_3, \quad \alpha \wedge \beta = x_1^2 x_2 dx_2 \wedge dx_3 + x_1 x_2 x_3 dx_1 \wedge dx_3.$$

Thus

$$d_\alpha(\beta) = (x_1^2 x_2 + x_1) dx_2 \wedge dx_3 + (x_1 x_2 x_3 + x_2) dx_1 \wedge dx_3.$$

From $d_\alpha(\beta) = 0$ we find the two equations

$$x_1^2 x_2 + x_1 = 0, \quad x_1 x_2 x_3 + x_2 = 0.$$

A solution is $x_1 = x_2 = 0$ and x_3 arbitrary. Another solution is $x_1 = 1$, $x_2 = -1$, $x_3 = -1$.

Problem 35.   Consider the smooth manifold $M = \mathbb{R}^3$ with coordinates (x, p, u) and the differential one-form

$$\alpha = du - pdx.$$

(i) Show that $\alpha \wedge d\alpha \neq 0$. Consider the vector fields

$$V = \frac{\partial}{\partial p}, \quad W = \frac{\partial}{\partial x} + p\frac{\partial}{\partial u}.$$

Find the functions $V \rfloor \alpha$, $W \rfloor \alpha$.
(ii) Consider the manifold $M = \mathbb{R}^5$ with coordinates (x_1, x_2, p_1, p_2, u) and the differential one-form

$$\alpha = du - \sum_{j=1}^{2} p_j dx_j.$$

Show that $\alpha \wedge d\alpha \wedge d\alpha \neq 0$. Consider the vector fields

$$V_1 = \frac{\partial}{\partial p_1}, \quad V_2 = \frac{\partial}{\partial p_2}, \quad W_1 = \frac{\partial}{\partial x_1} + p_1 \frac{\partial}{\partial u}, \quad W_2 = \frac{\partial}{\partial x_2} + p_2 \frac{\partial}{\partial u}.$$

Find the contractions $V_1 \rfloor \alpha$, $V_2 \rfloor \alpha$, $W_1 \rfloor \alpha$, $W_2 \rfloor \alpha$.

Solution 35.   (i) Since $d(du) = 0$ we have

$$d\alpha = -dp \wedge du.$$

Thus

$$\alpha \wedge d\alpha = -pdx \wedge dp \wedge du \neq 0.$$

We obtain $V \rfloor \alpha = 0$, $W \rfloor \alpha = 0$.
(ii) We have

$$d\alpha = -\sum_{j=1}^{2} dp_j \wedge dx_j.$$

Thus $\alpha \wedge d\alpha \wedge d\alpha = -2dx_1 \wedge dx_2 \wedge dp_1 \wedge dp_2 \neq 0$. We obtain

$$V_1 \rfloor \alpha = 0, \quad V_2 \rfloor \alpha = 0, \quad W_1 \rfloor \alpha = 0, \quad W_2 \rfloor \alpha = 0.$$

Problem 36.   (i) Consider the differential two-form in $\mathbb{R}^3$

$$\beta = x_1 dx_2 \wedge dx_3 + x_2 dx_3 \wedge dx_1 + x_3 dx_1 \wedge dx_2$$

and the vector field

$$V = x_1 \frac{\partial}{\partial x_2} + x_2 \frac{\partial}{\partial x_3} + x_3 \frac{\partial}{\partial x_1}.$$

Find $V \rfloor \beta$ and $V \rfloor d\beta$.
(ii) Consider the manifold $\mathbb{R}^3$ and the smooth vector fields

$$V = \sum_{j=1}^{3} V_j(\mathbf{x}) \frac{\partial}{\partial x_j}, \quad W = \sum_{k=1}^{3} W_k(\mathbf{x}) \frac{\partial}{\partial x_k}.$$

Consider the differential two-form in $\mathbb{R}^3$

$$\beta = b_1 dx_2 \wedge dx_3 + b_2 dx_3 \wedge dx_1 + b_3 dx_1 \wedge dx_2$$

with b_1, b_2, b_3 are smooth functions. Find the function

$$\beta(V, W) \equiv W \rfloor (V \rfloor \beta).$$

Solution 36. (i) We have

$$V \rfloor \beta = (x_2^2 - x_1 x_3)dx_1 + (x_3^2 - x_1 x_2)dx_2 + (x_1^2 - x_2 x_3)dx_3$$

and with $d\beta = 3dx_1 \wedge dx_2 \wedge dx_3$ we obtain

$$V \rfloor d\beta = 3(x_1 dx_3 \wedge dx_1 + x_2 dx_1 \wedge dx_2 + x_3 dx_2 \wedge dx_3).$$

(ii) We obtain

$$W \rfloor (V \rfloor \beta) = (b_2 V_3 - b_3 V_2)W_1 + (b_3 V_1 - b_1 V_3)W_2 + (b_1 V_2 - b_2 V_1)W_3.$$

Problem 37. Consider the differential one-forms in $\mathbb{R}^3$

$$\alpha_1 = \frac{dx_3 - x_1 dx_2 + x_2 dx_1}{1 + x_1^2 + x_2^2 + x_3^2}$$

$$\alpha_2 = \frac{dx_1 - x_2 dx_3 + x_3 dx_2}{1 + x_1^2 + x_2^2 + x_3^2}$$

$$\alpha_3 = \frac{dx_2 + x_1 dx_3 - x_3 dx_1}{1 + x_1^2 + x_2^2 + x_3^2}.$$

Find the *dual basis* of the vector fields V_1, V_2, V_3.

Solution 37. With

$$\frac{\partial}{\partial x_j} \rfloor dx_k = \delta_{jk}, \quad j, k = 1, 2, 3$$

we obtain

$$V_1 = (x_2 + x_1 x_3)\frac{\partial}{\partial x_1} + (-x_1 + x_2 x_3)\frac{\partial}{\partial x_2} + (1 + x_3^2)\frac{\partial}{\partial x_3}$$

$$V_2 = (1 + x_1^2)\frac{\partial}{\partial x_1} + (x_1 x_2 + x_3)\frac{\partial}{\partial x_2} + (-x_2 + x_1 x_3)\frac{\partial}{\partial x_3}$$

$$V_3 = (x_1 x_2 - x_3)\frac{\partial}{\partial x_1} + (1 + x_2^2)\frac{\partial}{\partial x_2} + (x_1 + x_2 x_3)\frac{\partial}{\partial x_3}.$$

These vector fields form a basis of the Lie algebra $so(3, \mathbb{R})$, i.e. we have

$$[V_1, V_2] = V_3, \quad [V_2, V_3] = V_1, \quad [V_3, V_1] = V_2.$$

Problem 38. Consider the differentiable manifold

$$\mathbb{S}^3 = \{ (x_1, x_2, x_3, x_4) : x_1^2 + x_2^2 + x_3^2 + x_4^2 = 1 \}.$$

(i) Show that the matrix

$$U(x_1, x_2, x_3, x_4) = -i \begin{pmatrix} x_3 + ix_4 & x_1 - ix_2 \\ x_1 + ix_2 & -x_3 + ix_4 \end{pmatrix}$$

is unitary. Show that the matrix is an element of the Lie group $SU(2)$.
(ii) Consider the parameters (θ, ψ, ϕ) with $0 \le \theta < \pi$, $0 \le \psi < 4\pi$, $0 \le \phi < 2\pi$. Show that

$$x_1(\theta, \psi, \phi) + ix_2(\theta, \psi, \phi) = \cos(\theta/2)e^{i(\psi+\phi)/2}$$
$$x_3(\theta, \psi, \phi) + ix_4(\theta, \psi, \phi) = \sin(\theta/2)e^{i(\psi-\phi)/2}$$

is a parametrization. Thus the matrix given in (i) takes the form

$$-i \begin{pmatrix} \sin(\theta/2)e^{i(\psi-\phi)/2} & \cos(\theta/2)e^{-i(\psi+\phi)/2} \\ \cos(\theta/2)e^{i(\psi+\phi)/2} & -\sin(\theta/2)e^{-i(\psi-\phi)/2} \end{pmatrix}.$$

(iii) Let $(\xi_1, \xi_2, \xi_3) = (\theta, \psi, \phi)$ with $0 \le \theta < \pi$, $0 \le \psi < 4\pi$, $0 \le \phi < 2\pi$. Show that

$$\frac{1}{24\pi^2} \int_0^\pi d\theta \int_0^{4\pi} d\psi \int_0^{2\pi} d\phi \sum_{j,k,\ell=1}^3 \epsilon_{jk\ell} \mathrm{tr}\left(U^{-1} \frac{\partial U}{\partial \xi_j} U^{-1} \frac{\partial U}{\partial \xi_k} U^{-1} \frac{\partial U}{\partial \xi_\ell} \right) = 1$$

where $\epsilon_{123} = \epsilon_{321} = \epsilon_{132} = +1$, $\epsilon_{213} = \epsilon_{321} = \epsilon_{132} = -1$ and 0 otherwise.
(iv) Consider the metric tensor field

$$g = dx_1 \otimes dx_1 + dx_2 \otimes dx_2 + dx_3 \otimes dx_3 + dx_4 \otimes dx_4.$$

Using the parametrization show that

$$g_{\mathbb{S}^3} = \frac{1}{4}(d\theta \otimes d\theta + d\psi \otimes d\psi + d\phi \otimes d\phi + \cos(\theta)d\psi \otimes d\phi + \cos(\theta)d\phi \otimes d\psi).$$

(v) Consider the differential one-forms e_1, e_2, e_3 defined by

$$\begin{pmatrix} e_1 \\ e_2 \\ e_3 \end{pmatrix} = \begin{pmatrix} -x_4 & -x_3 & x_2 & x_1 \\ x_3 & -x_4 & -x_1 & x_2 \\ -x_2 & x_1 & -x_4 & x_3 \end{pmatrix} \begin{pmatrix} dx_1 \\ dx_2 \\ dx_3 \\ dx_4 \end{pmatrix}.$$

Show that

$$g_{\mathbb{S}^3} = de_1 \otimes de_1 + de_2 \otimes de_2 + de_3 \otimes de_3.$$

(vi) Show that

$$de_j = \sum_{k,\ell=1}^3 \epsilon_{jk\ell} e_k \wedge e_\ell$$

i.e.

$$de_1 = 2e_2 \wedge e_3, \quad de_2 = 2e_3 \wedge e_1, \quad de_3 = 2e_1 \wedge e_2.$$

Solution 38.   (i) We have $UU^* = I_2$. Thus the matrix is unitary. We have $\det(U) = 1$. It follows that U is an element of $SU(2)$.
(ii) Since

$$x_1 = \cos(\theta/2)\cos((\psi + \phi)/2), \quad x_2 = \cos(\theta/2)\sin((\psi + \phi)/2),$$

$$x_3 = \sin(\theta/2)\cos((\psi - \phi)/2), \quad x_4 = \sin(\theta/2)\sin((\psi - \phi)/2)$$

we obtain

$$x_1^2 + x_2^2 = \cos^2(\theta/2), \qquad x_3^2 + x_4^2 = \sin^2(\theta/2).$$

Thus $x_1^2 + x_2^2 + x_3^2 + x_4^2 = 1$.
(iii) We have

$$\frac{\partial U}{\partial \xi_1} = -\frac{i}{2}\begin{pmatrix} \cos(\theta/2)e^{i(\psi-\phi)} & -\sin(\theta/2)e^{-i(\psi+\phi)/2} \\ -\sin(\theta/2)e^{i(\psi+\phi)/2} & -\cos(\theta/2)e^{-i(\psi-\phi)/2} \end{pmatrix}$$

$$\frac{\partial U}{\partial \xi_2} = -i\begin{pmatrix} \sin(\theta/2)e^{i(\psi-\phi)/2} & \cos(\theta/2)e^{-i(\psi+\phi)/2} \\ \cos(\theta/2)e^{i(\psi+\phi)/2} & -\sin(\theta/2)e^{-i(\psi-\phi)/2} \end{pmatrix}$$

$$\frac{\partial U}{\partial \xi_3} = -i\begin{pmatrix} \sin(\theta/2)e^{i(\psi-\phi)/2} & \cos(\theta/2)e^{-i(\psi+\phi)/2} \\ \cos(\theta/2)e^{i(\psi+\phi)/2} & -\sin(\theta/2)e^{-i(\psi-\phi)/2} \end{pmatrix}.$$

(vi) We show that $de_3 = 2e_1 \wedge e_2$. We have $de_3 = 2(dx_1 \wedge dx_2 + dx_3 \wedge dx_4)$. Now

$$e_1 \wedge e_2 = (x_3^2 + x_4^2)dx_1 \wedge dx_2 + (x_1x_4 - x_2x_3)dx_1 \wedge dx_3$$
$$+ (-x_1x_3 - x_2x_4)dx_1 \wedge dx_4 + (x_1x_3 + x_2x_4)dx_2 \wedge dx_3$$
$$+ (-x_2x_3 + x_1x_4)dx_2 \wedge dx_4 + (x_1^2 + x_2^2)dx_3 \wedge dx_4.$$

Using

$$x_1dx_1 = -x_2dx_2 - x_3dx_3 - x_4dx_4, \quad x_3dx_3 = -x_1dx_1 - x_2dx_2 - x_4dx_4$$

we have

$$(x_1x_4 - x_2x_3)dx_1 \wedge dx_3 =$$
$$x_2^2dx_1 \wedge dx_2 + x_4^2dx_3 \wedge dx_4 - x_2x_4dx_2 \wedge dx_3 + x_2x_4dx_1 \wedge dx_4.$$

Using

$$x_2dx_2 = -x_1dx_1 - x_3dx_3 - x_4dx_4, \quad x_4dx_4 = -x_1dx_1 - x_2dx_2 - x_3dx_3$$

we have

$$(-x_2 dx_3 + x_1 x_4)dx_2 \wedge dx_4 =$$

$$x_1^2 dx_1 \wedge dx_2 + x_3^2 dx_3 \wedge dx_4 + x_1 x_3 dx_1 \wedge dx_4 - x_1 x_3 dx_2 \wedge dx_3.$$

Inserting these terms into $e_1 \wedge e_2$ we arrive at

$$e_1 \wedge e_2 = (x_1^2 + x_2^2 + x_3^2 + x_4^2)dx_1 \wedge dx_2 + (x_1^2 + x_2^2 + x_3^2 + x_4^2)dx_3 \wedge dx_4$$
$$= dx_1 \wedge dx_2 + dx_3 \wedge dx_4.$$

Analogously we show that $de_1 = e_2 \wedge e_3$ and $de_3 = e_2 \wedge e_1$.

Problem 39. Find the maximum and minimum of the analytic function $f : \mathbb{R}^2 \to \mathbb{R}$

$$f(x_1, x_2) = x_1^2 - x_2^2$$

subject to the constraint $g(x_1, x_2) = x_1^2 + x_2^2 - 1 = 0$.
(i) Apply the *Lagrange multiplier method*.
(ii) Apply differential forms.
(iii) Apply the transformation $x_1(r, \phi) = r \cos(\phi)$, $x_2(r, \phi) = r \sin(\phi)$.

Solution 39. (i) From the Lagrange function

$$L(x_1, x_2, \lambda) = f(x_1, x_2) + \lambda g(x_1, x_2) = x_1^2 - x_2^2 + \lambda(x_1^2 + x_2^2 - 1)$$

where λ is the Lagrange multiplier we find

$$\frac{\partial L}{\partial x_1} = 2x_1(1 + \lambda) = 0, \qquad \frac{\partial L}{\partial x_2} = 2x_2(-1 + \lambda) = 0$$

together with the constraint $x_1^2 + x_2^2 = 1$.
Case 1. $\lambda = -1$ and therefore $x_2 = 0$. It follows that $x_1 = \pm 1$. For $(\pm 1, 0)$ we have a maximum $f(\pm 1, 0) = 1$.
Case 2. $\lambda = 1$ and therefore $x_1 = 0$ and $x_2 \pm 1$. For $(0, \pm 1)$ we have a minimum $f(0, \pm 1) = -1$.
(ii) We have

$$df = 2x_1 dx_1 - 2x_2 dx_2, \qquad dg = 2x_1 dx_1 + 2x_2 dx_2.$$

Thus $df \wedge dg = 8x_1 x_2 dx_1 \wedge dx_2$. We have to solve

$$x_1 x_2 = 0, \qquad x_1^2 + x_2^2 = 1.$$

This provides two solutions $x_1 = \pm 1$, $x_2 = 0$ and $x_1 = 0$, $x_2 \pm 1$ with $f(\pm 1, 0) = 1$ and $f(0, \pm 1) = -1$.

(iii) The transformation $x_1(r, \phi) = r\cos(\phi)$, $x_2(r, \phi) = r\sin(\phi)$ satisfies the constraint with $r = 1$. Then f takes the form

$$f(\phi) = \cos^2(\phi) - \sin^2(\phi) \equiv \cos(2\phi)$$

with $df/d\phi = -2\sin(2\phi) = 0$. The solutions are $\phi = 0$ with $f(0) = 1$ and $\phi = \pi$ with $f(\pi) = -1$.

Problem 40. Find the minimum of the function $f : \mathbb{R}^2 \to \mathbb{R}$, $f(x_1, x_2) = x_1$ subject to the constraint $g(x_1, x_2) = x_2^2 + x_1^4 - x_1^3 = 0$.
(i) Show that the Lagrange multiplier method fails. Explain why.
(ii) Apply differential forms and show that there is a solution.

Solution 40. (i) The Lagrange function is

$$L(x_1, x_2) = f(x_1, x_2) + \lambda g(x_1, x_2) = x_1 + \lambda(x_2^2 + x_1^4 - x_1^3).$$

It follows that

$$\frac{\partial L}{\partial x_1} = 0 \Rightarrow 1 + \lambda(4x_1^3 - 3x_1^3) = 0$$

$$\frac{\partial L}{\partial x_2} = 0 \Rightarrow 2x_2 = 0$$

$$\frac{\partial L}{\partial \lambda} = 0 \Rightarrow x_2^2 + x_1^4 - x_1^3 = 0.$$

From the second equation we find $x_2 = 0$. Thus

$$x_1^3(x_1 - 1) = 0$$

with the two solutions $x_1 = 0$ and $x_1 = 1$. For the case $x_1 = 0$ we obtain from the first equation $1 = 0$, i.e. a contradiction. For the case $x_1 = 1$ we find $\lambda = -1$ and $f(x_1 = 1, x_2 = 0) = 1$ which is not a minimum. The minimum is at $(x_1, x_2) = (0, 0)$ with $f(0, 0) = 0$. The curve $y^2 + x^4 - x^3 = 0$ has a *singular point* at the origin $(0, 0)$. At any such singular point we have $\nabla g = \mathbf{0}$ (*Implicit Function Theorem*). If the gradient were nonzero, the level set must be locally a smooth curve. The function f's minimum occurs at $(0, 0)$. This point does not satisfy the *Lagrange condition*

$$\nabla f(0, 0) = \lambda \nabla g(0, 0)$$

for any value λ.
(ii) Using differential forms we have

$$df = dx_1, \qquad dg = 2x_2 dx_2 + 4x_1^3 dx_1 - 3x_1^2 dx_1.$$

Thus $df \wedge dg = 2x_2 dx_2$. From $df \wedge dg = 0$ we obtain $x_2 = 0$. Inserting this into $g(x_1, x_2) = 0$ yields $x_1^4 - x_1^3 = x_1^3(x_1 - 1) = 0$. Thus $x_1 = 0$ and $x_1 = 1$. Thus $f(x_1 = 0, x_2 = 0) = 0$ is a minimum.

Problem 41.   The two planes in $\mathbb{R}^3$

$$x_1 + x_2 + x_3 = 4, \qquad x_1 + x_2 + 2x_3 = 6$$

intersect and create a line. Find the shortest distance from the origin $(0, 0, 0)$ to this line. Apply differential forms.

Solution 41.   The square of the distance D is

$$D^2 = x_1^2 + x_2^2 + x_3^2.$$

Thus

$$d(D^2) = 2(x_1 dx_1 + x_2 dx_2 + x_3 dx_3).$$

From the two constraints we find

$$d(x_1 + x_2 + x_3 - 4) = dx_1 + dx_2 + dx_3, \quad d(x_1 + x_2 + 2x_3 - 6) = dx_1 + dx_2 + 2dx_3.$$

Now

$$2(x_1 dx_1 + x_2 dx_2 + x_3 dx_3) \wedge (dx_1 + dx_2 + dx_3) \wedge (dx_1 + dx_2 + 2dx_3) = 0$$

provides $2(x_1 - x_2)dx_1 \wedge dx_2 \wedge dx_3 = 0$ and $x_1 = x_2$. Inserting this expression into the two constraints yields $2x_1 + x_3 = 4$, $2x_1 + 2x_3 = 6$ with the solution $x_1 = x_2 = 1$, $x_3 = 2$. Thus $D^2 = 6$ and $D = \sqrt{6}$.

Problem 42.   Let $a, b \in \mathbb{R}$ and $a \neq b$. Consider the *hyperplanes* in the Euclidean space $\mathbb{E}^4$, $x_1 + x_2 + x_3 + x_4 = a$, $y_1 + y_2 + y_3 + y_4 = b$. Find the shortest distance between the hyperplanes. Apply differential forms.

Solution 42.   The square of the distance D between the hyperplanes is given by

$$D^2 = (x_1 - y_1)^2 + (x_2 - y_2)^2 + (x_3 - y_3)^2 + (x_4 - y_4)^2.$$

Now

$$(x_1 - y_1) + (x_2 - y_2) + (x_3 - y_3) + (x_4 - y_4) = a - b.$$

Thus we introduce the quantities

$$c_j := x_j - y_j, \quad j = 1, 2, 3, 4.$$

This leads to $D^2 = c_1^2 + c_2^2 + c_3^2 + c_4^2$ and $c_1 + c_2 + c_3 + c_4 = a - b$. Now the exterior derivative of D^2 is

$$d(D^2) = 2(c_1 dc_1 + c_2 dc_2 + c_3 dc_3 + c_4 dc_4)$$

and $d(c_1 + c_2 + c_3 + c_4) = 0$. It follows that

$$\frac{1}{2}(d(D^2) \wedge d(c_1+c_2+c_3+c_4)) = (c_1-c_2)dc_1 \wedge dc_2 + (c_1-c_3)dc_1 \wedge dc_3$$
$$+ (c_1-c_4)dc_1 \wedge dc_4 + (c_2-c_3)dc_2 \wedge dc_3$$
$$+ (c_2-c_4)dc_2 \wedge dc_4 + (c_3-c_4)dc_3 \wedge dc_4$$
$$= 0$$

and $c_1 = c_2 = c_3 = c_4 = c$. Hence $4c = a - b$ or $c = \frac{1}{4}(a - b)$. Finally $D^2 = 4c^2 = \frac{1}{4}(a - b)^2$ and $D = \frac{1}{2}|a - b|$.

Problem 43.   Let $x_1 > 0$, $x_2 > 0$, $x_3 > 0$. Consider the surface

$$x_1 x_2 x_3 = 2$$

in $\mathbb{R}^3$. Find the shortest distance from the origin $(0,0,0)$ to the surface.
(i) Apply the *Lagrange multiplier method*.
(ii) Apply differential forms.
(iii) Apply symmetry consideration (permutation group).

Solution 43.   (i) The Lagrange function is given by

$$L(x_1, x_2, x_3, \lambda) = x_1^2 + x_2^2 + x_3^2 + \lambda(x_1 x_2 x_3 - 2).$$

From

$$\frac{\partial L}{\partial x_1} = 0, \qquad \frac{\partial L}{\partial x_2} = 0, \qquad \frac{\partial L}{\partial x_3} = 0$$

we obtain the three nonlinear equations

$$2x_1 + \lambda x_2 x_3 = 0, \quad 2x_2 + \lambda x_1 x_3 = 0, \quad 2x_3 + \lambda x_1 x_2 = 0.$$

Multiplying the first equation with x_1, the second with x_2 and the third with x_3 and taking into account the constraint $x_1 x_2 x_3 = 2$ we find that $x_1^2 = x_2^2 = x_3^2$ and thus $x_1 = x_2 = x_3$. This leads to $x_1 = x_2 = x_3 = 2^{1/3}$.
(ii) We have

$$f(x_1, x_2, x_3) = x_1^2 + x_2^2 + x_3^2 \Rightarrow df = 2x_1 dx_1 + 2x_2 dx_2 + 2x_3 dx_3$$

and

$$g(x_1, x_2, x_3) = x_1 x_2 x_3 - 2 \Rightarrow dg = x_2 x_3 dx_1 + x_3 x_1 dx_2 + x_1 x_2 dx_3.$$

Thus

$$df \wedge dg = 2(x_1^2 - x_2^2)x_3 dx_1 \wedge dx_2 + 2(x_3^2 - x_1^2)x_2 dx_3 \wedge dx_1 + 2(x_2^2 - x_3^2)x_1 dx_2 \wedge dx_3.$$

Consequently from $df \wedge dg = 0$ it follows that $x_1 = x_2 = x_3$. This leads to $x_1 = x_2 = x_3 = 2^{1/3}$.

(iii) The equation $x_1 x_2 x_3 = 2$ is invariant under all 6 permutations of 123. This implies one could look for a solution with $x_1 = x_2 = x_3 = x$.

Problem 44. Consider the differential $n + 1$ form

$$\alpha = df \wedge \omega + dt \wedge df \wedge \sum_{j=1}^{n} (-1)^{j+1} V_j(\mathbf{x}, t) dx_1 \wedge \cdots \wedge \widehat{dx_j} \wedge \cdots \wedge dx_n$$

$$+ (\mathrm{div}(V)) f dt \wedge \omega$$

where the circumflex indicates omission and $\omega = dx_1 \wedge \cdots \wedge dx_n$. Here $f : \mathbb{R}^{n+1} \to \mathbb{R}$ is a smooth function of $\mathbf{x}$, t and V is a smooth vector field.

(i) Show that the sectioned form

$$\widetilde{\alpha} = df(\mathbf{x}, t) \wedge \omega$$

$$+ dt \wedge df(\mathbf{x}, t) \wedge \left(\sum_{j=1}^{n} (-1)^{j+1} V_j(\mathbf{x}, t) dx_1 \wedge \cdots \wedge \widehat{dx_j} \wedge \cdots \wedge dx_n \right)$$

$$+ (\mathrm{div}(V)(\mathbf{x}, t)) f(\mathbf{x}, t) dt \wedge \omega$$

where we distinguish between the independent variables $x_1, \ldots, x_n$, t and the dependent variable f leads using the requirement that $\widetilde{\alpha} = 0$ to the generalized *Liouville equation*.

(ii) Show that the differential form α is closed, i.e. $d\alpha = 0$.

Solution 44. Since

$$df = \sum_{j=1}^{n} \frac{\partial f}{\partial x_j} dx_j + \frac{\partial f}{\partial t} dt$$

$dt \wedge dt = 0$ and $dx_j \wedge dx_j = 0$ we arrive at

$$\widetilde{\alpha} = \left(\frac{\partial f}{\partial t} + \sum_{j=1}^{n} V_j(\mathbf{x}, t) \frac{\partial f}{\partial x_j} + (\mathrm{div}(V)) f \right) dt \wedge \omega.$$

If we put $\tilde{\alpha} = 0$ we obtain the generalized Liouville equation.
(ii) Since

$$d(df \wedge \omega) = 0$$

$$d\left(\sum_{j=1}^{n}(-1)^{j+1}V_j(\mathbf{x})dx_1 \wedge \cdots \wedge \widehat{dx_j} \wedge \cdots \wedge dx_n\right) = (\mathrm{div}(V))\omega$$

$$d((\mathrm{div}(V))f dt \wedge \omega) = (\mathrm{div}(V))df \wedge dt \wedge \omega$$

we find that $d\alpha = 0$.

Problem 45.   Consider *Burgers equation*

$$\frac{\partial u}{\partial t} = u\frac{\partial u}{\partial x} + k\frac{\partial^2 u}{\partial x^2}$$

where u is the velocity field and k is a positive constant with dimension $meter^2/sec$. To describe Burgers equation within the jet bundle formalism with $(x, t; u_0, u_x) \in J^1(\mathbb{R}^2, \mathbb{R})$ one introduces the differential two-forms

$$\beta_1 = du_0 \wedge dt - u_x dx \wedge dt$$
$$\beta_2 = du_0 \wedge dx + u_0 du_0 \wedge dt + k du_x \wedge dt.$$

Show that $k d\beta_1 = dx \wedge \beta_2 - u_0 dx \wedge \beta_1$.

Solution 45.   For the left-hand side we have

$$k d\beta_1 = -k du_x \wedge dx \wedge dt$$

and for the right-hand side we find

$$dx \wedge \beta_2 - u_0 dx \wedge \beta_1 = u_0 dx \wedge du_0 \wedge dt + k dx \wedge du_x \wedge dt - u_0 dx \wedge du_0 \wedge dt$$
$$= k dx \wedge du_x \wedge dt.$$

Problem 46.   *Bessel's differential equation* of order zero is given by

$$x^2\frac{d^2y}{dx^2} + x\frac{dy}{dx} + x^2 y = 0$$

which is a second order linear differential equation. Apply the transformation

$$u = x, \qquad v = \frac{1}{y}\frac{dy}{dx}$$

and find dv/du.

Solution 46. We have

$$du = dx, \qquad dv = \frac{1}{y}\frac{d^2y}{dx^2}dx - \frac{1}{y^2}\left(\frac{dy}{dx}\right)^2 dx.$$

Consequently we obtain

$$\frac{dv}{du} = \frac{1}{y}\frac{d^2y}{dx^2} - \frac{1}{y^2}\left(\frac{dy}{dx}\right)^2 = -\frac{v}{u} - 1 - v^2.$$

Thus we find a nonlinear first order differential equation.

Problem 47. Consider the Euclidean space $\mathbb{E}^3$ with the metric tensor field $g = dx_1 \otimes dx_1 + dx_2 \otimes dx_2 + dx_3 \otimes dx_3$ and the sphere $\mathbb{S}^2 \subset \mathbb{E}^3$

$$\mathbb{S}^2 = \{\,(x_1, x_2, x_3) \,:\, x_1^2 + x_2^2 + x_3^2 = 1\,\}.$$

A parametrization (leaving out one longitude) is given by $\mathbf{f} : (0, 2\pi) \times (-\pi, \pi) \mapsto \mathbb{R}^3$

$$\mathbf{f}(\phi, \theta) = \begin{pmatrix} f_1(\phi, \theta) \\ f_2(\phi, \theta) \\ f_3(\phi, \theta) \end{pmatrix} = \begin{pmatrix} \cos(\phi)\cos(\theta) \\ \sin(\phi)\cos(\theta) \\ \sin(\theta) \end{pmatrix}.$$

Then

$$\mathbf{f}^*(g) = \mathbf{f}^*\left(\sum_{j=1}^{3} dx_j \otimes dx_j\right) = \sum_{j=1}^{3}(df_j \otimes df_j) = \cos^2(\theta)d\phi \otimes d\phi + d\theta \otimes d\theta.$$

The orthonormal *coframe* are the differential one-forms

$$\alpha_1 = d\theta, \qquad \alpha_2 = \cos(\theta)d\phi$$

and the corresponding orthonormal *frame* are the vector fields

$$V_1 = \frac{\partial}{\partial\theta}, \qquad V_2 = \frac{1}{\cos(\theta)}\frac{\partial}{\partial\phi}$$

i.e. $V_1 \rfloor \alpha_1 = 1$, $V_2 \rfloor \alpha_2 = 1$.
(i) Find $d\alpha_1$, $d\alpha_2$. Show that $d\alpha_2 = -\tan(\theta)\alpha_1 \wedge \alpha_2$.
(ii) Find the commutator $[V_1, V_2]$.

Solution 47.   (i) We have $d\alpha_1 = 0$, $d\alpha_2 = -\sin(\theta)d\theta \wedge d\phi$ and thus

$$d\alpha_2 = -\tan(\theta)\alpha_1 \wedge \alpha_2.$$

(ii) We obtain the vector field

$$[V_1, V_2] = \frac{\sin(\theta)}{\cos^2(\theta)} \frac{\partial}{\partial\phi} = \tan(\theta)V_2.$$

Problem 48.   Let $f : \mathbb{R}^4 \to \mathbb{R}$ be a smooth function $f(p_1, p_2, q_1, q_2)$ and consider the differential two-form in $\mathbb{R}^4$

$$\omega = dp_1 \wedge dq_1 + dp_2 \wedge dq_2.$$

We define a vector field

$$V_f = V_{q1}\frac{\partial}{\partial q_1} + V_{q2}\frac{\partial}{\partial q_2} + V_{p1}\frac{\partial}{\partial p_1} + V_{p2}\frac{\partial}{\partial p_2}$$

as solutions of the equation $V_f \rfloor \omega + df = 0$. Find the corresponding system of ordinary differential equations.

Solution 48.   We have

$$df = \sum_{j=1}^{2} \left(\frac{\partial f}{\partial q_j}dq_j + \frac{\partial f}{\partial p_j}dp_j \right)$$

and

$$V_f \rfloor \omega = -V_{q_1}dp_1 - V_{q_2}dp_2 + V_{p_1}dq_1 + V_{p_2}dq_2.$$

Thus from $V_f \rfloor \omega + df = 0$ it follows that

$$V_{q_1} = \frac{\partial f}{\partial p_1}, \quad V_{q_2} = \frac{\partial f}{\partial p_2}, \quad V_{p_1} = -\frac{\partial f}{\partial q_1}, \quad V_{p_2} = -\frac{\partial f}{\partial q_2}$$

with the vector field

$$V_f = \sum_{j=1}^{2} \left(\frac{\partial f}{\partial p_j}\frac{\partial}{\partial q_j} - \frac{\partial f}{\partial q_j}\frac{\partial}{\partial p_j} \right)$$

and the autonomous system of differential equations

$$\frac{dq_j}{dt} = \frac{\partial f}{\partial p_j}, \qquad \frac{dp_j}{dt} = -\frac{\partial f}{\partial q_j}$$

where $j = 1, 2$.

Problem 49.   Consider the system of partial differential equations (continuity equation and *Euler equation* of hydrodynamics in one space dimension)

$$\frac{\partial u}{\partial t} + u\frac{\partial u}{\partial x} + 2c\frac{\partial c}{\partial x} = \frac{\partial H}{\partial x}, \qquad \frac{\partial c}{\partial t} + u\frac{\partial c}{\partial x} + \frac{1}{2}c\frac{\partial u}{\partial x} = 0$$

where u and c are the velocities of the fluid and of the disturbance with respect to the fluid, respectively. H the depth is a given function of x. Show that the partial differential equations can be written in the forms $d\alpha = 0$ and $d\omega = 0$, where α and β are differential one-forms. Owing to $d\alpha = 0$ and $d\omega = 0$ one can find locally (*Poincaré lemma*) zero-forms (functions) (also called potentials) such that

$$\alpha = d\Phi, \qquad \beta = d\Psi.$$

Solution 49.   We have

$$\alpha = udx - \left(\frac{1}{2}u^2 + c^2 - H\right)dt, \qquad \omega = c^2 dx - c^2 udt.$$

Then

$$d\alpha = \frac{\partial u}{\partial t}dt \wedge dx - \left(u\frac{\partial u}{\partial x} + 2c\frac{\partial c}{\partial x} - \frac{\partial H}{\partial x}\right)dx \wedge dt$$

$$= \left(\frac{\partial u}{\partial t} + u\frac{\partial u}{\partial x} + 2c\frac{\partial c}{\partial x} - \frac{\partial H}{\partial x}\right)dt \wedge dx$$

and

$$d\omega = 2c\frac{\partial c}{\partial t}dt \wedge dx - \left(2c\frac{\partial c}{\partial x}u + c^2\frac{\partial u}{\partial x}\right)dx \wedge dt$$

$$= 2c\left(\frac{\partial c}{\partial t} + \frac{\partial c}{\partial x}u + \frac{c}{2}\frac{\partial u}{\partial x}\right)dt \wedge dx.$$

For the potentials Φ and Ψ we have

$$\frac{\partial \Phi}{\partial x} = u, \qquad \frac{\partial \Psi}{\partial t} = -\frac{1}{2}u^2 - c^2 + H,$$

$$\frac{\partial \Psi}{\partial x} = c^2, \qquad \frac{\partial \Psi}{\partial t} = -uc^2.$$

Thus one can write

$$\frac{\partial \Psi}{\partial t} + \frac{1}{2}\left(\frac{\partial \Phi}{\partial x}\right)^2 + \frac{\partial \Psi}{\partial x} = H, \qquad \frac{\partial \Psi}{\partial t} + \frac{\partial \Psi}{\partial x}\frac{\partial \Psi}{\partial x} = 0.$$

Problem 50. In the two-dimensional Euclidean space with metric tensor field

$$g = dx_1 \otimes dx_1 + dx_2 \otimes dx_2$$

the Hodge duality operator $\star$ is given by

$$\star(f_1(\mathbf{x})dx_1 + f_2(\mathbf{x})dx_2) = f_1(\mathbf{x})(\star dx_1) + f_2(\mathbf{x})(\star dx_2) = f_1(\mathbf{x})dx_2 - f_2(\mathbf{x})dx_1$$

i.e. $\star dx_1 = dx_2$, $\star dx_2 = -dx_1$. Hence $dx_1 \wedge (\star dx_1) = dx_1 \wedge dx_2$. Consider the one-form $\alpha = x_1 dx_2 - x_2 dx_1$. Find $d(\star\alpha)$ and $\star(d\alpha)$.

Solution 50. We have

$$d(\star\alpha) = d(-x_1 dx_1 - x_2 dx_2) = 0$$

and with $\star(dx_1 \wedge dx_2) = 1$ and $d\alpha = 2dx_1 \wedge dx_2$ we obtain

$$\star(d\alpha) = \star(2dx_1 \wedge dx_2) = 2.$$

Problem 51. Consider the manifold $M = \mathbb{R}^3$ with the metric tensor field

$$g = dx_1 \otimes dx_1 + dx_2 \otimes dx_2 + dx_3 \otimes dx_3$$

and the smooth differential one-form

$$\alpha = f_1(\mathbf{x})dx_1 + f_2(\mathbf{x})dx_2 + f_3(\mathbf{x})dx_3.$$

Find solutions of $d\alpha = \star\alpha$, where $\star$ is the Hodge duality operator. Note that

$$\star dx_1 = dx_2 \wedge dx_3, \quad \star dx_2 = dx_3 \wedge dx_1, \quad \star dx_3 = dx_1 \wedge dx_2.$$

Solution 51. We have

$$d\alpha = \left(-\frac{\partial f_1}{\partial x_2} + \frac{\partial f_2}{\partial x_1}\right)dx_1 \wedge dx_2 + \left(-\frac{\partial f_2}{\partial x_3} + \frac{\partial f_3}{\partial x_2}\right)dx_2 \wedge dx_3$$
$$+ \left(\frac{\partial f_1}{\partial x_3} - \frac{\partial f_3}{\partial x_1}\right)dx_3 \wedge dx_1$$

and

$$\star\alpha = f_1 dx_2 \wedge dx_3 + f_2 dx_3 \wedge dx_1 + f_3 dx_1 \wedge dx_2.$$

Hence we obtain the three equations

$$\frac{\partial f_2}{\partial x_1} - \frac{\partial f_1}{\partial x_2} = f_3, \quad \frac{\partial f_3}{\partial x_2} - \frac{\partial f_2}{\partial x_3} = f_1, \quad \frac{\partial f_1}{\partial x_3} - \frac{\partial f_3}{\partial f_1} = f_2.$$

Note that since $dd\alpha = 0$ we have $d(\star\alpha) = 0$, i.e.

$$\left(\frac{\partial f_1}{\partial x_1} + \frac{\partial f_2}{\partial x_2} + \frac{\partial f_3}{\partial x_3} \right) dx_1 \wedge dx_2 \wedge dx_3 = 0 \;\Rightarrow\; \frac{\partial f_1}{\partial x_1} + \frac{\partial f_2}{\partial x_2} + \frac{\partial f_3}{\partial x_3} = 0.$$

Find $d(\star(d\alpha)) = 0$.

Problem 52. Consider the vector fields

$$V_{12} = x_2 \frac{\partial}{\partial x_1} - x_1 \frac{\partial}{\partial x_2}, \quad V_{23} = x_3 \frac{\partial}{\partial x_2} - x_2 \frac{\partial}{\partial x_3}, \quad V_{31} = x_1 \frac{\partial}{\partial x_3} - x_3 \frac{\partial}{\partial x_1}$$

in $\mathbb{R}^3$ and the volume form $\Omega = dx_1 \wedge dx_2 \wedge dx_3$.
(i) Find the commutators $[V_{12}, V_{23}]$, $[V_{23}, V_{31}]$, $[V_{31}, V_{12}]$. Discuss.
(ii) Find $V_{12}\rfloor\Omega$, $V_{23}\rfloor\Omega$, $V_{31}\rfloor\Omega$.
(iii) Let $\star$ be the Hodge duality operator in $\mathbb{R}^3$ with metric tensor field

$$g = dx_1 \otimes dx_1 + dx_2 \otimes dx_2 + dx_3 \otimes dx_3.$$

Find $\star(V_{12}\rfloor\Omega)$, $\star(V_{23}\rfloor\Omega)$, $\star(V_{31}\rfloor\Omega)$.
(iv) Find $d(\star(V_{12}\rfloor\Omega))$, $d(\star(V_{23}\rfloor\Omega))$, $d(\star(V_{31}\rfloor\Omega))$.

Solution 52. (i) We obtain

$$[V_{12}, V_{23}] = V_{31}, \quad [V_{23}, V_{31}] = V_{12}, \quad [V_{31}, V_{12}] = V_{23}.$$

Thus we have a basis of the Lie algebra $so(3, \mathbb{R})$.
(ii) We find

$$V_{12}\rfloor\Omega = x_2 dx_2 \wedge dx_3 + x_1 dx_1 \wedge dx_3,$$
$$V_{23}\rfloor\Omega = x_3 dx_3 \wedge dx_1 + x_2 dx_2 \wedge dx_1,$$
$$V_{31}\rfloor\Omega = x_1 dx_1 \wedge dx_2 + x_3 dx_3 \wedge dx_2.$$

(iii) We find

$$\star(V_{12}\rfloor\Omega) = x_2 dx_1 - x_1 dx_2, \quad \star(V_{23}\rfloor\Omega) = x_3 dx_2 - x_2 dx_3,$$

$$\star(V_{31}\rfloor\Omega) = x_1 dx_3 - x_3 dx_1.$$

(iv) The exterior derivative yields

$$d(\star(V_{12}\rfloor\Omega)) = -2dx_1 \wedge dx_2, \quad d(\star(V_{23}\rfloor\Omega)) = -2dx_2 \wedge dx_3,$$

$$d(\star(V_{31}\rfloor\Omega)) = -2dx_3 \wedge dx_1.$$

Problem 53. A transformation $(\mathbf{q}, \mathbf{p}) \to (\mathbf{Q}, \mathbf{P})$ is called *symplectic* if it preserves the differential two-form

$$\omega = \sum_{j=1}^{n}(dq_j \wedge dp_j).$$

Consider the Hamilton function

$$H(\mathbf{q}, \mathbf{p}) = \frac{|\mathbf{p}|^2}{2\mu} - \frac{\mu M}{|\mathbf{q}|}, \qquad \mathbf{p} := \mu\frac{d\mathbf{q}}{dt}$$

where μ and M are positive constants and $\mathbf{p} = (p_1, p_2)^T$, $\mathbf{q} = (q_1, q_2)^T$, i.e. $n = 2$. The phase space is $\mathbb{R}^2 \setminus \{0\} \times \mathbb{R}^2$. The parameter μ is the reduced mass $m_1 m_2/M$. The *symplectic two-form* is

$$\omega = dq_1 \wedge dp_1 + dq_2 \wedge dp_2.$$

Show that ω is invariant under the transformation

$$\mathbf{f} : ((r, \phi), (R, \Phi)) \to (q_1, q_2, p_1, p_2)$$

with

$$q_1(r, \phi, R, \Phi) = r\cos(\phi)$$
$$q_2(r, \phi, R, \Phi) = r\sin(\phi)$$
$$p_1(r, \phi, R, \Phi) = R\cos(\phi) - \frac{\Phi}{r}\sin(\phi)$$
$$p_2(r, \phi, R, \Phi) = R\sin(\phi) + \frac{\Phi}{r}\cos(\phi).$$

Find the Hamilton function in this new symplectic variables.

Solution 53. We have

$$dq_1 = \cos(\phi)dr - r\sin(\phi)d\phi$$
$$dq_2 = \sin(\phi)dr + r\cos(\phi)d\phi$$
$$dp_1 = \cos(\phi)dR - R\sin(\phi) - \frac{\sin(\phi)}{r}d\Phi + \frac{\Phi}{r^2}\sin(\phi)dr - \frac{\Phi}{r}\cos(\phi)d\phi$$
$$dp_2 = \sin(\phi)dR + R\cos(\phi)d\phi + \frac{\cos(\phi)}{r}d\Phi - \frac{\Phi}{r^2}\cos(\phi)dr - \frac{\Phi}{r}\sin(\phi)d\phi.$$

Using $dr \wedge dr = 0$, $dr \wedge d\phi = -d\phi \wedge dr$ etc we obtain

$$dq_1 \wedge dp_1 + dq_2 \wedge dp_2 = dr \wedge dR + d\phi \wedge d\Phi.$$

The Hamilton function takes the form

$$H_{pc}(r, \phi, R, \Phi) = H_{Kep} \circ \mathbf{f}(r, \phi, R, \Phi) = \frac{1}{2\mu} \left(R^2 + \frac{\Phi^2}{r^2} \right) - \frac{\mu M}{r}.$$

Problem 54. *Poincaré upper half plane* is given by the set

$$\mathbb{H}_+^2 = \{ (x_1, x_2) \in \mathbb{R}^2 : x_2 > 0 \}$$

with the metric tensor field

$$g = \frac{1}{x_2} dx_1 \otimes \frac{1}{x_2} dx_1 + \frac{1}{x_2} dx_2 \otimes \frac{1}{x_2} dx_2$$

which is conformal with the standard inner product. The orthonormal *coframe* and *frame* are given by

$$\alpha_1 = \frac{1}{x_2} dx_1, \quad \alpha_2 = \frac{1}{x_2} dx_2, \quad V_1 = x_2 \frac{\partial}{\partial x_1}, \quad V_2 = x_2 \frac{\partial}{\partial x_2}$$

i.e. we have $V_1 \rfloor \alpha_1 = 1$, $V_2 \rfloor \alpha_2 = 1$, $V_1 \rfloor \alpha_2 = 0$, $V_2 \rfloor \alpha_1 = 0$.
(i) Find $d\alpha_1$ and $d\alpha_2$ and show that $d\alpha_1 = \alpha_1 \wedge \alpha_2$.
(ii) Find the commutator $[V_1, V_2]$.
(iii) Find the curvature forms.

Solution 54. (i) We have $d\alpha_1 = \frac{1}{x_2} dx_1 \wedge dx_2$, $d\alpha_2 = 0$ and thus

$$d\alpha_1 = \frac{1}{x_2} dx_1 \wedge dx_2 = \alpha_1 \wedge \alpha_2.$$

(ii) For the commutator we find

$$[V_1, V_2] = x_2 \frac{\partial}{\partial x_1} = V_1.$$

(iii) For the exterior derivative of α_1 and α_2 we have

$$d\alpha_1 = d\left(\frac{1}{x_2} dx_1\right) = \frac{1}{x_2^2} dx_1 \wedge dx_2 = \alpha_1 \wedge \alpha_2, \quad d\alpha_2 = 0.$$

The connection forms we compute from

$$d\alpha_k + \sum_{j=1}^{2} \omega_j^k \wedge \alpha_j = 0.$$

We find

$$-d\alpha_1 = 0 + \omega_2^1 \wedge \alpha_2 = -\alpha_1 \wedge \alpha_2, \quad -d\alpha_2 = \omega_1^2 \wedge \alpha_1 + 0 = 0.$$

Therefore $\omega_2^1 = -\alpha_1 = -x_2^{-1}dx_1$. Then

$$\omega = \begin{pmatrix} 0 & -\alpha_1 \\ \alpha_1 & 0 \end{pmatrix}.$$

For the curvature forms we obtain

$$\Omega_2^1 = d\omega_2^1 + \omega_1^1 \wedge \omega_2^1 + \omega_2^1 \wedge \omega_2^2 = d(-x_2^{-1}dx_1) = -\alpha_1 \wedge \alpha_2$$

and therefore

$$\Omega = \begin{pmatrix} 0 & -\alpha_1 \wedge \alpha_2 \\ \alpha_1 \wedge \alpha_2 & 0 \end{pmatrix}.$$

Now we have $V_2\rfloor(V_1\rfloor(-\alpha_1 \wedge \alpha_2)) = -1$.

Problem 55.   Let $r > 0$ and fixed. Consider the curve (circle)

$$x_1(\tau) = r\cos(\tau), \quad x_2(\tau) = r\sin(\tau), \quad \tau \in [0, 2\pi)$$

and the differential one-form $\alpha = \frac{1}{2}(x_1 dx_2 - x_2 dx_1)$. Find

$$\alpha(\tau) = \frac{1}{2}(x_1(\tau)dx_2(\tau) - x_2(\tau)dx_1(\tau))$$

and

$$\oint \alpha(\tau) = \int_0^{2\pi} \frac{1}{2}(x_1(\tau)dx_2(\tau) - x_2(\tau)dx_1(\tau)).$$

Find $d\alpha$.

Solution 55.   With

$$dx_1(\tau) = -r\sin(\tau)d\tau, \quad dx_2(\tau) = r\cos(\tau)d\tau$$

we find

$$x_1(\tau)dx_2(\tau) - x_2(\tau)dx_1(\tau) = r^2\cos^2(\tau)d\tau + r^2\sin^2(\tau)d\tau = r^2 d\tau.$$

Then

$$\oint \alpha(\tau) = \frac{1}{2}r^2 \int_0^{2\pi} d\tau = r^2\pi.$$

This is the area of a circle with radius r. With $dx_1 \wedge dx_2 = -dx_2 \wedge dx_1$ we obtain

$$d\alpha = \frac{1}{2}(dx_1 \wedge dx_2 - dx_2 \wedge dx_1) = dx_1 \wedge dx_2.$$

Problem 56. Consider the manifold $M = \mathbb{R}^2$ and the differential one-form

$$\alpha = \frac{1}{2}(x_1 dx_2 - x_2 dx_1).$$

(i) Find the differential two-form $d\alpha$.

(ii) Consider the domains in $\mathbb{R}^2$

$$D = \{\, (x_1, x_2) : x_1^2 + x_2^2 \leq 1 \,\}, \quad \partial D = \{\, (x_1, x_2) : x_1^2 + x_2^2 = 1 \,\}$$

i.e. ∂D is the boundary of D. Show that

$$\int_D d\alpha = \int_{\partial D} \alpha.$$

Apply *polar coordinates*, i.e. $x_1(r, \phi) = r\cos(\phi)$, $x_2(r, \phi) = r\sin(\phi)$.

Solution 56. (i) With $dx_1 \wedge dx_2 = -dx_2 \wedge dx_1$ we have

$$d\alpha = \frac{1}{2}(dx_1 \wedge dx_2 - dx_2 \wedge dx_1) = dx_1 \wedge dx_2.$$

(ii) With $r > 0$ and constant we obtain

$$dx_1 = -r\sin(\phi)d\phi, \qquad dx_2 = r\cos(\phi)d\phi$$

and therefore with $\cos^2(\phi) + \sin^2(\phi) = 1$ we find

$$\frac{1}{2}(x_1 dx_2 - x_2 dx_1) = \frac{1}{2}r^2 d\phi.$$

Consequently

$$\frac{1}{2}\oint r^2 d\phi = \frac{1}{2}r^2 \oint d\phi = \frac{1}{2}r^2 \int_0^{2\pi} d\phi = r^2\pi.$$

Problem 57. Consider the differential one-form in the plane

$$\alpha = x_2^2 dx_1 + x_1^2 dx_2.$$

Calculate the integral

$$\oint_C \alpha$$

where C is the closed curve which the boundary of a triangle with vertices $(0,0)$, $(1,1)$, $(1,0)$ and counterclockwise orientation. Apply *Green's theorem*

$$\oint_C (f(x_1,x_2)dx_1 + g(x_1,x_2)dx_2) = \int\int_D \left(\frac{\partial g}{\partial x_1} - \frac{\partial f}{\partial x_2}\right) dx_1 dx_2.$$

D is the domain inside the triangle given by $0 \le x_1 \le 1$, $0 \le x_2 \le x_1$.

Solution 57. Since $f(x_1,x_2) = x_2^2$, $g(x_1,x_2) = x_1^2$ we have

$$\frac{\partial g}{\partial x_1} - \frac{\partial f}{\partial x_2} = 2(x_1 - x_2).$$

Thus

$$\int_0^1 \int_0^{x_1} 2(x_1 - x_2)dx_2 dx_1 = \frac{1}{3}.$$

Problem 58. The *lemniscate of Gerono* is described by the equation

$$x^4 = x^2 - y^2.$$

A parametrization is given by $x(\tau) = \sin(\tau)$, $y(\tau) = \sin(\tau)\cos(\tau)$ with $\tau \in [0, \pi]$. Consider the differential one-form $\alpha = x dy$ in the plane $\mathbb{R}^2$. Let $x(\tau) = \sin(\tau)$, $y(\tau) = \sin(\tau)\cos(\tau)$. Find $\alpha(\tau)$. Calculate

$$-\int_0^\pi x(\tau)dy(\tau).$$

Discuss.

Solution 58. We have

$$dy(\tau) = \cos^2(\tau)d\tau - \sin^2(\tau)d\tau = \cos(2\tau)d\tau.$$

Thus

$$\alpha(\tau) = x(\tau)dy(\tau) = \sin(\tau)\cos(2\tau)d\tau = \frac{1}{2}(-\sin(\tau)d\tau + \sin(3\tau)d\tau).$$

Hence

$$-\int_0^\pi x(\tau)dy(\tau) = \frac{1}{2}\int_0^\pi \sin(\tau)d\tau - \frac{1}{2}\int_0^\pi \sin(3\tau)d\tau = 1 - \frac{1}{3} = \frac{2}{3}.$$

This is an application of *Green's Theorem*. Let C be a simple closed curve in the xy-plane enclosing a region D with positive orientation. The area of D is then given by

$$\oint_C xdy.$$

Thus we calculated the area enclosed by the lemniscate of Gerono (actually half of it).

Problem 59. Let $\alpha(\tau) = (x(\tau), y(\tau))$ be a positive oriented simple closed curve, i.e. $x(b) = x(a)$, $y(b) = y(a)$. Show that

$$A = -\int_a^b y(\tau)x'(\tau)d\tau = \int_a^b x(\tau)y'(\tau)d\tau = \frac{1}{2}\int_a^b (x(\tau)y'(\tau) - y(\tau)x'(\tau))d\tau.$$

Solution 59. Using the product rule for differentiation we have

$$\int_a^b x(\tau)y'(\tau)d\tau = \int_a^b (x(\tau)y(\tau))'d\tau - \int_a^b x'(\tau)y(\tau)d\tau$$

$$= (x(b)y(b) - x(a)y(a)) - \int_a^b x'(\tau)y(\tau)d\tau$$

$$= -\int_a^b x'(\tau)y(\tau)d\tau.$$

Problem 60. Consider the differential one-form

$$\alpha = (2x_1x_2 - x_1^2)dx_1 + (x_1 + x_2^2)dx_2.$$

We can write $\alpha = \alpha_1 + \alpha_2$ with

$$\alpha_1 = 2x_1x_2dx_1 + x_1dx_2, \qquad \alpha_2 = -x_1^2dx_1 + x_2^2dx_2$$

with $d\alpha_2 = 0$, $d\alpha_1 = (1 - 2x_1)dx_1 \wedge dx_2$.
(i) Calculate $d\alpha$.
(ii) Calculate

$$\oint \alpha$$

with the closed path $C_1 - C_2$ starting from $(0,0)$ moving along via the curve $C_1 : x_2 = x_1^2$ to $(1,1)$ and back to $(0,0)$ via the curve $C_2 : x_2 = \sqrt{x_1}$. Let D be the (convex) domain enclosed by the two curves C_1 and C_2.
(iii) Calculate the double integral

$$\int\int_D d\alpha$$

where D is the domain given in (i), i.e. $C_1 - C_2$ is the boundary of D. Thus verify the theorem of Gauss-Green.

Solution 60.   (i) Since $dx_1 \wedge dx_1 = dx_2 \wedge dx_2 = 0$ and $dx_1 \wedge dx_2 = -dx_2 \wedge dx_1$ we have

$$d\alpha = (1 - 2x_1)dx_1 \wedge dx_2.$$

(ii) For the curve C_1 we set $x_1(\tau) = \tau$, $x_2(\tau) = \tau^2$ and therefore $dx_1 = d\tau$, $dx_2 = 2\tau d\tau$. For the curve C_2 we set $x_1(\tau) = \tau$, $x_2(\tau) = \sqrt{\tau}$ and therefore $dx_1 = d\tau$, $dx_2 = (1/(2\sqrt{\tau}))d\tau$. It follows that

$$\oint \alpha = \int_{C_1} \alpha + \int_{C_2} \alpha$$

with

$$\int_{C_1} \alpha = \int_{\tau=0}^1 (2\tau^3 - \tau^2)dt + \int_{\tau=0}^1 (\tau + \tau^4)2\tau d\tau = \frac{7}{6}$$

$$\int_{C_2} \alpha = \int_{\tau=1}^0 (2\tau\sqrt{\tau} - \tau^2)dt + \int_{\tau=1}^0 (\tau + \tau)\frac{1}{2\sqrt{\tau}}d\tau = -\frac{17}{15}.$$

Finally $\oint \alpha = 1/30$.
(iii) Since $d\alpha = (1 - 2x_1)dx_1 \wedge dx_2$ we have

$$\int\int_D d\alpha = \int_{y=0}^1 \int_{x_1=x_2^2}^{x_1=\sqrt{x_2}} (1 - 2x_1)dx_1 dx_2$$

$$= \int_{x_2=0}^1 \left(\int_{x_1=x_2^2}^{x_1=\sqrt{x_2}} dx_1 \right) dx_2 - 2\int_{x_2=0}^1 \left(\int_{x_1=x_2^2}^{x_1=\sqrt{x_2}} x_1 dx_1 \right) dx_2$$

$$= \int_{x_2=0}^1 (\sqrt{x_2} - x_2^2)dy - \int_{x_2=0}^1 (x_2 - x_2^4)dy = 1/30.$$

Problem 61.   Show that

$$dg = \frac{1}{16\pi^2} \sin(\beta)d\alpha d\beta d\gamma$$

is a normalized left-invariant measure on the compact Lie group $SU(2)$, where α, β, γ are the *Euler angles*.

Solution 61. Let $B \in SU(2)$. Under left multiplication $A' = BA$ is expressed in terms of the new variables x_0', x_1', x_2' and x_3'. By regarding $x_0 + ix_3$ and $x_2 - ix_1$ as two independent complex variables, we find that the Jacobian determinant of changing variables from (x_0, x_1, x_2, x_3) to (x_0', x_1', x_2', x_3') is $\det(B) = 1$. It follows that dg is invariant. Let $r^2 = x_1^2 + x_2^2 + x_3^2$. We have

$$\int dg = \frac{1}{2\pi^2} \int_{|r| \le 1} \frac{dx_1 dx_2 dx_3}{\sqrt{1 - r^2}} = \frac{2}{\pi} \int_0^1 \frac{r^2 dr}{\sqrt{1 - r^2}} = 1$$

where we used spherical coordinates for the angle integration which provides 4π and

$$\int \frac{x^2 dx}{\sqrt{a^2 - x^2}} = -\frac{x\sqrt{a^2 - x^2}}{2} + \frac{a^2}{2} \arcsin\left(\frac{x}{a}\right).$$

We find

$$x_1(\alpha, \beta, \gamma) = \sin\left(\frac{\beta}{2}\right) \sin\left(\frac{\alpha - \gamma}{2}\right)$$

$$x_2(\alpha, \beta, \gamma) = \sin\left(\frac{\beta}{2}\right) \cos\left(\frac{\alpha - \gamma}{2}\right)$$

$$x_3(\alpha, \beta, \gamma) = -\cos\left(\frac{\beta}{2}\right) \sin\left(\frac{\alpha + \gamma}{2}\right).$$

For the Jacobian determinant we obtain

$$\left| \frac{\partial(x_1, x_2, x_3)}{\partial(\alpha, \beta, \gamma)} \right| = -\frac{1}{4} \sin\left(\frac{\beta}{2}\right) \cos^2\left(\frac{\beta}{2}\right) \cos\left(\frac{\alpha + \gamma}{2}\right).$$

Since

$$r^2 = x_1^2 + x_2^2 + x_3^2 = 1 - \cos^2\left(\frac{\alpha + \gamma}{2}\right) \cos^2\left(\frac{\beta}{2}\right)$$

we have

$$\sqrt{1 - r^2} = \sqrt{1 - (x_1^2 + x_2^2 + x_3^2)} = \cos\left(\frac{\alpha + \gamma}{2}\right) \cos\left(\frac{\beta}{2}\right).$$

Furthermore $\sin(\beta/2) \cos(\beta/2) = \frac{1}{2} \sin(\beta)$. Using these results and the results from above we obtain

$$dg = \frac{1}{2\pi^2} \frac{dx_1 dx_2 dx_3}{\sqrt{1 - r^2}} = \frac{1}{16\pi^2} \sin(\beta) d\alpha d\beta d\gamma.$$

Problem 62.   Let M be a smooth, compact, and oriented n-manifold. Let $f : M \to \mathbb{R}^{n+1} \setminus \{\mathbf{0}\}$ be a smooth map. The *Kronecker characteristic* is given by the following integral

$$K(f) := (\text{vol}(\mathbb{S}^n))^{-1} \int_M \|f(\mathbf{x})\|^{-(n+1)} \det \left(f(\mathbf{x}), \frac{\partial f}{\partial x_1}(\mathbf{x}), \dots, \frac{\partial f}{\partial x_n}(\mathbf{x}) \right) d\mathbf{x}$$

where $(\mathbf{x} = (x_1, \dots, x_n))$ are local coordinates of M and $d\mathbf{x} = dx_1 \cdots dx_n$. Express this integral in terms of differential forms.

Solution 62.   In terms of differential forms we have

$$K(f) = (\text{vol}(\mathbb{S}^n))^{-1} \int_M (g \circ f)^* \omega$$

where $g : \mathbb{R}^{n+1} \setminus \{\mathbf{0}\}$ is the map $g(\mathbf{x}) = \mathbf{x}/\|\mathbf{x}\|$ and ω is the n-differential form

$$\omega := \left. \sum_{j=1}^{n+1} (-1)^{j+1} x_j dx_1 \wedge \cdots dx_{j-1} \wedge \widehat{dx_j} \wedge dx_{j+1} \wedge \cdots \wedge dx_{n+1} \right|_{\mathbb{S}^n} .$$

i.e. the volume form of the compact smooth manifold $\mathbb{S}^n$. Here $\widehat{dx_j}$ indicates omission. $K(f)$ is also called the *Kronecker integral*.

Problem 63.   Let C be the unit circle centered at the origin $(0,0)$. Calculate

$$\frac{1}{2\pi} \oint_C \frac{PdQ - QdP}{P^2 + Q^2}$$

where $P(x,y) = -y$, $Q(x,y) = x$.

Solution 63.   Since

$$PdQ - QdP = -ydx + xdy = xdy - ydx$$

and $P^2 + Q^2 = x^2 + y^2$ we have

$$\frac{1}{2\pi} \oint_C \frac{PdQ - QdP}{P^2 + Q^2} = \frac{1}{2\pi} \oint_C \frac{xdy - ydx}{x^2 + y^2}.$$

Introducing polar coordinates $x = r\cos(\phi)$, $y = r\sin(\phi)$ with $r = 1$ we find

$$\frac{1}{2\pi} \int_0^{2\pi} d\phi = 1.$$

Problem 64. The parameter representation for the *torus* is given by

$$x_1(u_1, u_2) = (R + r\cos(u_1))\cos(u_2)$$
$$x_2(u_1, u_2) = (R + r\cos(u_1))\sin(u_2)$$
$$x_3(u_1, u_2) = r\sin(u_1)$$

where $u_1 \in [0, 2\pi]$ and $u_2 \in [0, 2\pi]$ and $R > r$. Let

$$\mathbf{t}_1(u_1, u_2) := \begin{pmatrix} \partial x_1/\partial u_1 \\ \partial x_2/\partial u_1 \\ \partial x_3/\partial u_1 \end{pmatrix}, \qquad \mathbf{t}_2(u_1, u_2) := \begin{pmatrix} \partial x_1/\partial u_2 \\ \partial x_2/\partial u_2 \\ \partial x_3/\partial u_2 \end{pmatrix}.$$

The *surface element* of the torus is given by

$$do = \sqrt{g}\, du_1 du_2$$

where $g = g_{11}g_{22} - g_{12}g_{21}$ and

$$g_{jk}(u_1, u_2) := \mathbf{t}_j(u_1, u_2) \cdot \mathbf{t}_k(u_1, u_2)$$

with $\cdot$ denoting the scalar product. Calculate the surface area of the torus.

Solution 64. Since

$$\mathbf{t}_1(u_1, u_2) = \begin{pmatrix} -r\sin(u_1)\cos(u_2) \\ -r\sin(u_1)\sin(u_2) \\ r\cos(u_1) \end{pmatrix},$$

$$\mathbf{t}_2(u_1, u_2) = \begin{pmatrix} -(R + r\cos(u_1))\sin(u_2) \\ (R + r\cos(u_1))\cos(u_2) \\ 0 \end{pmatrix}$$

we obtain that

$$\text{rank}(\mathbf{t}_1, \mathbf{t}_2) = \text{rank} \begin{pmatrix} -r\sin(u_1)\cos(u_2) & -(R + r\cos(u_1))\sin(u_2) \\ -r\sin u_1 \sin u_2 & (R + r\cos(u_1))\cos(u_2) \\ r\cos(u_1) & 0 \end{pmatrix} = 2.$$

Now

$$g_{11}(u_1, u_2) = \mathbf{t}_1 \cdot \mathbf{t}_1 = r^2, \qquad g_{22}(u_1, u_2) = \mathbf{t}_2 \cdot \mathbf{t}_2 = (R + r\cos(u_1))^2$$

$$g_{12}(u_1, u_2) = \mathbf{t}_1 \cdot \mathbf{t}_2 = 0, \qquad g_{21}(u_1, u_2) = \mathbf{t}_2 \cdot \mathbf{t}_1 = 0.$$

Therefore

$$\sqrt{g} = \sqrt{g_{11}g_{22} - g_{12}g_{21}} = \sqrt{r^2(R + r\cos(u_1))^2} = r(R + r\cos(u_1))$$

and

$$do = \sqrt{g}du_1 du_2 = r(R + r\cos(u_1))du_1 du_2.$$

The integration yields for the surface of the torus

$$O = \int_0^{2\pi} \int_0^{2\pi} r(R + r\cos(u_1))du_1 du_2 = 4\pi^2 rR.$$

Problem 65. Let $x_0 = ct$. Consider the metric tensor field

$$g = -e^{2u_1(x_1)}dx_0 \otimes dx_0 + dx_1 \otimes dx_1 + e^{2u_2(x_1)}dx_2 \otimes dx_2 + e^{2u_3(x_1)}dx_3 \otimes dx_3$$

and the "corresponding" differential one-forms

$$\alpha_0 = e^{u_1(x_1)}dx_0, \quad \alpha_1 = dx_1, \quad \alpha_2 = e^{u_2(x_1)}dx_2, \quad \alpha_3 = e^{u_3(x_1)}dx_3$$

i.e. g can be written as

$$g = -\alpha_0 \otimes \alpha_0 + \alpha_1 \otimes \alpha_1 + \alpha_2 \otimes \alpha_2 + \alpha_3 \otimes \alpha_3.$$

Find $d\alpha_0$, $d\alpha_1$, $d\alpha_2$, $d\alpha_3$ and express them as a sum of exterior products of α_0, α_1, α_2, α_3.

Solution 65. We have

$$d\alpha_0 = e^{u_1(x_1)}\frac{du_1}{dx_1}dx_1 \wedge dx_0, \quad d\alpha_1 = 0,$$

$$d\alpha_2 = e^{u_2(x_1)}\frac{du_2}{dx_1}dx_1 \wedge dx_2, \quad d\alpha_3 = e^{u_3(x_1)}\frac{du_3}{dx_1}dx_1 \wedge dx_3.$$

Since

$$dx_0 = e^{-u_1(x_1)}\alpha_0, \quad dx_1 = \alpha_1, \quad dx_2 = e^{-u_2(x_1)}\alpha_2, \quad dx_3 = e^{-u_3(x_1)}\alpha_3.$$

It follows that

$$d\alpha_0 = -\frac{du_1}{dx_1}\alpha_0 \wedge \alpha_1, \quad d\alpha_1 = 0, \quad d\alpha_2 = \frac{du_2}{dx_1}\alpha_1 \wedge \alpha_2, \quad d\alpha_3 = \frac{du_3}{dx_1}\alpha_1 \wedge \alpha_3.$$

Problem 66. Consider the metric tensor field

$$g = -e^{2f(r)}dx_0 \otimes dx_0 + e^{2h(r)}dr \otimes dr + r^2 d\theta \otimes d\theta + r^2\sin^2(\theta)d\phi \otimes d\phi$$
$$= -e^{f(r)}dx_0 \otimes e^{f(r)}dx_0 + e^{h(r)}dr \otimes e^{h(r)}dr + rd\theta \otimes rd\theta$$
$$+ r\sin(\theta)d\phi \otimes r\sin(\theta)d\phi.$$

The differentiable functions $f(r)$ and $g(r)$ tend to 0 for $r \to \infty$. The corresponding 4×4 matrix is given by

$$(g_{\mu\nu}) = \begin{pmatrix} -e^{2f(r)} & 0 & 0 & 0 \\ 0 & e^{2h(r)} & 0 & 0 \\ 0 & 0 & r^2 & 0 \\ 0 & 0 & 0 & r^2 \sin^2(\theta) \end{pmatrix}, \quad \mu, \nu = 0, 1, 2, 3$$

with the inverse matrix given by

$$(g^{\mu\nu}) = \begin{pmatrix} -e^{-2f(r)} & 0 & 0 & 0 \\ 0 & e^{-2h(r)} & 0 & 0 \\ 0 & 0 & 1/r^2 & 0 \\ 0 & 0 & 0 & 1/(r^2 \sin^2(\theta)) \end{pmatrix}.$$

Within the *Cartan approach* we can form a basis of differential one-forms (*dual tetrad*)

$$\alpha_0 = e^{f(r)} dx_0, \quad \alpha_1 = e^{h(r)} dr, \quad \alpha_2 = r d\theta, \quad \alpha_3 = r\sin(\theta) d\phi.$$

Hence the metric tensor field can be written as

$$g = \sum_{\mu=0}^{3} \sum_{\nu=0}^{3} \widetilde{g}_{\mu\nu} \alpha_\mu \otimes \alpha_\nu, \quad (\widetilde{g}_{\mu\nu}) = \text{diag}(-1, 1, 1, 1)$$

where $\text{diag}(-1, 1, 1, 1)$ is called the *signature*.

(i) Find the differential two-forms $d\alpha_0$, $d\alpha_1$, $d\alpha_2$, $d\alpha_3$. Express them using the differential one-forms α_0, α_1, α_2, α_3 and the exterior product.

(ii) Find ω^μ_ν (*connection forms*) from the *first structure equation*

$$d\alpha_\mu = -\sum_{\nu=0}^{3} \omega^\mu_\nu \wedge \alpha_\nu.$$

Note that $\omega_{\mu\nu} + \omega_{\nu\mu} = 0$.

(iii) Find the *curvature two-forms* Ω^μ_ν given by

$$\Omega^\mu_\nu = d\omega^\mu_\nu + \sum_{\rho=0}^{3} \omega^\mu_\rho \wedge \omega^\rho_\nu.$$

(iv) Find the *Ricci tensor*

$$R_{\mu\nu} = \Omega^\nu_\mu(E_\rho, E_\nu) \equiv \sum_{\rho=0}^{3} E_\nu \rfloor (E_\rho \rfloor \Omega^\rho_\mu)$$

where E_0, E_1, E_2, E_3 is the *tetrad*

$$E_0 = e^{-f(r)}\frac{\partial}{\partial x_0}, \quad E_1 = e^{-h(r)}\frac{\partial}{\partial r}, \quad E_2 = \frac{1}{r}\frac{\partial}{\partial \theta}, \quad E_3 = \frac{1}{r\sin(\theta)}\frac{\partial}{\partial \phi}$$

and $\rfloor$ is the *contraction (interior product)*.
(v) The *Ricci scalar* R is given by

$$R = \sum_{\mu=0}^{3} R_\mu^\mu, \qquad R_\mu^\nu = \sum_{\rho=0}^{3} \tilde{g}^{\nu\rho} R_{\rho\mu}.$$

Find R and the *Einstein tensor* $G_{\mu\nu}$ given by

$$G_{\mu\nu} := R_{\mu\nu} - \frac{1}{2}\tilde{g}_{\mu\nu}R.$$

Solution 66.   (i) We have

$$d\alpha_0 = e^{f(r)}\frac{df}{dr}dr \wedge dx_0, \quad d\alpha_1 = 0,$$

$$d\alpha_2 = dr \wedge d\theta, \quad d\alpha_3 = \sin(\theta)dr \wedge d\phi + r\cos(\theta)d\theta \wedge d\phi.$$

Then using the basis α_0, α_1, α_2, α_3 we can write

$$d\alpha_0 = \frac{df}{dr}e^{-h(r)}\alpha_1 \wedge \alpha_0, \quad d\alpha_1 = 0, \quad d\alpha_2 = \frac{1}{r}e^{-h(r)}\alpha_1 \wedge \alpha_2$$

$$d\alpha_3 = \frac{1}{r}e^{-h(r)}\alpha_1 \wedge \alpha_3 + \frac{1}{r}\cot(\theta)\alpha_2 \wedge \alpha_3.$$

(ii) Note that $\omega_{00} = \omega_{11} = \omega_{22} = \omega_{33} = 0$ and $\omega_0^0 = \omega_1^1 = \omega_2^2 = \omega_3^3 = 0$. The system of equations

$$d\alpha_\mu = -\sum_{\nu=0}^{3} \omega_\nu^\mu \wedge \alpha_\nu$$

provides the solution

$$\omega_0^0 = \omega_1^1 = \omega_2^2 = \omega_3^3 = 0, \quad \omega_2^0 = \omega_0^2 = \omega_0^3 = \omega_3^0 = 0$$

and

$$\omega_1^0 = \omega_0^1 = \frac{df}{dr}e^{-h(r)}\alpha_0, \quad \omega_1^2 = -\omega_2^1 = \frac{1}{r}e^{-h(r)}\alpha_2,$$

$$\omega_1^3 = -\omega_3^1 = \frac{1}{r}e^{-h(r)}\alpha_3, \quad \omega_2^3 = -\omega_3^2 = \frac{1}{r}\cot(\theta)\alpha_3.$$

For example we have

$$d\alpha_0 = -\omega_1^0 \wedge \alpha_1 = -\frac{df}{dr}e^{-h(r)}\alpha_0 \wedge \alpha_1.$$

(iii) We find

$$\Omega_1^0 = d\omega_1^0 + \sum_{\rho=0}^{3} \omega_\rho^0 \wedge \omega_0^\rho = d\omega_1^0$$

$$= e^{-2h(r)}\left(-\left(\frac{df}{dr}\right)^2 + \left(\frac{df}{dr}\right)\left(\frac{dh}{dr}\right) - \frac{d^2f}{dr^2}\right)\alpha_0 \wedge \alpha_1 = \Omega_0^1$$

$$\Omega_2^0 = d\omega_2^0 + \omega_0^0 \wedge \omega_2^0 + \omega_1^0 \wedge \omega_2^1 + \omega_2^0 \wedge \omega_2^2 + \omega_3^0 \wedge \omega_2^3 = \omega_1^0 \wedge \omega_2^1$$

$$= -\frac{1}{r}e^{-2h(r)}\frac{df}{dr}\alpha_0 \wedge \alpha_2 = \Omega_0^2.$$

Analogously we find

$$\Omega_3^0 = -\frac{e^{-2h(r)}}{r}\frac{df}{dr}\alpha_0 \wedge \alpha_3 = \Omega_0^3, \quad \Omega_2^1 = \frac{e^{-2h(r)}}{r}\frac{dh}{dr}\alpha_1 \wedge \alpha_2 = -\Omega_1^2,$$

$$\Omega_3^1 = \frac{e^{-2h(r)}}{r}\frac{dh}{dr}\alpha_1 \wedge \alpha_3 = -\Omega_1^3, \quad \Omega_3^2 = \frac{1-e^{-2h(r)}}{r^2}\alpha_2 \wedge \alpha_3 = -\Omega_2^3.$$

The other curvature two-forms follow from

$$\Omega_{\mu\nu} = \sum_{\lambda=0}^{3} \widetilde{g}_{\mu\lambda}\Omega_\nu^\lambda = -\Omega_{\nu\mu}.$$

(iv) Note that

$$R_{\mu\nu} = \sum_{\lambda=0}^{3} R_{\mu\lambda\nu}^\lambda = \sum_{\lambda=0}^{3}(E_\nu \lrcorner (E_\lambda \lrcorner \Omega_\mu^\lambda)).$$

Utilizing $E_0 \lrcorner \alpha_0 = 1$, $E_1 \lrcorner \alpha_1 = 1$, $E_2 \lrcorner \alpha_2 = 1$, $E_3 \lrcorner \alpha_3 = 1$ we have

$$R_{00} = E_0 \lrcorner E_1 \lrcorner \Omega_0^1 + E_0 \lrcorner E_2 \lrcorner \Omega_0^2 + E_0 \lrcorner E_3 \lrcorner \Omega_0^3$$

$$= e^{-2h(r)}\left(\frac{d^2f}{dr^2} - \frac{df}{dr}\frac{dh}{dr} + \left(\frac{df}{dr}\right)^2 + \frac{2}{r}\frac{df}{dr}\right).$$

Analogously

$$R_{11} = -e^{-2h}\left(\frac{d^2f}{dr^2} - \frac{df}{dr}\frac{dh}{dr} + \left(\frac{df}{dr}\right)^2 - \frac{2}{r}\frac{dh}{dr}\right)$$

$$R_{22} = \frac{e^{-2h}}{r}\left(-\frac{df}{dr} + \frac{dh}{dr} + \frac{e^{2h}-1}{r}\right)$$

$$R_{33} = R_{22}.$$

(v) With the signature diag$(-1, 1, 1, 1)$ we obtain for the scalar curvature

$$R = e^{-2h}\left(-2\frac{d^2f}{dr^2} + 2\frac{df}{dr}\frac{dh}{dr} - 2\left(\frac{df}{dr}\right)^2 + \frac{4}{r}\left(\frac{dh}{dr} - \frac{df}{dr}\right) + \frac{2}{r^2}\left(e^{2h} - 1\right)\right)$$

and the nonzero components of the Einstein tensor $G_{\mu\nu}$ are given by

$$G_{00} = R_{00} - \frac{1}{2}R\tilde{g}_{00} = \frac{1}{r^2} - e^{-2h}\left(\frac{1}{r^2} - \frac{2}{r}\frac{dh}{dr}\right)$$

$$G_{11} = -\frac{1}{r^2} + e^{-2h}\left(\frac{1}{r^2} + \frac{2}{r}\frac{df}{dr}\right)$$

$$G_{22} = e^{-2h}\left(\frac{d^2f}{dr^2} - \frac{df}{dr}\frac{dh}{dr} + \left(\frac{df}{dr}\right)^2 + \frac{1}{r}\left(\frac{df}{dr} - \frac{dh}{dr}\right)\right) = G_{33}.$$

The vacuum *Einstein field equation* is $G_{00} = G_{11} = G_{22} = G_{33} = 0$. They admit the *Schwarzschild solution*

$$g = \left(1 - \frac{r_S}{r}\right)dx_0 \otimes dx_0 - \left(1 - \frac{r_S}{r}\right)^{-1}dr \otimes dr$$
$$-r^2(d\theta \otimes d\theta + \sin^2(\theta)d\phi \otimes d\phi)$$

with

$$e^{2f(r)} = \left(1 - \frac{r_S}{r}\right), \quad e^{2h(r)} = \left(1 - \frac{r_S}{r}\right)^{-1}, \quad r_S = \frac{2GM}{c^2}.$$

Problem 67.   Let A be an $n \times n$ matrix. Assume that the entries are analytic functions of x and that A is invertible for all x. Let d be the exterior derivative. We have the identity

$$d(\det(A)) \equiv \det(A)\mathrm{tr}(A^{-1}dA).$$

Let

$$A(x) = \begin{pmatrix} \cos(x) & \sin(x) \\ -\sin(x) & \cos(x) \end{pmatrix}.$$

Calculate the left-hand side and right-hand side of the identity.

Solution 67.   Since $\det(A) = 1$ we have $d1 = 0$. With

$$A^{-1}(x) = \begin{pmatrix} \cos(x) & -\sin(x) \\ \sin(x) & \cos(x) \end{pmatrix}$$

and $\sin^2(x) + \cos^2(x) = 1$ we obtain

$$A^{-1}dA = \begin{pmatrix} \cos(x) & -\sin(x) \\ \sin(x) & \cos(x) \end{pmatrix} \begin{pmatrix} -\sin(x)dx & \cos(x)dx \\ -\cos(x)dx & -\sin(x)dx \end{pmatrix}$$

$$= \begin{pmatrix} 0 & dx \\ -dx & 0 \end{pmatrix}.$$

Thus $\text{tr}(A^{-1}dA) = 0$ and therefore $\det(A)\text{tr}(A^{-1}dA) = 0$.

Problem 68.   (i) Let $\alpha \in \mathbb{R}$. Consider the *rotation matrix*

$$R(\alpha) = \begin{pmatrix} \cos(\alpha) & -\sin(\alpha) \\ \sin(\alpha) & \cos(\alpha) \end{pmatrix}.$$

Hence $R(\alpha) \in SO(2, \mathbb{R})$. Find $R^{-1}(\alpha)dR(\alpha)$ and $(dR(\alpha))R^{-1}(\alpha)$.
(ii) Let $\alpha \in \mathbb{R}$. Consider the matrix

$$S(\alpha) = \begin{pmatrix} \cos(\alpha) & \sin(\alpha) \\ \sin(\alpha) & -\cos(\alpha) \end{pmatrix}.$$

Hence $S(\alpha) \in O(2, \mathbb{R})$, but $S(\alpha) \notin SO(2, \mathbb{R})$ since $\det(S(\alpha)) = -1$. Find $S^{-1}(\alpha)dS(\alpha)$.
(iii) Let $\phi \in \mathbb{R}$. Consider the matrix

$$W(\phi) = \begin{pmatrix} 0 & e^{-i\phi} \\ e^{i\phi} & 0 \end{pmatrix}.$$

Hence $W(\phi) \in U(2)$. Find $W^{-1}(\phi)dW(\phi)$.
(iv) Consider the matrix

$$U(\theta, \phi) = \begin{pmatrix} \cos(\theta/2) & -e^{-i\phi}\sin(\theta/2) \\ e^{i\phi}\sin(\theta/2) & \cos(\theta/2) \end{pmatrix}.$$

Hence $U(\theta, \phi) \in SU(2)$. Find $U^{-1}(\theta, \phi)dU(\theta, \phi)$.
(v) Consider the matrix

$$V(\theta, \phi) = \begin{pmatrix} \cos(\theta/2) & e^{-i\phi}\sin(\theta/2) \\ e^{i\phi}\sin(\theta/2) & -\cos(\theta/2) \end{pmatrix}.$$

Hence $V(\theta, \phi) \in U(2)$, but $V(\theta, \phi) \notin SU(2)$ since $\det(V(\theta, \phi)) = -1$.
Find $V^{-1}(\theta, \phi)dV(\theta, \phi)$. Utilize

$$\sin(\theta/2)\cos(\theta/2) \equiv \frac{1}{2}\sin(\theta), \quad \sin^2(\theta/2) \equiv \frac{1}{2}(1 - \cos(\theta)),$$

$$\cos^2(\theta/2) \equiv \frac{1}{2}(1 + \cos(\theta)).$$

Solution 68.   (i) We have

$$R^{-1}(\alpha) = \begin{pmatrix} \cos(\alpha) & \sin(\alpha) \\ -\sin(\alpha) & \cos(\alpha) \end{pmatrix}, \quad dR(\alpha) = \begin{pmatrix} -\sin(\alpha)d\alpha & -\cos(\alpha)d\alpha \\ \cos(\alpha)d\alpha & -\sin(\alpha)d\alpha \end{pmatrix}.$$

Then

$$R^{-1}(\alpha)dR(\alpha) = \begin{pmatrix} 0 & -d\alpha \\ d\alpha & 0 \end{pmatrix} = \begin{pmatrix} 0 & -1 \\ 1 & 0 \end{pmatrix} d\alpha$$

and $R^{-1}(\alpha)dR(\alpha) = (dR(\alpha))R^{-1}(\alpha)$. Thus the *left-invariant* matrix differential one-form is equal to the *right-invariant* matrix differential one-form.

(ii) We have

$$S^{-1}(\alpha) = \begin{pmatrix} \cos(\alpha) & \sin(\alpha) \\ \sin(\alpha) & -\cos(\alpha) \end{pmatrix} = S(\alpha)$$

$$dS(\alpha) = \begin{pmatrix} -\sin(\alpha)d\alpha & \cos(\alpha)d\alpha \\ \cos(\alpha)d\alpha & \sin(\alpha)d\alpha \end{pmatrix}.$$

Then

$$S^{-1}(\alpha)dS(\alpha) = \begin{pmatrix} 0 & d\alpha \\ -d\alpha & 0 \end{pmatrix} = \begin{pmatrix} 0 & 1 \\ -1 & 0 \end{pmatrix} d\alpha.$$

Note that $(dS(\alpha))S^{-1}(\alpha) = -S^{-1}dS(\alpha)$.

(iii) We have $W^{-1}(\phi) = W(\phi)$ and

$$W^{-1}(\phi)dW(\phi) = \begin{pmatrix} 0 & e^{-i\phi} \\ e^{i\phi} & 0 \end{pmatrix} \begin{pmatrix} 0 & -ie^{-i\phi}d\phi \\ ie^{i\phi}d\phi & 0 \end{pmatrix} = \begin{pmatrix} id\phi & 0 \\ 0 & -id\phi \end{pmatrix}$$

$$= \begin{pmatrix} i & 0 \\ 0 & -i \end{pmatrix} d\phi.$$

(iv) We have

$$U^{-1}(\theta, \phi) = \begin{pmatrix} \cos(\theta/2) & -e^{-i\phi}\sin(\theta/2) \\ -e^{i\phi}\sin(\theta/2) & \cos(\theta/2) \end{pmatrix}$$

$$dU(\theta, \phi) =$$

$$\begin{pmatrix} -\frac{1}{2}\sin(\theta/2)d\theta & ie^{-i\phi}\sin(\theta/2)d\phi - \frac{1}{2}e^{-i\phi}\cos(\theta/2)d\theta \\ ie^{i\phi}\sin(\theta/2)d\phi + \frac{1}{2}e^{i\phi}\cos(\theta/2)d\theta & -\frac{1}{2}\sin(\theta/2)d\theta \end{pmatrix}.$$

Then

$$U^{-1}(\theta,\phi)dU(\theta,\phi) = \frac{i}{2}\begin{pmatrix} (1-\cos(\theta))d\phi & e^{-i\phi}(id\theta+\sin(\theta)d\phi) \\ e^{i\phi}(-id\theta+\sin(\theta)d\phi) & -(1-\cos(\theta))d\phi \end{pmatrix}.$$

(v) We have

$$V^{-1}(\theta,\phi) = \begin{pmatrix} \cos(\theta/2) & e^{-i\phi}\sin(\theta/2) \\ e^{i\phi}\sin(\theta/2) & -\cos(\theta/2) \end{pmatrix} = V(\theta,\phi)$$

$$dV(\theta,\phi) =$$

$$\begin{pmatrix} -\frac{1}{2}\sin(\theta/2)d\theta & -ie^{-i\phi}\sin(\theta/2)d\phi+\frac{1}{2}e^{-i\phi}\cos(\theta/2)d\theta \\ ie^{i\phi}\sin(\theta/2)d\phi+\frac{1}{2}e^{i\phi}\cos(\theta/2)d\theta & \frac{1}{2}\sin(\theta/2)d\theta \end{pmatrix}.$$

Thus

$$V^{-1}(\theta,\phi)dV(\theta,\phi) = \frac{i}{2}\begin{pmatrix} (1-\cos(\theta))d\phi & e^{-i\phi}(-id\theta-\sin(\theta)d\phi) \\ e^{i\phi}(id\theta-\sin(\theta)d\phi) & -(1-\cos(\theta))d\phi \end{pmatrix}.$$

Programming Problem

Problem 69. Consider the optimization of the function $f(x,y)$ subject to the constraint $g(x,y) = 0$. We assume that $f(x,y)$ and $g(x,y)$ are smooth functions. The *Lagrange multiplier* method provides critical points for this problem when the partial derivatives of f and g with respect to x and y exist and

$$\frac{\partial g}{\partial x} \neq 0, \qquad \frac{\partial g}{\partial y} \neq 0.$$

The method consists of solving the equations

$$\frac{\partial f}{\partial x} = \lambda\frac{\partial g}{\partial x}, \qquad \frac{\partial f}{\partial y} = \lambda\frac{\partial g}{\partial y}, \qquad g(x,y) = 0$$

to obtain the critical points where λ is the Lagrange multiplier. Since

$$df = \frac{\partial f}{\partial x}dx + \frac{\partial f}{\partial y}dy, \qquad dg = \frac{\partial g}{\partial x}dx + \frac{\partial g}{\partial y}dy$$

we find

$$df \wedge dg = \left(\frac{\partial f}{\partial x}\frac{\partial g}{\partial y} - \frac{\partial f}{\partial y}\frac{\partial g}{\partial x}\right) dx \wedge dy$$

where we used $dx \wedge dx = dy \wedge dy = 0$ and $dy \wedge dx = -dx \wedge dy$. At the critical points we find

$$df \wedge dg = 0.$$

Thus to find the critical points it is sufficient to solve

$$df \wedge dg = 0, \qquad g(x, y) = 0.$$

Since $dx \wedge dy = -dy \wedge dx$ we treat dx and dy as noncommuting variables in the exterior product. Give a SymbolicC++ implementation to determine the equations to find the critical points of $f(x, y) = 2x^2 + y^2$ subject to the constraint $x + y = 1$.

Solution 69.   In SymbolicC++ `df(.)` denotes differentiation. The quantities dx and dy declared noncommutative and the rules $dx \wedge dx = 0$, $dy \wedge dy = 0$ and $dy \wedge dx = -dx \wedge dy$ are implemented.

```
// lagrange.cpp
#include <iostream>
#include "symbolicc++.h"
using namespace std;

int main(void)
{
 Symbolic x("x"), y("y"), dx("dx"), dy("dy");
 Symbolic f = 2*x*x+y*y;
 Symbolic g = x+y-1;
 dx = ~dx; dy = ~dy; // noncommutative
 cout << "f = " << f << endl;
 cout << "g = " << g << endl;
 Symbolic d_f = df(f,x)*dx + df(f,y)*dy;
 Symbolic d_g = df(g,x)*dx + df(g,y)*dy;
 cout << "d_f = " << d_f << endl;
 cout << "d_g = " << d_g << endl;
 Symbolic w = (d_f*d_g).subst_all((dx*dx==0,dy*dy==0,dy*dx==-dx*dy));
 cout << "w = " << w << endl;
 cout << (w.coeff(dx*dy)==0) << endl;
 return 0;
}
```

The output is `(4*x-2*y)==0`. Hence $y = 2x$.

4.3   Supplementary Problems

Problem 1.   Let $f_1 : \mathbb{R}^2 \to \mathbb{R}$, $f_2 : \mathbb{R}^2 \to \mathbb{R}$ given by

$$f_1(x_1, x_2) = x_1 x_2, \qquad f_2(x_1, x_2) = \frac{1}{2}(-x_1^2 + x_2^2).$$

Show that $df_1 = x_1 dx_2 + x_2 dx_1$, $df_2 = -x_1 dx_1 + x_2 dx_2$ and therefore with $dx_1 \wedge dx_2 = -dx_2 \wedge dx_1$ we have $df_1 \wedge df_2 = (x_1^2 + x_2^2) dx_1 \wedge dx_2$.

Problem 2.   Consider the differential one-form in $\mathbb{R}^2$

$$\alpha = (e^{2x_2} - x_2 \cos(x_1 x_2)) dx_1 + (2x_1 e^{2x_2} - x_1 \cos(x_1 x_2) + 2x_2) dx_2.$$

Show that $d\alpha = 0$. Find a smooth function $f : \mathbb{R}^2 \to \mathbb{R}$ such that $df = \alpha$.

Problem 3.   Consider the differential one-form in $\mathbb{R}^2$

$$\alpha = dx_1 - (x_1 + x_2^2) dx_2.$$

Then $d\alpha = -dx_1 \wedge dx_2$. Show that the integrating factor is e^{-x_2}, i.e. show that $d(e^{-x_2}\alpha) = 0$.

Problem 4.   Consider the differential one-forms

$$\alpha_1 = \cos(u_1 + u_2) du_1 + 2\sin^2(u_1) du_2, \qquad \alpha_2 = 3\sin^2(u_2) du_1 - 4du_2.$$

Find $d\alpha_1$, $d\alpha_2$, $\alpha_1 \wedge \alpha_2$ and $d(\alpha_1 \wedge \alpha_2)$.

Problem 5.   Let $M = \mathbb{R}^2$ and $\alpha = x_1 dx_2 - x_2 dx_1$. Then

$$d\alpha = 2dx_1 \wedge dx_2.$$

Now let $M = \mathbb{R}^2 \setminus \{(0,0)\}$. Consider the differential one-form

$$\beta = \frac{1}{x_1^2 + x_2^2}(x_1 dx_2 - x_2 dx_1).$$

Find $d\beta$ and show that

$$d(\arctan(x_2/x_1)) = \frac{1}{x_1^2 + x_2^2}(x_1 dx_2 - x_2 dx_1).$$

Problem 6.   Let $M = \mathbb{R}^2$ and $c > 0$. Consider the *elliptical coordinates*

$$x_1(\alpha, \beta) = c \cosh(\alpha) \cos(\beta), \quad x_2(\alpha, \beta) = c \sinh(\alpha) \sin(\beta).$$

Find the differential two-form $\Omega = dx_1 \wedge dx_2$ in this coordinate system. Note that

$$dx_1 = c(\sinh(\alpha)\cos(\beta)d\alpha - \cosh(\alpha)\sin(\beta)d\beta)$$

$$dx_2 = c(\cosh(\alpha)\sin(\beta)d\alpha + \sinh(\alpha)\cos(\beta)d\beta).$$

Problem 7.   (i) Consider the differential one-forms in $\mathbb{R}^3$

$$\alpha_1 = (x_2 + x_3)dx_1 + (x_3 + x_1)dx_2 + (x_1 + x_2)dx_3$$
$$\alpha_2 = x_2 x_3 dx_1 + x_1 x_3 dx_2 + x_1 x_2 dx_3.$$

Show that $d\alpha_1 = 0$, $d\alpha_2 = 0$ and

$$\alpha_1 \wedge \alpha_2 = x_3^2(x_1 - x_2)dx_1 \wedge dx_2 + x_1^2(x_2 - x_3)dx_2 \wedge dx_3 + x_2^2(x_3 - x_1)dx_3 \wedge dx_1.$$

(ii) Let a_1, a_2, a_3 be real constants. Consider the differential one-form in $\mathbb{R}^3$

$$\alpha = (a_2 \cos(x_2) + a_3 \sin(x_3))dx_1 + (a_1 \sin(x_1) + a_3 \cos(x_3))dx_2$$
$$+ (a_1 \cos(x_1) + a_2 \sin(x_2))dx_3.$$

Find $d\alpha$ and solve the equation $d\alpha = 0$.

Problem 8.   (i) Find nonzero differential one-forms α_1, α_2, α_3 in $\mathbb{R}^3$ with $d\alpha_1 = \alpha_2 \wedge \alpha_3$, $d\alpha_2 = \alpha_3 \wedge \alpha_1$, $d\alpha_3 = \alpha_1 \wedge \alpha_2$.
(ii) Consider the three differential one-forms in $\mathbb{R}^3$

$$\alpha_1 = f_{11}(\mathbf{x})dx_1 + f_{12}(\mathbf{x})dx_2 + f_{13}(\mathbf{x})dx_3$$
$$\alpha_2 = f_{21}(\mathbf{x})dx_1 + f_{22}(\mathbf{x})dx_2 + f_{23}(\mathbf{x})dx_3$$
$$\alpha_3 = f_{31}(\mathbf{x})dx_1 + f_{32}(\mathbf{x})dx_2 + f_{33}(\mathbf{x})dx_3.$$

Find the conditions on the functions f_{jk} such that

$$d\alpha_j = \sum_{k=1}^{3} \sum_{\ell=1}^{3} c_{jk\ell} dx_k \wedge dx_\ell$$

where $c_{123} = c_{321} = c_{132} = 1$, $c_{213} = c_{321} = c_{132} = -1$ and 0 otherwise.

Problem 9.   Consider the smooth one-forms in $\mathbb{R}^3$

$$\alpha_1 = f_1(\mathbf{x})dx_1 + f_2(\mathbf{x})dx_2 + f_3(\mathbf{x})dx_3$$
$$\alpha_2 = g_1(\mathbf{x})dx_1 + g_2(\mathbf{x})dx_2 + g_3(\mathbf{x})dx_3.$$

Find the differential equation from the condition $d(\alpha_1 \wedge \alpha_2) = 0$ and provide solution of it.

Problem 10. Consider the differential two-form in $\mathbb{R}^3$

$$\beta = b_{12}(\mathbf{x})dx_1 \wedge dx_2 + b_{23}(\mathbf{x})dx_2 \wedge dx_3 + b_{31}(\mathbf{x})dx_3 \wedge dx_1$$

and the three differential one-forms in $\mathbb{R}^3$

$$\alpha_j = \sum_{k=1}^{3} a_{jk}(\mathbf{x})dx_k, \quad j = 1, 2, 3.$$

Find the conditions on the functions b_{jk} and a_{jk} such that

$$d\beta = \alpha_1 \wedge \alpha_2 \wedge \alpha_3.$$

Problem 11. (i) Let $\mathbf{B}$ be a vector field in $\mathbb{R}^3$. Calculate

$$(\text{curl}(\mathbf{B})) \times \mathbf{B}$$

where $\times$ denotes the vector product. Formulate the problem with differential forms.

(ii) Find the curl of the vector field

$$V(x_1, x_2, x_3) = \begin{pmatrix} x_2 x_3 \\ x_1 x_3 \\ x_1 x_2 \end{pmatrix}.$$

Express the problem using differential forms in $\mathbb{R}^3$. Let $f : \mathbb{R}^3 \to \mathbb{R}$ be a smooth function and V be a vector field in $\mathbb{R}^3$. Show that

$$\text{curl}(fV) \equiv \nabla(f) \times V + f\text{curl}(V).$$

Let $\alpha = a_1(\mathbf{x})dx_1 + a_2(\mathbf{x})dx_2 + a_3(\mathbf{x})dx_3$ be a smooth differential form in $\mathbb{R}^3$. Find $d(f\alpha)$.

Problem 12. Consider the differential two-forms in $\mathbb{R}^3$

$$\beta_1 = x_1 dx_2 \wedge dx_3 + x_2 dx_3 \wedge dx_1 + x_3 dx_1 \wedge dx_2$$

$$\beta_2 = \frac{1}{1 + x_1^2 + x_2^2 + x_3^2}\beta_1.$$

Find $d\beta_1$ and $d\beta_2$.

Problem 13.   Consider the three differential one-forms in $\mathbb{R}^3$

$$\alpha_k = \sum_{\ell=1}^{3} f_{k\ell}(\mathbf{x})dx_\ell$$

with $k = 1, 2, 3$ and the nine differential one-forms

$$\Theta_{jk} = \sum_{n=1}^{3} g_{jkn}(\mathbf{x})dx_n.$$

Find the conditions on the functions $f_{k\ell}$ and g_{jkn} such that

$$d\alpha_j = \sum_{k=1}^{3} \Theta_{jk} \wedge \alpha_k.$$

Note that

$$d\alpha_j = \left(\frac{\partial f_{j2}}{\partial x_1} - \frac{\partial f_{j1}}{\partial x_2} \right) dx_1 \wedge dx_2 + \left(\frac{\partial f_{j3}}{\partial x_2} - \frac{\partial f_{j2}}{\partial x_3} \right) dx_2 \wedge dx_3$$
$$+ \left(\frac{\partial f_{j1}}{\partial x_3} - \frac{\partial f_{j3}}{\partial x_1} \right) dx_3 \wedge dx_1.$$

Problem 14.   Consider the *torus* in $\mathbb{R}^3$ given by

$$x_1(u_1, u_2) = (3 + \cos(u_1)) \cos(u_2)$$
$$x_2(u_1, u_2) = (3 + \cos(u_1)) \sin(u_2)$$
$$x_3(u_1, u_2) = \sin(u_1)$$

where $(u_1, u_2) \in (0, 2\pi) \times (0, 2\pi)$ and the differential one-forms on the torus

$$\alpha_1 = \cos(u_1 + u_2)du_1 + 2\sin^2(u_1)du_2, \qquad \alpha_2 = 3\sin^2(u_2)du_1 - 4du_2.$$

Find $\alpha_1 \wedge \alpha_2$ and $d\alpha_1, d\alpha_2$.

Problem 15.   Consider the smooth map $\mathbf{f} : \mathbb{R}^3 \to \mathbb{R}^3$

$$f_1(x_1, x_2, x_3) = x_1x_2 - x_3, \quad f_2(x_1, x_2, x_3) = x_1, \quad f_3(x_1, x_2, x_3) = x_2.$$

(i) Show that

$$\mathbf{f}^*(dx_1 \wedge dx_2 \wedge dx_3) = df_1 \wedge df_2 \wedge df_3 = -dx_1 \wedge dx_2 \wedge dx_3.$$

(ii) Show that the inverse of the map is

$$f_1^{-1}(x_1, x_2, x_3) = x_2$$
$$f_2^{-1}(x_1, x_2, x_3) = x_3$$
$$f_3^{-1}(x_1, x_2, x_3) = x_2 x_3 - x_1.$$

(iii) Find $\mathbf{f}^*(dx_1 \wedge dx_2)$, $\mathbf{f}^*(dx_2 \wedge dx_3)$, $\mathbf{f}^*(dx_3 \wedge dx_1)$. Discuss.

Problem 16. Consider a four-dimensional vector space (z_0, z_1, z_2, z_3) and the differential three-form

$$\gamma = \sum_{\alpha, \beta, \gamma, \delta = 0}^{3} \epsilon_{\alpha\beta\gamma\delta} z_\alpha dz_\beta \wedge dz_\gamma \wedge dz_\delta$$

where $\epsilon_{\alpha,\beta,\gamma,\delta}$ is the *Levi-Civita symbol* with

$$\epsilon_{\alpha,\beta,\gamma,\delta} := \begin{cases} +1 & \text{if} & (\alpha, \beta, \gamma, \delta) \text{ even permutation of } (0,1,2,3) \\ -1 & \text{if} & (\alpha, \beta, \gamma, \delta) \text{ odd permutation of } (0,1,2,3) \\ 0 & \text{otherwise} \end{cases}$$

Find $d\gamma$.

Problem 17. (i) Consider three-dimensional *spherical coordinates*

$$x_1(r, \theta, \phi) = r \sin(\theta) \cos(\phi)$$
$$x_2(r, \theta, \phi) = r \sin(\theta) \sin(\phi)$$
$$x_3(r, \theta, \phi) = r \cos(\theta)$$

where $0 \leq r < \infty$, $0 \leq \phi < 2\pi$, $0 \leq \theta \leq \pi$. Find

$$dx_1(r, \theta, \phi) \wedge dx_2(r, \theta, \phi) \wedge dx_3(r, \theta, \phi).$$

(ii) Consider four-dimensional *spherical coordinates*

$$x_1(r, \theta_1, \theta_2, \theta_3) = r \cos(\theta_1)$$
$$x_2(r, \theta_1, \theta_2, \theta_3) = r \sin(\theta_1) \cos(\theta_2)$$
$$x_3(r, \theta_1, \theta_2, \theta_3) = r \sin(\theta_1) \sin(\theta_2) \cos(\theta_3)$$
$$x_4(r, \theta_1, \theta_2, \theta_3) = r \sin(\theta_1) \sin(\theta_2) \sin(\theta_3)$$

where $0 \leq r < \infty$, $0 \leq \theta_1 < \pi$, $0 \leq \theta_2 < \pi$, $0 \leq \theta_3 < 2\pi$. Find

$$dx_1(r, \theta_1, \theta_2, \theta_3) \wedge dx_2(r, \theta_1, \theta_2, \theta_3) \wedge dx_3(r, \theta_1, \theta_2, \theta_3) \wedge dx_4(r, \theta_1, \theta_2, \theta_3).$$

Problem 18. (i) Consider the differential one-forms

$$\alpha_{12} := \frac{dz_1 - dz_2}{z_1 - z_2}, \quad \alpha_{23} := \frac{dz_2 - dz_3}{z_2 - z_3}, \quad \alpha_{31} := \frac{dz_3 - dz_1}{z_3 - z_1}.$$

Show that $\alpha_{12} \wedge \alpha_{23} + \alpha_{23} \wedge \alpha_{31} + \alpha_{31} \wedge \alpha_{12} = 0$.
(ii) Let $z_1, z_2 \in \mathbb{C}$. Consider the differential one-form

$$\omega = \frac{1}{2\pi i} \frac{dz_1 - dz_2}{z_1 - z_2}.$$

Find $d\omega$ and $\omega \wedge \omega$.

Problem 19. Let $\mathbf{f} : \mathbb{R}^2 \to \mathbb{R}^2$ be a two-dimensional analytic map.
(i) Find the condition on $\mathbf{f}$ such that $dx_1 \wedge dx_2$ is invariant, i.e. $\mathbf{f}$ should be area preserving.
(ii) Find the condition on $\mathbf{f}$ such that $x_1 dx_1 + x_2 dx_2$ is invariant.
(iii) Find the condition on $\mathbf{f}$ such that $x_1 dx_1 - x_2 dx_2$ is invariant.
(iv) Find the condition on $\mathbf{f}$ such that $x_1 dx_2 + x_2 dx_1$ is invariant.
(v) Find the condition on $\mathbf{f}$ such that $x_1 dx_2 - x_2 dx_1$ is invariant.

Problem 20. Consider the vector space $\mathbb{R}^3$. Find a differential one-form α such that $d\alpha \neq 0$ but $\alpha \wedge d\alpha = 0$.

Problem 21. Consider the differential one-form

$$\alpha = f_1(\mathbf{x})dx_1 + f_2(\mathbf{x})dx_2 + f_3(\mathbf{x})dx_3$$

in $\mathbb{R}^3$ with $f_j : \mathbb{R}^3 \to \mathbb{R}$ be analytic functions and the metric tensor field

$$g = dx_1 \otimes dx_1 + dx_2 \otimes dx_2 + dx_3 \otimes dx_3.$$

Note that with the given metric tensor field we have

$$\star dx_1 = dx_2 \wedge dx_3, \quad \star dx_2 = dx_3 \wedge dx_1, \quad \star dx_3 = dx_1 \wedge dx_2.$$

Furthermore

$$d\alpha = \left(\frac{\partial f_2}{\partial x_1} - \frac{\partial f_1}{\partial x_2}\right) dx_1 \wedge dx_2 + \left(\frac{\partial f_3}{\partial x_2} - \frac{\partial f_2}{\partial x_3}\right) dx_2 \wedge dx_3$$
$$+ \left(\frac{\partial f_1}{\partial x_3} - \frac{\partial f_3}{\partial x_1}\right) dx_3 \wedge dx_1$$

and

$$\star(d\alpha) = \left(-\frac{\partial f_1}{\partial x_2} + \frac{\partial f_2}{\partial x_1}\right) dx_3 + \left(-\frac{\partial f_2}{\partial x_3} + \frac{\partial f_3}{\partial x_2}\right) dx_1 + \left(-\frac{\partial f_1}{\partial x_3} - \frac{\partial f_3}{\partial x_1}\right) dx_2.$$

Find all α's such that $d\alpha = c(\star\alpha)$, where c is a nonzero constant. Find $d(\star(d\alpha))$.

Problem 22. Find smooth maps $\mathbf{f} : \mathbb{R}^3 \to \mathbb{R}^3$ such that

$$\mathbf{f}^*(dx_1 \wedge dx_2) = dx_2 \wedge dx_3$$
$$\mathbf{f}^*(dx_2 \wedge dx_3) = dx_3 \wedge dx_1$$
$$\mathbf{f}^*(dx_3 \wedge dx_1) = dx_1 \wedge dx_2.$$

Note that

$$df_1 = \left(\frac{\partial f_1}{\partial x_1} dx_1 + \frac{\partial f_1}{\partial x_2} dx_2 + \frac{\partial f_1}{\partial x_3} dx_3 \right)$$

$$df_2 = \left(\frac{\partial f_2}{\partial x_1} dx_1 + \frac{\partial f_2}{\partial x_2} dx_2 + \frac{\partial f_2}{\partial x_3} dx_3 \right)$$

$$df_3 = \left(\frac{\partial f_3}{\partial x_1} dx_1 + \frac{\partial f_3}{\partial x_2} dx_2 + \frac{\partial f_3}{\partial x_3} dx_3 \right)$$

and

$$\mathbf{f}^*(dx_1 \wedge dx_2) = df_1 \wedge df_2$$
$$= \left(\frac{\partial f_1}{\partial x_1} \frac{\partial f_2}{\partial x_2} - \frac{\partial f_1}{\partial x_2} \frac{\partial f_2}{\partial x_1} \right) dx_1 \wedge dx_2$$
$$+ \left(\frac{\partial f_1}{\partial x_1} \frac{\partial f_2}{\partial x_2} - \frac{\partial f_1}{\partial x_2} \frac{\partial f_2}{\partial x_1} \right) dx_1 \wedge dx_2$$
$$+ \left(\frac{\partial f_1}{\partial x_1} \frac{\partial f_2}{\partial x_2} - \frac{\partial f_1}{\partial x_2} \frac{\partial f_2}{\partial x_1} \right) dx_1 \wedge dx_2.$$

Problem 23. Consider the differential two-form in $\mathbb{R}^5$

$$\beta = f_{12} dx_1 \wedge dx_2 + f_{23} dx_2 \wedge dx_3 + f_{34} dx_3 \wedge dx_4 + f_{45} dx_4 \wedge dx_5 + f_{51} dx_5 \wedge dx_1$$

with the metric tensor field

$$g = \sum_{j=1}^{5} (dx_j \otimes dx_j)$$

and the $f_{jk} : \mathbb{R}^5 \to \mathbb{R}$ are smooth functions. Find the conditions of the f_{jk}'s such that

$$d\beta = c(\star\beta)$$

where c is a nonzero constant.

Problem 24.   Consider the Euclidean space $\mathbb{E}^4$ and the smooth differential two-form

$$\beta = f_{12}(\mathbf{x})dx_1 \wedge dx_2 + f_{13}(\mathbf{x})dx_1 \wedge dx_3 + f_{14}(\mathbf{x})dx_1 \wedge dx_4$$
$$+ f_{23}(\mathbf{x})dx_2 \wedge dx_3 + f_{24}(\mathbf{x})dx_2 \wedge dx_4 + f_{34}(\mathbf{x})dx_3 \wedge dx_4.$$

Let $\star$ be the Hodge duality operator.
(i) Find solutions of the system of partial differential equations

$$d(\star(d\beta)) = \beta.$$

(ii) Find solutions of the system of partial differential equations

$$d(\star(d\beta)) = \star\beta.$$

Problem 25.   Let α be the $(n-1)$ differential form on $\mathbb{R}^n$ given by

$$\alpha = \sum_{j=1}^{n}(-1)^{j-1}x_j dx_1 \wedge \cdots \wedge \widehat{dx_j} \wedge \cdots \wedge dx_n$$

where $\widehat{}$ indicates omission. Show that α is invariant under the orthogonal group of $\mathbb{R}^n$. Show that the volume form $\Omega = dx_1 \wedge \cdots \wedge dx_n$ is invariant under the orthogonal group.

Problem 26.   Let $n \geq 2$ and Ω be the volume form in $\mathbb{R}^n$

$$\Omega = dx_1 \wedge dx_2 \wedge \cdots \wedge dx_n.$$

(i) Find the condition on the smooth vector V in $\mathbb{R}^n$ such $V \rfloor \Omega = 0$.
(ii) Let V, W be two smooth vector fields in $\mathbb{R}^n$. Find the conditions on V, W such that

$$W \rfloor (V \rfloor \Omega) = 0.$$

Problem 27.   A *symplectic structure* on a $2n$-dimensional manifold M is a closed non-degenerate differential two-form ω such that $d\omega = 0$ and ω^n does not vanish. Every symplectic form is locally diffeomorphic to the standard differential form

$$\omega_0 = dx_1 \wedge dx_2 + dx_3 \wedge dx_4 + \cdots + dx_{2n-1} \wedge dx_{2n}$$

on $\mathbb{R}^{2n}$. Consider the vector field

$$V = x_1\frac{\partial}{\partial x_1} + x_2\frac{\partial}{\partial x_2} + \cdots + x_{2n}\frac{\partial}{\partial x_{2n}}$$

in $\mathbb{R}^{2n}$. Find $V\rfloor\omega_0$ and $L_V\omega_0$.

Problem 28. Let A be a differential one-form in space-time with the metric tensor field

$$g = -dx_0 \otimes dx_0 + dx_1 \otimes dx_1 + dx_2 \otimes dx_2 + dx_3 \otimes dx_3$$

with $x_0 = ct$. Let $F = dA$. Find $F \wedge \star F$, where $\star$ is the Hodge duality operator.

Problem 29. Given a *Lagrange function L*. Show that the *Cartan form* for a Lagrange function is given by

$$\alpha = L(\mathbf{x}, \mathbf{v}, t)dt + \sum_{j=1}^{n} \left(\frac{\partial L}{\partial v_j}(dx_j - v_j dt) \right). \tag{1}$$

Let

$$H = \sum_{j=1}^{n} v_j \frac{\partial L}{\partial v_j} - L, \qquad p_j = \frac{\partial L}{\partial v_j}. \tag{2}$$

Find the Cartan form for the Hamilton function.

Problem 30. Let $a > 0$. *Toroidal coordinates* are given by

$$x_1(\mu, \theta, \phi) = \frac{a \sinh(\mu) \cos(\phi)}{\cosh(\mu) - \cos(\theta)}$$

$$x_2(\mu, \theta, \phi) = \frac{a \sinh(\mu) \sin(\phi)}{\cosh(\mu) - \cos(\theta)}$$

$$x_3(\mu, \theta, \phi) = \frac{a \sin(\theta)}{\cosh(\mu) - \cos(\theta)}$$

where $0 < \mu < \infty$, $-\pi < \theta < \pi$, $0 < \phi < 2\pi$. Show that the volume element $dx_1 \wedge dx_2 \wedge dx_3$ using toroidal coordinates is given by

$$\frac{a^3}{(\cosh(\mu) - \cos(\theta))^3} \sinh(\mu)d\mu \wedge d\theta \wedge d\phi.$$

Problem 31. Let G be a Lie group with Lie algebra L. A differential form ω on G is called *left-invariant* if

$$f(x)^*\omega = \omega \tag{1}$$

for all $x \in G$, $f(x)$ denoting the left translation $g \to xg$ on G. Let X_1, ..., X_n be a basis of L and $\omega_1, \dots \omega_n$ the one-forms on G determined by

$$\omega_i(\widetilde{X}_j) \equiv \omega_i \rfloor \widetilde{X}_j = \delta_{ij} \tag{2}$$

where $\widetilde{X}_i$ are the corresponding left invariant vector fields on G and δ_{ij} is the Kronecker delta. Show that

$$d\omega_i = -\frac{1}{2} \sum_{j,k=1}^{n} c^i_{jk} \omega_j \wedge \omega_k, \qquad i = 1, \dots, n \tag{3}$$

where the *structure constants* c^i_{jk} are given by

$$[X_j, X_k] = \sum_{i=1}^{n} c^i_{jk} X_i. \tag{4}$$

System (3) is known as the *Maurer-Cartan equations*.

Problem 32.   Consider the non-compact Lie group $SU(1,1)$ and the compact Lie group $U(1)$. Let $z \in \mathbb{C}$ and $|z| < 1$. Consider the coset space $SU(1,1)/U(1)$ with the element ($\alpha \in \mathbb{R}$)

$$U(z, \alpha) = \frac{1}{\sqrt{1 - |z|^2}} \begin{pmatrix} 1 & -z \\ -\bar{z} & 1 \end{pmatrix} \begin{pmatrix} e^{i\alpha} & 0 \\ 0 & e^{-i\alpha} \end{pmatrix}.$$

The coset space $SU(1,1)/U(1)$ can be viewed as an open unit disc in the complex plane. Consider the Cartan differential one-forms

$$\mu = i\frac{\bar{z}dz - zd\bar{z}}{1 - |z|^2}, \quad \omega_+ = \frac{idz}{1 - |z|^2}, \quad \omega_- = -\frac{id\bar{z}}{1 - |z|^2}.$$

Show that (*Maurer-Cartan equations*)

$$d\mu = 2i\omega_- \wedge \omega_+, \quad d\omega_+ = i\mu \wedge \omega_+, \quad d\omega_- = -i\mu \wedge \omega_-.$$

Show that

$$\omega_+ \wedge \omega_- = \frac{1}{(1 - |z|^2)^2} dz \wedge d\bar{z}.$$

Problem 33.   Consider the compact Lie group $SU(3)$. Show that the group volume $V_{SU(3)}$ of $SU(3)$ with respect to the Cartan-Killing metric volume element is given by

$$V_{SU(3)} = \int_{SU(3)} (*1)$$

$$= 8\sqrt{3} \int_0^{2\pi} \int_0^{\pi/2} \int_0^{2\pi} \int_0^{\pi} \int_0^{\pi/2} \int_0^{2\pi} \int_0^{\pi/2} \int_0^{2\pi}$$
$$\sin(2\theta_2)\sin(2\theta_5)\sin^2(\theta_5)\sin(2\theta_7)d\theta_1 d\theta_2 d\theta_3 d\theta_4 d\theta_5 d\theta_6 d\theta_7 d\theta_8$$
$$= 64\sqrt{3}\pi^5.$$

Problem 34. Let $\mathbb{S}^n \subset \mathbb{R}^{n+1}$ be given by

$$\mathbb{S}^n := \{\, (x_1, \ldots, x_{n+1}) \,:\, x_1^2 + \cdots + x_{n+1}^2 = 1 \,\}.$$

Show that the invariant normalized n-differential form on $\mathbb{S}^n$ is given by

$$\omega = \frac{1}{2}\pi^{-n/2}\Gamma\left(\frac{n}{2}\right)\frac{dx_1 \wedge \cdots \wedge dx_n}{|x_{n+1}|}$$

where Γ denotes the gamma function.

Problem 35. A volume differential form on a manifold M of dimension n is an n-form Ω such that $\Omega(p) \neq 0$ at each point $p \in M$. Consider $M = \mathbb{R}^3$ (or an open set here) with coordinate system (x_1, x, x_3) with respect to the usual right-handed orthonormal frame. The volume differential form is defined as $\Omega = dx_1 \wedge dx_2 \wedge dx_3$. Any differential three-form can be written as $\gamma = f(x_1, x_2, x_3)dx_1 \wedge dx_2 \wedge dx_3$ for some function f. The integral of η is (if it exists)

$$\int_{\mathbb{R}^3} \gamma = \int_{\mathbb{R}^3} f(x_1, x_2, x_3)dx_1 dx_2 dx_3.$$

(i) Express Ω in terms of *spherical coordinates* (r, θ, ϕ) with $r \geq 0$, $0 \leq \phi < 2\pi$, $0 \leq \theta \leq \pi$

$$x_1(r, \theta, \phi) = r\sin(\theta)\cos(\phi)$$
$$x_2(r, \theta, \phi) = r\sin(\theta)\sin(\phi)$$
$$x_3(r, \theta, \phi) = r\cos(\theta).$$

(ii) Express Ω in terms of *prolate spherical coordinates* (ξ, η, ϕ) $(a > 0)$

$$x_1(\xi, \eta, \phi) = a\sinh(\xi)\sin(\eta)\cos(\phi)$$
$$x_2(\xi, \eta, \phi) = a\sinh(\xi)\sin(\eta)\sin(\phi)$$
$$x_3(\xi, \eta, \phi) = a\cosh(\xi)\cos(\eta).$$

Problem 36.   Let A_1, A_2 be two $n \times n$ matrices ($n \geq 2$) over $\mathbb{R}$. Consider the differential one-forms

$$\alpha_1 = (\, x_1 \quad \cdots \quad x_n \,) A_1 \begin{pmatrix} dx_1 \\ \vdots \\ dx_n \end{pmatrix}, \quad \alpha_2 = (\, x_1 \quad \cdots \quad x_n \,) A_2 \begin{pmatrix} dx_1 \\ \vdots \\ dx_n \end{pmatrix}.$$

(i) Assume that $\alpha_1 \wedge \alpha_2 = 0$. Can we conclude that $[A_1, A_2] = 0_n$?
(ii) Assume that $[A_1, A_2] = 0_n$. Can we conclude that $\alpha_1 \wedge \alpha_2 = 0$?
(iii) Let A be an $n \times n$ matrix over $\mathbb{R}$. Consider the differential one-form and vector field

$$\alpha = (\, x_1 \quad \cdots \quad x_n \,) A \begin{pmatrix} dx_1 \\ \vdots \\ dx_n \end{pmatrix}, \quad V = (\, x_1 \quad \cdots \quad x_n \,) A \begin{pmatrix} \partial/\partial x_1 \\ \vdots \\ \partial/\partial x_n \end{pmatrix}.$$

Find the function $V \rfloor \alpha$.

Problem 37.   Let $\star$ be the Hodge duality operator.
(i) Consider the three-dimensional Euclidean space and the differential one-form

$$\alpha = \sum_{j=1}^{3} a_j(\mathbf{x}) dx_j.$$

Find all solutions of $d\alpha = c(\star\alpha)$, where c is a nonzero constant.
(ii) Consider the five-dimensional Euclidean space and the differential two-form

$$\beta = \sum_{j=1}^{5} \sum_{k=1}^{5} b_{jk}(\mathbf{x}) dx_j \wedge dx_k.$$

Find all solutions of $d\beta = c(\star\beta)$, where c is a nonzero constant.
(iii) Consider the seven-dimensional Euclidean space and the differential three-form

$$\gamma = \sum_{j=1}^{7} \sum_{k=1}^{7} \sum_{\ell=1}^{7} \gamma_{jk\ell}(\mathbf{x}) dx_j \wedge dx_k \wedge dx_\ell.$$

Find all solutions of $d\gamma = c(\star\gamma)$, where c is a nonzero constant.

Problem 38.   Let α be smooth differential form in $\mathbb{R}^n$ with $n \geq 3$. Can we conclude that $\alpha \wedge d\alpha = -d\alpha \wedge \alpha$?

Problem 39.   Consider the differential one-form in $\mathbb{R}^2$

$$\alpha = \frac{1}{2}(x_1 dx_2 - x_2 dx_1).$$

Then $d\alpha = dx_1 \wedge dx_2$. Find

$$\frac{1}{2}\int_P (x_1(\tau)dx_2(\tau) - x_2(\tau)dx_1(\tau)) = \oint \alpha$$

with the closed path P given by the four linear segments

1. $x_1(\tau) = \tau, \quad x_2(\tau) = \frac{1}{2}\tau, \quad 0 \le \tau \le 1$

2. $x_1(\tau) = 1, \quad x_2(\tau) = \tau, \quad 1/2 \le \tau \le 1$

3. $x_1(\tau) = \tau. \; x_2(\tau) = 1, \quad 1 \ge \tau \ge 1/2$

4. $x_1(\tau) = \frac{1}{2}\tau, \quad x_2(\tau) = \tau, \quad 1 \ge \tau \ge 0.$

Problem 40. Consider the smooth vector fields and differential one-forms in $\mathbb{R}^n$

$$\alpha_1 = \sum_{j=1}^{n} a_{1,j}(\mathbf{x})dx_j, \quad \alpha_2 = \sum_{k=1}^{n} a_{2,k}(\mathbf{x})dx_k, \quad \alpha = \sum_{j=1}^{n} a_j(\mathbf{x})dx_j$$

$$V_1 = \sum_{\ell=1}^{n} v_{1,\ell}(\mathbf{x})\frac{\partial}{\partial x_\ell}, \quad V_2 = \sum_{m=1}^{n} v_{2,m}(\mathbf{x})\frac{\partial}{\partial x_m}.$$

(i) Find $\alpha_1 \wedge \alpha_2$, $d\alpha_1$, $d\alpha_2$, $[V_1, V_2]$.
(ii) Show that $V_2\rfloor(V_1\rfloor(\alpha_1 \wedge \alpha_2)) \equiv (V_1\rfloor\alpha_1)(V_2\rfloor\alpha_2) - (V_2\rfloor\alpha_1)(V_1\rfloor\alpha_2)$.
(iii) Show that $V_2\rfloor(V_1\rfloor d\alpha) \equiv V_1(V_2\rfloor\alpha) - V_2(V_1\rfloor\alpha) - ([V_1, V_2])\rfloor\alpha$. Note that one also finds the notation

$$d\alpha(V_1, V_2) \equiv V_1(V_2(\alpha)) - V_2(V_1(\alpha)) - \alpha([V_1, V_2]).$$

Problem 41. Let α be a smooth differential one-form in $\mathbb{R}^n$ and V, W be smooth vector fields in $\mathbb{R}^n$. Find the condition such that

$$[V, W]\rfloor\alpha = W\rfloor(V\rfloor d\alpha).$$

Problem 42. Consider the Euclidean space $\mathbb{E}^7$ with the metric tensor field

$$g = \sum_{j=1}^{7}(dx_j \otimes dx_j).$$

Let γ be a smooth differential three-form in $\mathbb{E}^7$. Find solutions of the differential equations $d\gamma = \star\gamma$, where $\star$ is Hodge duality operator.

Problem 43.   Find solutions of the *Pfaffian system*

$$dx_3 = x_1 dx_2 - x_2 dx_1, \quad dx_4 = x_2 dx_3 - x_3 dx_2, \quad dx_5 = x_3 dx_1 - x_1 dx_3.$$

Problem 44.   The sphere $\mathbb{S}^3$ can be considered as the $SU(2)$ Lie group manifold. Let $U \in SU(2)$ with

$$U = \begin{pmatrix} u & -\overline{v} \\ v & \overline{v} \end{pmatrix}, \quad u\overline{u} + v\overline{v} = 1.$$

A basis T_j $(j = 1, 2, 3)$ of the Lie algebra $su(2)$ $[T_j, T_k] = \sum_{\ell=1}^{3} \epsilon_{jk}^{\ell} T_\ell$ with $\epsilon_{12}^3 = 1$. Consider the *Maurer-Cartan differential one-form*

$$U^{-1} dU = \sum_{j=1}^{3} (T_j \otimes \alpha_j)$$

with

$$\alpha_1 = i(udv - vdu + \overline{v}d\overline{u} - \overline{u}d\overline{v})$$
$$\alpha_2 = udv - vdu + \overline{v}d\overline{u} + \overline{u}d\overline{v}$$
$$\alpha_3 = 2i(\overline{u}du + \overline{v}dv) = -2i(ud\overline{u} + vd\overline{v}).$$

Let

$$T_1 = \frac{i}{2}\left(\overline{v}\frac{\partial}{\partial u} - v\frac{\partial}{\partial \overline{u}} - \overline{u}\frac{\partial}{\partial v} + u\frac{\partial}{\partial \overline{v}}\right)$$

$$T_2 = \frac{1}{2}\left(-\overline{v}\frac{\partial}{\partial u} - v\frac{\partial}{\partial \overline{v}} + \overline{u}\frac{\partial}{\partial v} + u\frac{\partial}{\partial \overline{v}}\right)$$

$$T_3 = -\frac{i}{2}\left(u\frac{\partial}{\partial u} - \overline{u}\frac{\partial}{\partial \overline{u}} + v\frac{\partial}{\partial v} - \overline{v}\frac{\partial}{\partial \overline{v}}\right).$$

Show that

$$T_1 \lrcorner \alpha_1 = 1, \quad T_1 \lrcorner \alpha_2 = 0, \quad T_1 \lrcorner \alpha_3 = 0,$$
$$T_2 \lrcorner \alpha_1 = 0, \quad T_2 \lrcorner \alpha_2 = 1, \quad T_2 \lrcorner \alpha_3 = 0,$$
$$T_3 \lrcorner \alpha_1 = 0, \quad T_3 \lrcorner \alpha_2 = 0, \quad T_3 \lrcorner \alpha_3 = 1.$$

From the differential one-forms α_1, α_2, α_3 we can form the Cartan-Killing metric tensor field

$$g = \alpha_1 \otimes \alpha_1 + \alpha_2 \otimes \alpha_2 + \alpha_3 \otimes \alpha_3.$$

Find the Killing vector fields of g.

Problem 45. Consider the four-dimensional Euclidean space $\mathbb{R}^4$ with the metric tensor field

$$g = dx_1 \otimes dx_1 + dx_2 \otimes dx_2 + dx_3 \otimes dx_3 + dx_4 \otimes dx_4.$$

Let $\star$ be the Hodge duality operator. Find

$$\star(dx_1 \wedge dx_2), \quad \star(dx_1 \wedge dx_3), \quad \star(dx_1 \wedge dx_4),$$

$$\star(dx_2 \wedge dx_3), \quad \star(dx_2 \wedge dx_4), \quad \star(dx_3 \wedge dx_4).$$

Problem 46. Let $R > 0$ and $x_0 = ct$. Consider the *anti-de Sitter* metric tensor field

$$g = -\alpha_t \otimes \alpha_t + \alpha_r \otimes \alpha_r + \alpha_\theta \otimes \alpha_\theta + \alpha_\phi \otimes \alpha_\phi$$

with the differential one-forms (spherical orthonormal coframe)

$$\alpha_t = e^{\Phi(r)}dx_0, \quad \alpha_r = e^{-\Phi(r)}dr, \quad \alpha_\theta = rd\theta, \quad \alpha_\phi = r\sin(\theta)d\phi$$

where $e^{2\Phi_0(r)} = 1 + (r/R)^2$ and r, θ, ϕ are spherical coordinates. Note that

$$d\alpha_t = e^{\Phi(r)}\frac{d\Phi(r)}{dr}dr \wedge dx_0.$$

Show that the Riemannian curvature two-form

$$\Omega_{\alpha,\beta} = -\frac{1}{R^2}\omega_\beta \wedge \omega_\alpha, \quad \alpha, \beta \in \{x_0, r, \theta, \phi\}$$

is that of a constant negative curvature space with radius of curvature R. Show that the underlying Riemannian geometry has vanishing torsion.

Problem 47. Consider the differential one-forms

$$\alpha_{x_0} = e^{\mu(x_0, R)}dx_0, \quad \alpha_R = e^{\nu(x_0, R)}dR$$
$$\alpha_\theta = r(x_0, R)d\theta, \quad \alpha_\phi = r(x_0.R)\sin(\theta)d\phi$$

where θ and ϕ are coordinates on the sphere $\mathbb{S}^2$. The corresponding metric tensor field (with signature $(-, +, +, +)$) is

$$g = -\alpha_{x_0} \otimes \alpha_{x_0} + \alpha_R \otimes \alpha_R + \alpha_\theta \otimes \alpha_\theta + \alpha_\phi \otimes \alpha_\phi.$$

Find $d\alpha_{x_0}$, $d\alpha_R$, $d\alpha_\theta$, $d\alpha_\phi$.

Problem 48.   Consider the differential two-form in $\mathbb{R}^4$

$$\beta = a_{12}(\mathbf{x})dx_1 \wedge dx_2 + a_{13}(\mathbf{x})dx_1 \wedge dx_3 + a_{14}(\mathbf{x})dx_1 \wedge dx_4$$
$$+ a_{23}(\mathbf{x})dx_2 \wedge dx_3 + a_{24}(\mathbf{x})dx_2 \wedge dx_4 + a_{34}(\mathbf{x})dx_3 \wedge dx_4$$

where $a_{jk} : \mathbb{R}^4 \to \mathbb{R}$ are smooth functions. Find $d\beta$ and the conditions from $d\beta = 0$.

Problem 49.   (i) Consider the differential one-forms in $\mathbb{R}^n$

$$\alpha_1 = \sum_{j=1}^{n} x_j dx_j$$
$$\alpha_2 = x_2 dx_1 + x_3 dx_2 + \cdots + x_n dx_{n-1} + x_1 dx_n.$$

Find the differential two-forms $d\alpha_1$ and $d\alpha_2$. Find $\alpha_1 \wedge \alpha_2$ and then $d(\alpha_1 \wedge d\alpha_2)$.
(ii) Consider the manifold $M = \mathbb{R}^n$, $f : \mathbb{R}^n \to \mathbb{R}$

$$f(\mathbf{x}) = \frac{1}{2}(x_1^2 + \cdots + x_n^2)$$

$$\alpha_1 = \sum_{j=1}^{n} x_j dx_j, \qquad \alpha_2 = x_2 dx_1 + x_3 dx_2 + \cdots + x_n dx_{n-1} + x_1 dx_n.$$

Find df, $d\alpha_1$, $d\alpha_2$, $\alpha_1 \wedge \alpha_2$, $d(\alpha_1 \wedge \alpha_2)$.

Problem 50.   Consider the differential one-form in space-time

$$\alpha = a_0(\mathbf{x})dx_0 + a_1(\mathbf{x})dx_1 + a_2(\mathbf{x})dx_2 + a_3(\mathbf{x})dx_3$$

with $\mathbf{x} = (x_0, x_1, x_2.x_3)$ $(x_0 = ct)$.
(i) Find the conditions on the a_j's such that $d\alpha = 0$.
(ii) Find the conditions on the a_j's such that $d\alpha \neq 0$ and $\alpha \wedge d\alpha = 0$.
(iii) Find the conditions on the a_j's such that $\alpha \wedge d\alpha \neq 0$ and $d\alpha \wedge d\alpha = 0$.
(iv) Find the conditions on the a_j's such that $d\alpha \wedge d\alpha \neq 0$.
(v) Consider the metric tensor field

$$g = -dx_0 \otimes dx_0 + dx_1 \otimes dx_1 + dx_2 \otimes dx_2 + dx_3 \otimes dx_3.$$

Find the condition such that

$$d(\star\alpha) = 0$$

where $\star$ denotes the Hodge duality operator.

Problem 51. We consider the case where $M = \mathbb{R}^2$ and $L = s\ell(2, \mathbb{R})$. In local coordinates (x, t) a Lie algebra-valued differential one-form is given by

$$\tilde{\alpha} = \sum_{j=1}^{3} (\alpha_j \otimes T_j)$$

where

$$\alpha_j := a_j(x, t)dx + A_j(x, t)dt$$

and $\{T_1, T_2, T_3\}$ is a basis of the semi-simple Lie algebra $s\ell(2, \mathbb{R})$. A choice is

$$T_1 = \begin{pmatrix} 1 & 0 \\ 0 & -1 \end{pmatrix}, \quad T_2 = \begin{pmatrix} 0 & 1 \\ 0 & 0 \end{pmatrix}, \quad T_3 = \begin{pmatrix} 0 & 0 \\ 1 & 0 \end{pmatrix}.$$

(i) Show that the condition that the covariant derivative vanishes

$$D_{\tilde{\alpha}}\tilde{\alpha} = 0$$

leads to the following systems of partial differential equations of first order

$$-\frac{\partial a_1}{\partial t} + \frac{\partial A_1}{\partial x} + a_2 A_3 - a_3 A_2 = 0$$

$$-\frac{\partial a_2}{\partial t} + \frac{\partial A_2}{\partial x} + 2(a_1 A_2 - a_2 A_1) = 0$$

$$\frac{\partial a_3}{\partial t} + \frac{\partial A_3}{\partial x} - 2(a_1 A_3 - a_3 A_1) = 0.$$

(ii) Show that the *sine-Gordon equation*

$$\frac{\partial^2 u}{\partial t^2} - \frac{\partial^2 u}{\partial x^2} + \sin(u) = 0$$

can be represented as follows

$$a_2 = -\frac{1}{4}(\cos(u) + 1), \quad A_1 = \frac{1}{4}(\cos(u) - 1)$$

$$a_2 = \frac{1}{4}\left(\frac{\partial u}{\partial x} + \frac{\partial u}{\partial t} - \sin u\right), \quad A_2 = \frac{1}{4}\left(\frac{\partial u}{\partial x} + \frac{\partial u}{\partial t} + \sin u\right)$$

$$a_3 = -\frac{1}{4}\left(\frac{\partial u}{\partial x} + \frac{\partial u}{\partial t} + \sin(u)\right), \quad A_4 = -\frac{1}{4}\left(\frac{\partial u}{\partial x} + \frac{\partial u}{\partial t} - \sin(u)\right).$$

Problem 52. Let $f_{jk\ell} : \mathbb{R}^3 \to \mathbb{R}$ be smooth functions. Consider the matrix valued differential one-form

$$A = \begin{pmatrix} f_{111}dx_1 + f_{112}dx_2 + f_{113}dx_3 & f_{121}dx_1 + f_{122}dx_2 + f_{123}dx_3 \\ f_{211}dx_1 + f_{212}dx_2 + f_{213}dx_3 & f_{221}dx_1 + f_{222}dx_2 + f_{223}dx_3 \end{pmatrix}.$$

Find dA, $A \wedge dA$, $A \wedge A \wedge A$ and then

$$\operatorname{tr}(A \wedge dA + \frac{2}{3} A \wedge A \wedge A).$$

Find solutions of the partial differential equation

$$\operatorname{tr}(A \wedge dA + \frac{2}{3} A \wedge A \wedge A) = 0.$$

Chapter 5

Lie Derivative and Applications

5.1 Notations and Definitions

The concept of the Lie derivative $L_V(.)$ of functions, vector fields, differential forms, tensor fields and currents with respect to a smooth vector field V plays an important role in many domains in physics such as classical mechanics, hydrodynamics, optics, quantum mechanics, theory of relativity, supergravity and string theory. The Lie derivative is also the core in the study of symmetries of ordinary and partial differential equations.

Let M be an n-dimensional C^∞ differentiable manifold with local coordinates x_j, $j = 1, \ldots, n$. Real valued C^∞ vector fields and real valued C^∞ differential forms on M can be considered. The components of the vector field V are denoted by $V_j(\mathbf{x})\partial/\partial x_j$. This means

$$V := V_1(\mathbf{x})\frac{\partial}{\partial x_1} + V_2(\mathbf{x})\frac{\partial}{\partial x_2} + \cdots + V_n(\mathbf{x})\frac{\partial}{\partial x_n} = \sum_{j=1}^{n} V_j(\mathbf{x})\frac{\partial}{\partial x_j}$$

in local coordinates. The *Lie derivative* of a differential form γ with respect to V is defined by the derivative of γ along the integral curve

$\tau \mapsto \Phi_\tau$ of V, i.e.

$$L_V\gamma := \lim_{\tau \to 0} \frac{\Phi_\tau^*\gamma - \gamma}{\tau}.$$

It can be shown that this can be written as

$$L_V\gamma := d(V\rfloor\gamma) + V\rfloor(d\gamma)$$

where $d\gamma$ is the exterior derivative of the differential form γ and $V\rfloor\gamma$ is the *contraction* of γ by V (*interior product*). In local coordinates we have

$$\frac{\partial}{\partial x_j}\rfloor dx_k = \delta_{jk}$$

where δ_{jk} denotes the Kronecker delta and

$$V\rfloor df \equiv V(f) = \sum_{j=1}^n V_j \frac{\partial f}{\partial x_j}$$

where f is a smooth function. Furthermore, we have the *product rule*

$$V\rfloor(\alpha \wedge \beta) = (V\rfloor\alpha) \wedge \beta + (-1)^r \alpha \wedge (V\rfloor\beta)$$

where α is an r-form. The linear operators $d(.)$, $V\rfloor$ and $L_V(.)$ are coordinate-free operators. Note that d and L_V commute, i.e.

$$d(L_V\gamma) = L_V(d\gamma).$$

In local coordinates we have

$$L_V dx_j = dV_j = \sum_{k=1}^n \frac{\partial V_j}{\partial x_k} dx_k, \quad L_V\left(\frac{\partial}{\partial x_j}\right) = \left[V, \frac{\partial}{\partial x_j}\right] = -\sum_{k=1}^n \frac{\partial V_k}{\partial x_j}\frac{\partial}{\partial x_k}$$

where $j = 1, \ldots, n$.

The requirement that a smooth differential form γ be a *conformal invariant* with respect to a smooth vector field V is given by

$$L_V\gamma = f(\mathbf{x})\gamma$$

where $f \in C^\infty(M)$. If $f = 0$, the we call γ *invariant* with respect to V. Let g be a metric tensor field and V be a vector field. Then V is called a *Killing vector field* if

$$L_V g = 0$$

i.e. the Lie derivative of g with respect to V vanishes. If V and W are Killing vector fields, i.e. $L_V g = 0$ and $L_W g = 0$, then $[W, V]$ is also a Killing vector field, i.e. $L_{[V,W]} g = 0$. Hence the Killing vector fields provide a basis of a Lie algebra.

5.2 Solved Problems

Problem 1. (i) Let $f : \mathbb{R} \to \mathbb{R}$ be a smooth function. Consider the vector field $V = f(x)d/dx$. Find

$$L_V(dx \otimes dx), \quad L_V\left(\frac{d}{dx} \otimes \frac{d}{dx}\right), \quad L_V\left(\frac{d}{dx} \otimes dx\right), \quad L_V\left(dx \otimes \frac{d}{dx}\right).$$

(ii) Let $n \geq 2$. Consider the smooth vector field in $\mathbb{R}^n$

$$V = \sum_{j=1}^{n} V_j(\mathbf{x})\frac{\partial}{\partial x_j}.$$

Find the Lie derivative of the tensor fields

$$\frac{\partial}{\partial x_k} \otimes \frac{\partial}{\partial x_\ell}, \quad dx_k \otimes \frac{\partial}{\partial x_\ell}, \quad dx_k \otimes dx_\ell$$

with $j, k, \ell = 1, \dots, n$.

Solution 1. (i) We have

$$L_V dx = d(V \rfloor dx) = df = \frac{df}{dx}dx$$

$$L_V\left(\frac{d}{dx}\right) = [V, d/dx] = -\frac{df}{dx}\frac{d}{dx}$$

and applying the *product rule* provides

$$L_V(dx \otimes dx) = (L_V dx) \otimes dx + dx \otimes L_V dx = 2\frac{df}{dx}dx \otimes dx$$

$$L_V\left(\frac{d}{dx} \otimes \frac{d}{dx}\right) = [V, d/dx] \otimes \frac{d}{dx} = \frac{d}{dx} \otimes [V, d/dx] = -2\frac{df}{dx}\left(\frac{d}{dx} \otimes \frac{d}{dx}\right)$$

$$L_V\left(\frac{d}{dx} \otimes dx\right) = [V, d/dx] \otimes dx + \frac{d}{dx} \otimes L_V dx = 0$$

$$L_V\left(dx \otimes \frac{d}{dx}\right) = (L_V dx) \otimes \frac{d}{dx} + dx \otimes ([V, d/x]) = 0.$$

(ii) Utilizing

$$L_V(dx_k) = d(L_V x_k) = dV_k = \sum_{\ell=1}^{n} \frac{\partial V_k}{\partial x_\ell}dx_\ell$$

$$L_V\left(\frac{\partial}{\partial x_k}\right) = \sum_{j=1}^{n}\left[V_j\frac{\partial}{\partial x_j}, \frac{\partial}{\partial x_k}\right] = -\sum_{j=1}^{n}\frac{\partial V_j}{\partial x_k}\frac{\partial}{\partial x_j}$$

and the product rule

$$L_V\left(\frac{\partial}{\partial x_k} \otimes \frac{\partial}{\partial x_\ell}\right) = L_V\left(\frac{\partial}{\partial x_k}\right) \otimes \frac{\partial}{\partial x_\ell} + \frac{\partial}{\partial x_k} \otimes \left(L_V \frac{\partial}{\partial x_\ell}\right)$$

$$L_V\left(dx_k \otimes \frac{\partial}{\partial x_\ell}\right) = (L_V dx_k) \otimes \frac{\partial}{\partial x_\ell} + dx_k \otimes \left(L_V \frac{\partial}{\partial x_\ell}\right)$$

$$L_V(dx_k \otimes dx_\ell) = (L_V dx_k) \otimes dx_\ell + dx_k \otimes (L_V dx_\ell)$$

we obtain

$$L_V\left(\frac{\partial}{\partial x_k} \otimes \frac{\partial}{\partial x_\ell}\right) = \left(-\sum_{j=1}^n \frac{\partial V_j}{\partial x_k}\frac{\partial}{\partial x_j}\right) \otimes \frac{\partial}{\partial x_\ell} + \frac{\partial}{\partial x_k} \otimes \left(-\sum_{j=1}^n \frac{\partial V_j}{\partial x_\ell}\frac{\partial}{\partial x_j}\right)$$

$$L_V\left(dx_k \otimes \frac{\partial}{\partial x_\ell}\right) = \left(\sum_{j=1}^n \frac{\partial V_k}{x_j} dx_j\right) \otimes \frac{\partial}{\partial x_\ell} + dx_k \otimes \left(-\sum_{j=1}^n \frac{\partial V_j}{\partial x_\ell}\frac{\partial}{\partial x_j}\right)$$

$$L_V(dx_k \otimes dx_\ell) = \left(\sum_{j=1}^n \frac{\partial V_k}{\partial x_j} dx_j\right) \otimes dx_\ell + dx_k \otimes \left(\sum_{j=1}^n \frac{\partial V_\ell}{\partial x_j} dx_j\right).$$

Problem 2.   Consider the vector field in $\mathbb{R}^2$

$$V = x_1 \frac{\partial}{\partial x_2} - x_2 \frac{\partial}{\partial x_1}.$$

(i) Let $f : \mathbb{R}^2 \to \mathbb{R}$, $f(x_1, x_2) = x_1^2 + x_2^2$. Find $L_V f$.
(ii) Let $\alpha_1 = x_1 dx_2 + x_2 dx_1$. Find $L_V \alpha_1$.
(iii) Let $\alpha_2 = x_1 dx_1 + x_2 dx_2$. Find $L_V \alpha_2$.
(iv) Let $\Omega = dx_1 \wedge dx_2$. Find $L_V \Omega$.
(v) Let $W = x_1 \frac{\partial}{\partial x_1} + x_2 \frac{\partial}{\partial x_2}$. Find $L_V W \equiv [V, W]$.
(vi) Let $g = dx_1 \otimes dx_1 + dx_2 \otimes dx_2$. Find $L_V g$.

Solution 2.   (i) We find

$$L_V f = V(f) = \left(x_1 \frac{\partial}{\partial x_2} - x_2 \frac{\partial}{\partial x_1}\right)(x_1^2 + x_2^2)$$
$$= 2x_1 x_2 - 2x_2 x_1 = 0.$$

(ii) We find

$$
\begin{aligned}
L_V \alpha_1 &= L_V(x_1 dx_2) + L_V(x_2 dx_1) \\
&= (L_V x_1) dx_2 + x_1 L_V(dx_2) + (L_V x_2) dx_1 + x_2 L_V(dx_1) \\
&= -x_2 dx_2 + x_1 dx_1 + x_1 dx_1 - x_2 dx_2 \\
&= 2x_1 dx_1 - 2x_2 dx_2.
\end{aligned}
$$

(iii) We have

$$
\begin{aligned}
L_V \alpha_2 &= L_V(x_1 dx_1) + L_V(x_2 dx_2) \\
&= (L_V x_1) dx_1 + x_1 (L_V dx_1) + (L_V x_2) dx_2 + x_2 (L_V dx_2) \\
&= -x_2 dx_1 - x_1 dx_2 + x_1 dx_2 + x_2 dx_1 \\
&= 0
\end{aligned}
$$

i.e. α_2 is invariant under V.

(iv) Since $L_V dx_1 = -dx_2$ and $L_V dx_2 = dx_1$ we have applying the product rule

$$
L_V(dx_1 \wedge dx_2) = (L_V dx_1) \wedge dx_2 + dx_1 \wedge (L_V dx_2) = 0.
$$

(v) We obtain

$$
L_V W \equiv [V, W] = x_1 \frac{\partial}{\partial x_2} - x_2 \frac{\partial}{\partial x_1} - x_1 \frac{\partial}{\partial x_2} + x_2 \frac{\partial}{\partial x_1} = 0.
$$

(vi) Since

$$
L_V(dx_1) = d(L_V x_1) = -dx_2, \qquad L_V(dx_2) = d(L_V x_2) = dx_1
$$

we obtain

$$
\begin{aligned}
L_V g &= (L_V dx_1) \otimes dx_1 + dx_1 \otimes (L_V dx_1) + (L_V dx_2) \otimes dx_2 + dx_2 \otimes (dx_2) \\
&= -dx_2 \otimes dx_1 + dx_1 \otimes (-dx_2) + dx_1 \otimes dx_2 + dx_2 \otimes dx_1 \\
&= 0.
\end{aligned}
$$

Thus V is a Killing vector field for the given g.

Problem 3. Consider the autonomous system of first order differential equations

$$
\frac{dx_0}{d\tau} = x_1, \qquad \frac{dx_1}{d\tau} = x_0
$$

with the corresponding vector field

$$
V = x_1 \frac{\partial}{\partial x_0} + x_0 \frac{\partial}{\partial x_1}.
$$

(i) Find $L_V(x_0^2 - x_1^2)$.
(ii) Find

$$L_V\left(x_0\frac{\partial}{\partial x_0} + x_1\frac{\partial}{\partial x_1}\right).$$

(iii) Find $L_V(dx_0 \wedge dx_1)$.
(iv) Find $L_V(dx_0 \otimes dx_0 - dx_1 \otimes dx_1)$.
(v) Let $f_1, f_2 : \mathbb{R} \to \mathbb{R}$ be smooth functions. Consider the vector field

$$W = \left(x_0 f_2(x_0^2 - x_1^2) + x_1 f_1(x_0^2 - x_1^2)\right)\frac{\partial}{\partial x_0}$$

$$+ \left(x_1 f_2(x_0^2 - x_1^2) + x_0 f_1(x_0^2 - x_1^2)\right)\frac{\partial}{\partial x_1}$$

where $f_1, f_2 : \mathbb{R} \to \mathbb{R}$ are smooth functions. Find the commutator $[V, W]$.

Solution 3.   (i) We obtain $L_V(x_0^2 - x_1^2) = 0$.
(ii) We obtain

$$L_V\left(x_0\frac{\partial}{\partial x_0} + x_1\frac{\partial}{\partial x_1}\right) = 0.$$

(iii) We obtain $L_V(dx_0 \wedge dx_1) = 0$.
(iv) We obtain $L_V(dx_0 \otimes dx_0 - dx_1 \otimes dx_1) = 0$. Thus V is a Killing vector field of $g = dx_0 \otimes dx_0 - dx_1 \otimes dx_1$.
(v) We obtain $[V, W] = 0$. This condition is an *integrability condition*.

Problem 4.   Consider the metric tensor field

$$g = dx_1 \otimes dx_2 + dx_2 \otimes dx_1$$

of $\mathbb{E}^2$ and the vector field

$$V = x_1\frac{\partial}{\partial x_1} - x_2\frac{\partial}{\partial x_2}.$$

Find $L_V g$ and $L_V(dx_1 \wedge dx_2)$.

Solution 4.   With

$$L_{x_j \partial/\partial x_k}(dx_\ell \otimes dx_m) = \delta_{k\ell}dx_j \otimes dx_m + \delta_{km}dx_\ell \otimes dx_j$$

and linearity we obtain $L_V g = 0$. Hence $L_V(dx_1 \wedge dx_2) = 0$.

Problem 5.   (i) Find smooth vector fields in $\mathbb{R}^2$

$$V = f_1(x_1, x_2)\frac{\partial}{\partial x_1} + f_2(x_1, x_2)\frac{\partial}{\partial x_2}$$

such that $L_V dx_1 = dx_2$, $L_V dx_2 = dx_1$.

(ii) Apply these vector fields to find $L_V(dx_1 \wedge dx_2)$.

Solution 5.   (i) We have

$$L_V dx_1 = d(L_V x_1) = df_1 = \sum_{j=1}^{2} \frac{\partial f_1}{\partial x_j} dx_j$$

$$L_V dx_2 = d(L_V x_2) = df_2 = \sum_{j=1}^{2} \frac{\partial f_2}{\partial x_j} dx_j.$$

Thus

$$\frac{\partial f_1}{\partial x_1} = 0, \quad \frac{\partial f_1}{\partial x_2} = 1, \quad \frac{\partial f_2}{\partial x_1} = 1, \quad \frac{\partial f_2}{\partial x_2} = 0.$$

Hence

$$V = x_2 \frac{\partial}{\partial x_1} + x_1 \frac{\partial}{\partial x_2}.$$

(ii) Using the product rule we find

$$L_V(dx_1 \wedge dx_2) = (L_V dx_1) \wedge dx_2 + dx_1 \wedge (L_V dx_2) = 0.$$

Problem 6.   Consider the smooth vector field in $\mathbb{R}^2$

$$V = V_1(x_1, x_2) \frac{\partial}{\partial x_1} + V_2(x_1, x_2) \frac{\partial}{\partial x_2}$$

and the differential two-form $\Omega = dx_1 \wedge dx_2$. Find the equation $d(V \rfloor \Omega) = 0$. One also writes $d(\Omega(V)) = 0$. Calculate the Lie derivative $L_V \Omega$.

Solution 6.   We have $d\Omega = 0$. Thus $L_V \Omega = d(V \rfloor \Omega)$. Now

$$V \rfloor \Omega = V_1(x_1, x_2) dx_2 - V_2(x_1, x_2) dx_1.$$

Thus

$$L_V \Omega = d(V \rfloor \Omega) = \left(\frac{\partial V_1}{\partial x_1} + \frac{\partial V_2}{\partial x_2} \right) dx_1 \wedge dx_2 = \text{div}(V) dx_1 \wedge dx_2.$$

Problem 7.   Consider the two-dimensional Euclidean space $\mathbb{E}^2$ with the metric tensor field $g = dx_1 \otimes dx_1 + dx_2 \otimes dx_2$. Let $\star$ be the *Hodge duality operator*. Then $\star dx_1 = dx_2$, $\star dx_2 = -dx_1$.

(i) Find

$$L_V(\star dx_1) \quad \text{and} \quad \star(L_V dx_1)$$

and thus show that $L_V(\star dx_1) \neq \star L_V(dx_1)$ in general.

(ii) Let V be a smooth vector field in $\mathbb{E}^2$. Find the condition on V such that $L_V(\star dx_1) = \star(L_V dx_1)$.

Solution 7.   (i) With

$$V = V_1(x_1, x_2)\frac{\partial}{\partial x_1} + V_2(x_1, x_2)\frac{\partial}{\partial x_2}$$

we have

$$L_V(\star dx_1) = L_V(dx_2) = dV_2 = \frac{\partial V_2}{\partial x_1}dx_1 + \frac{\partial V_2}{\partial x_2}dx_2$$

and

$$\star(L_V dx_1) = \star(dV_1) = \frac{\partial V_1}{\partial x_1}dx_2 - \frac{\partial V_1}{\partial x_2}dx_1.$$

(ii) Since $d(dx_1) = 0$ and $d(dx_2) = 0$ we have

$$L_V(\star dx_1) = L_V dx_2 = dV_2 = \frac{\partial V_2}{\partial x_1}dx_1 + \frac{\partial V_2}{\partial x_2}dx_2$$

and

$$\star(L_V dx_1) = \star(d(V \rfloor dx_1)) = \star(dV_1) = \frac{\partial V_1}{\partial x_1}dx_2 - \frac{\partial V_1}{\partial x_2}dx_1.$$

Hence the condition is

$$\frac{\partial V_2}{\partial x_1} = \frac{\partial V_1}{\partial x_1}, \quad \frac{\partial V_2}{\partial x_2} = -\frac{\partial V_1}{\partial x_2}.$$

Problem 8.   Consider the two-dimensional Euclidean space with the metric tensor field $g = dx_1 \otimes dx_1 + dx_2 \otimes dx_2$. Find the Killing vector fields, i.e. the analytic vector fields

$$V = V_1(x_1, x_2)\frac{\partial}{\partial x_1} + V_2(x_1, x_2)\frac{\partial}{\partial x_2}$$

such that $L_V g = 0$. Show that the set of Killing vector fields form a basis of a Lie algebra under the commutator.

Solution 8. Using the sum rule and product rule for the Lie derivative provides

$$
\begin{aligned}
L_V g &= L_V(dx_1 \otimes dx_1 + dx_2 \otimes dx_2) \\
&= L_V(dx_1 \otimes dx_1) + L_V(dx_2 \otimes dx_2) \\
&= (L_V dx_1) \otimes dx_1 + dx_1 \otimes (L_V dx_1) + (L_V dx_2) \otimes dx_2 + dx_2 \otimes (L_V dx_2) \\
&= dV_1 \otimes dx_1 + dx_1 \otimes dV_1 + dV_2 \otimes dx_2 + dx_2 \otimes dV_2.
\end{aligned}
$$

Thus we obtain the equation

$$
2\frac{\partial V_1}{\partial x_1} dx_1 \otimes dx_1 + \left(\frac{\partial V_1}{\partial x_2} + \frac{\partial V_2}{\partial x_1} \right) dx_1 \otimes dx_2
$$
$$
+ \left(\frac{\partial V_1}{\partial x_2} + \frac{\partial V_2}{\partial x_1} \right) dx_2 \otimes dx_1 + 2\frac{\partial V_2}{\partial x_2} dx_2 \otimes dx_2 = 0.
$$

Separating out the terms with $dx_1 \otimes dx_1$, $dx_1 \otimes dx_2$, $dx_2 \otimes dx_1$, $dx_2 \otimes dx_2$ it follows that

$$
\frac{\partial V_1}{\partial x_1} = 0, \qquad \frac{\partial V_1}{\partial x_2} + \frac{\partial V_2}{\partial x_1} = 0, \qquad \frac{\partial V_2}{\partial x_2} = 0.
$$

From the first and third equations it follows that

$$
V_1(x_1, x_2) = f(x_2), \qquad V_2(x_1, x_2) = g(x_1)
$$

where f and g are analytic functions. Inserting these solutions into the second equation yields the following three solutions

$$
\begin{aligned}
(1) &\quad f(x_2) = 1, \quad g(x_1) = 0 \\
(2) &\quad f(x_2) = 0, \quad g(x_1) = 1 \\
(3) &\quad f(x_2) = x_2, \quad g(x_1) = -x_1.
\end{aligned}
$$

Therefore the metric tensor field g admits the three Killing vector fields

$$
\left\{ \frac{\partial}{\partial x_1}, \ \frac{\partial}{\partial x_2}, \ x_2\frac{\partial}{\partial x_1} - x_1\frac{\partial}{\partial x_2} \right\}.
$$

These three vector fields form a basis of a Lie algebra.

Problem 9. Consider the differential one-form in $\mathbb{R}^3$

$$
\alpha = dx_2 - x_3 dx_1.
$$

(i) Find $d\alpha$ and $\alpha \wedge d\alpha$.

(ii) Consider the smooth vector field

$$V = f_1(\mathbf{x})\frac{\partial}{\partial x_1} + f_2(\mathbf{x})\frac{\partial}{\partial x_2} + f_3(\mathbf{x})\frac{\partial}{\partial x_3}.$$

Find the conditions on f_1, f_2, f_3 such that $L_V\alpha = g(\mathbf{x})\alpha$, where $g : \mathbb{R}^3 \to \mathbb{R}$ is a smooth function.

Solution 9.   (i) We have $d\alpha = -dx_3 \wedge dx_1$ and

$$\alpha \wedge d\alpha = (dx_2 - x_3 dx_1) \wedge (-dx_3 \wedge dx_1) = -dx_1 \wedge dx_2 \wedge dx_3 \neq 0.$$

(ii) We have

$$\begin{aligned}
L_V\alpha &= L_V(dx_2 - x_3 dx_1) = L_V dx_2 - L_V(x_3 dx_1) \\
&= d(V \lfloor dx_2) - (L_V x_3)dx_1 - x_3 L_V dx_1 = d(f_2) - f_3 dx_1 - x_3 df_1 \\
&= \left(\frac{\partial f_2}{\partial x_1} - x_3\frac{\partial f_1}{\partial x_1} - f_3\right)dx_1 + \left(\frac{\partial f_2}{\partial x_2} - x_3\frac{\partial f_1}{\partial x_2}\right)dx_2 \\
&\quad + \left(\frac{\partial f_2}{\partial x_3} - x_3\frac{\partial f_1}{\partial x_3}\right)dx_3.
\end{aligned}$$

Thus $L_V\alpha = g(\mathbf{x})\alpha$ provides the system of differential equations

$$\left(\frac{\partial f_2}{\partial x_1} - x_3\frac{\partial f_1}{\partial x_1} - f_3\right) = -g(\mathbf{x})x_3$$

$$\left(\frac{\partial f_2}{\partial x_2} - x_3\frac{\partial f_1}{\partial x_2}\right) = g(\mathbf{x}), \qquad \left(\frac{\partial f_2}{\partial x_3} - x_3\frac{\partial f_1}{\partial x_3}\right) = 0.$$

Thus from the equation $L_V\alpha = g(\mathbf{x})\alpha$ and after eliminating g we obtain the system of partial differential equations

$$\frac{\partial f_2}{\partial x_3} - x_3\frac{\partial f_1}{\partial x_3} = 0, \quad f_3 = \frac{\partial f_2}{\partial x_1} + x_3\left(\frac{\partial f_2}{\partial\partial x_2} - \frac{\partial f_1}{\partial x_1}\right) - x_3^2\frac{\partial f_1}{\partial x_2}.$$

Problem 10.   Let V be a smooth vector field in $\mathbb{R}^3$ and

$$\alpha = x_1 dx_2 + x_2 dx_3 + x_3 dx_1.$$

Find the condition on V such that $L_V\alpha = 0$.

Solution 10.   Note that

$$d\alpha = dx_1 \wedge dx_2 + dx_2 \wedge dx_3 + dx_3 \wedge dx_1$$

$$V\rfloor\alpha = x_1 V_2(\mathbf{x}) + x_2 V_3(\mathbf{x}) + x_3 V_1(\mathbf{x})$$
$$V\rfloor d\alpha = V_1 dx_2 - V_1 dx_3 - V_2 dx_1 + V_2 dx_3 - V_3 dx_2 + V_3 dx_1$$
$$d(V\rfloor\alpha) = V_2 dx_1 + V_3 dx_2 + V_1 dx_3$$
$$+ x_1\left(\frac{\partial V_2}{\partial x_1}dx_1 + \frac{\partial V_2}{\partial x_2}dx_2 + \frac{\partial V_2}{\partial x_3}dx_3\right)$$
$$+ x_2\left(\frac{\partial V_3}{\partial x_1}dx_1 + \frac{\partial V_3}{\partial x_2}dx_2 + \frac{\partial V_3}{\partial x_3}dx_3\right)$$
$$+ x_3\left(\frac{\partial V_1}{\partial x_1}dx_1 + \frac{\partial V_1}{\partial x_2}dx_2 + \frac{\partial V_1}{\partial x_3}dx_3\right).$$

The Lie derivative follows as

$$L_V\alpha = V_1 dx_2 + V_2 dx_3 + V_3 dx_1$$
$$+ x_1\left(\frac{\partial V_2}{\partial x_1}dx_1 + \frac{\partial V_2}{\partial x_2}dx_2 + \frac{\partial V_2}{\partial x_3}dx_3\right)$$
$$+ x_2\left(\frac{\partial V_3}{\partial x_1}dx_1 + \frac{\partial V_3}{\partial x_2}dx_2 + \frac{\partial V_3}{\partial x_3}dx_3\right)$$
$$+ x_3\left(\frac{\partial V_1}{\partial x_1}dx_1 + \frac{\partial V_1}{\partial x_2}dx_2 + \frac{\partial V_1}{\partial x_3}dx_3\right).$$

Hence the three conditions are

$$V_3 + x_1\frac{\partial V_2}{\partial x_1} + x_2\frac{\partial V_3}{\partial x_1} + x_3\frac{\partial V_1}{\partial x_1} = 0$$
$$V_1 + x_1\frac{\partial V_2}{\partial x_2} + x_2\frac{\partial V_3}{\partial x_2} + x_3\frac{\partial V_1}{\partial x_2} = 0$$
$$V_2 + x_1\frac{\partial V_2}{\partial x_3} + x_2\frac{\partial V_3}{\partial x_3} + x_3\frac{\partial V_1}{\partial x_3} = 0.$$

Problem 11. (i) Let V, W be smooth vector fields in $\mathbb{R}^n$ and $f : \mathbb{R} \to \mathbb{R}$ be a smooth function. Find $L_V(fW)$.

(ii) Let V, W be vector fields and α be a differential form. Find the Lie derivative

$$L_V(W \otimes \alpha).$$

(iii) Let α_1, α_2 be smooth differential forms in $\mathbb{R}^n$ and V be a smooth vector field in $\mathbb{R}^n$. Assume that $L_V\alpha_1 = 0$ and $L_V\alpha_2 = 0$. Show that $L_V(\alpha_1 \wedge \alpha_2) = 0$.

Solution 11. (i) Applying the product rule we find

$$L_V(fW) = (V(f))W + f(L_V W) = (V(f))W + f[V, W].$$

(ii) Applying the product rule we find

$$L_V(W \otimes \alpha) = (L_V W) \otimes \alpha + W \otimes L_V \alpha = [V, W] + W \otimes L_V \alpha.$$

(iii) Applying the product rule provides

$$L_V(\alpha_1 \wedge \alpha_2) = (L_V \alpha_1) \wedge \alpha_2 + \alpha_1 \wedge (L_V \alpha_2) = 0.$$

Problem 12. Consider the unit ball $x_1^2 + x_2^2 + x_3^2 = 1$ and the vector field

$$V = (a_0 + a_1 x_1 + a_2 x_2 + a_3 x_3 + x_1(e_1 x_1 + e_2 x_2 + e_3 x_3)) \frac{\partial}{\partial x_1}$$

$$+ (b_0 + b_1 x_1 + b_2 x_2 + b_3 x_3 + x_2(e_1 x_1 + e_2 x_2 + e_3 x_3)) \frac{\partial}{\partial x_2}$$

$$+ (c_0 + c_1 x_1 + c_2 x_2 + c_3 x_3 + x_3(e_1 x_1 + e_2 x_2 + e_3 x_3)) \frac{\partial}{\partial x_3}.$$

Find the coefficients from the conditions

$$L_V(x_1^2 + x_2^2 + x_3^2) = 0, \qquad x_1^2 + x_2^2 + x_3^2 = 1.$$

Solution 12. Inserting $x_1^2 + x_2^2 + x_3^2 = 1$ into $L_V(x_1^2 + x_2^2 + x_3^2) = 0$ yields the equation

$$x_1(a_0 + a_1 x_1 + a_2 x_2 + a_3 x_3) + x_2(b_0 + b_1 x_1 + b_2 x_2 + b_3 x_3)$$
$$+ x_3(c_0 + c_1 x_1 + c_2 x_2 + c_3 x_3) + (e_1 x_1 + e_2 x_2 + e_3 x_3) = 0.$$

The solution is

$$a_1 = b_2 = c_3 = 0$$

$$a_2 + b_1 = a_3 + c_1 = b_3 + c_2 = a_0 + e_1 = b_0 + e_2 = c_0 + e_3 = 0.$$

Thus we obtain the vector fields

$$V_1 = (x_1^2 - 1) \frac{\partial}{\partial x_1} + x_1 x_2 \frac{\partial}{\partial x_2} + x_1 x_3 \frac{\partial}{\partial x_3}, \quad W_1 = x_2 \frac{\partial}{\partial x_3} - x_3 \frac{\partial}{\partial x_2}$$

$$V_2 = x_2 x_1 \frac{\partial}{\partial x_1} + (x_2^2 - 1) \frac{\partial}{\partial x_2} + x_2 x_3 \frac{\partial}{\partial x_3}, \quad W_2 = x_3 \frac{\partial}{\partial x_1} - x_1 \frac{\partial}{\partial x_3}$$

$$V_3 = x_3 x_1 \frac{\partial}{\partial x_1} + x_3 x_2 \frac{\partial}{\partial x_2} + (x_3^2 - 1) \frac{\partial}{\partial x_1}, \quad W_3 = x_1 \frac{\partial}{\partial x_2} - x_2 \frac{\partial}{\partial x_1}.$$

They form a basis of a six-dimensional Lie algebra under the commutator.

Problem 13. (i) Let V, W be vector fields. Let f, g be smooth functions and α be a differential form. Assume that $L_V \alpha = f\alpha$, $L_W \alpha = g\alpha$. Show that

$$L_{[V,W]}\alpha = (L_V f - L_W g)\alpha. \tag{1}$$

(ii) Let g be a smooth metric tensor field and V, W be smooth vector fields with $L_V g = \rho_1 g$, $L_W g = \rho_2 g$, where ρ_1 and ρ_2 are smooth functions. Find $L_{[V,W]}g$.

Solution 13. (i) We have

$$L_{[V,W]}\alpha = [L_V, L_W]\alpha = L_V(L_W \alpha) - L_W(L_V \alpha) = L_V(g\alpha) - L_W(f\alpha).$$

Applying the product rule we have

$$L_V(g\alpha) = (L_V g)\alpha + g L_V \alpha = (L_V g)\alpha + (gf)\alpha$$

$$L_W(f\alpha) = (L_W f)\alpha + f L_W \alpha = (L_W f)\alpha + (fg)\alpha.$$

Thus (1) follows.
(ii) We have

$$\begin{aligned}
L_{[V,W]}g &= L_V(L_W g) - L_W(L_V g) = L_V(\rho_2 g) - L_W(\rho_1 g) \\
&= (L_V \rho_2)g + \rho_2(L_V g) - (L_W \rho_1)g - \rho_1(L_W g) \\
&= (V(\rho_2) - W(\rho_1))g + (\rho_2 \rho_1 - \rho_1 \rho_2)g \\
&= (V(\rho_2) - W(\rho_1))g.
\end{aligned}$$

Problem 14. Consider the smooth vector fields V and W defined on $\mathbb{R}^n$. Let f and g be smooth functions. Assume that

$$L_V f \equiv V(f) = 0, \qquad L_W g \equiv W(g) = 0$$

i.e. f, g are first integrals of the dynamical system given by the vector fields V and W. Find $L_{[V,W]}(f + g)$, $L_{[V,W]}(fg)$.

Solution 14. Using the sum rule we find

$$\begin{aligned}
L_{[V,W]}(f + g) &= L_{[V,W]}f + L_{[V,W]}g \\
&= (L_V L_W - L_W L_V)f + (L_V L_W - L_W L_V)g \\
&= L_V(L_W f) - L_W(L_V g).
\end{aligned}$$

Using the product rule we obtain

$$
\begin{aligned}
L_{[V,W]}(fg) &= (L_{[V,W]}f)g + f(L_{[V,W]}g) \\
&= ((L_V L_W - L_W L_V)f)g + f((L_V L_W - L_W L_V)g) \\
&= g(L_V L_W f) - f(L_W L_V g).
\end{aligned}
$$

Problem 15. Let V, W be smooth vector fields in $\mathbb{R}^n$ ($n \geq 2$) and α, β be differential one-forms. Calculate $L_{[V,W]}(\alpha \wedge \beta)$. Then assume that $L_V \alpha = 0$ and $L_W \beta = 0$.

Solution 15. Applying the product rule we find

$$
\begin{aligned}
L_{[V,W]}(\alpha \wedge \beta) &= L_V L_W(\alpha \wedge \beta) - L_W L_V(\alpha \wedge \beta) \\
&= L_V((L_W \alpha) \wedge \beta) + L_V(\alpha \wedge (L_W \beta)) \\
&\quad - L_W((L_V \alpha) \wedge \beta)) - L_W(\alpha \wedge L_V \beta) \\
&= (L_V(L_W \alpha) - L_W(L_V \alpha)) \wedge \beta \\
&\quad + \alpha \wedge (L_V(L_W \beta) - L_W(L_V \beta)).
\end{aligned}
$$

For the case $L_V \alpha = 0$, $L_W \beta = 0$ the result simplifies to

$$
L_{[V,W]}(\alpha \wedge \beta) = (L_V(L_W \alpha)) \wedge \beta - \alpha \wedge (L_W(L_V \beta)).
$$

Problem 16. Let V_j ($j = 1, \ldots, n$) be smooth vector fields and α a smooth differential one-form. Assume that

$$
L_{V_j} \alpha = d\phi_j, \quad j = 1, \ldots, n
$$

where ϕ_j are smooth functions.
(i) Find $L_{[V_j, V_k]} \alpha$.
(ii) Assume that the vector fields V_j ($j = 1, \ldots, n$) form basis of a Lie algebra, i.e.

$$
[V_j, V_k] = \sum_{\ell=1}^{n} c_{jk}^{\ell} V_\ell
$$

where c_{jk}^{ℓ} are the *structure constants*. Find the conditions on the functions ϕ_j.

Solution 16. (i) We have

$$
L_{[V_j, V_k]} \alpha = (L_{V_j} L_{V_k} - L_{V_k} L_{V_j})\alpha = d(L_{V_j} \phi_k - L_{V_k} \phi_j).
$$

(ii) We have

$$L_{[V_j,V_k]}\alpha = L_{\sum_{\ell=1}^n c_{jk}^\ell V_\ell}\alpha = \sum_{\ell=1}^n c_{jk}^\ell d\phi_\ell = d\sum_{\ell=1}^n c_{jk}^\ell \phi_\ell.$$

Using the result from (i) we obtain

$$L_{V_j}\phi_k - L_{V_k}\phi_j - \sum_{\ell=1}^n c_{jk}^\ell \phi_\ell = 0$$

with $j, k = 1, \ldots, n$ and $j \neq k$. Note that $c_{jk}^\ell = -c_{kj}^\ell$. Thus we have $(n^2 - n)/2$ equations.

Problem 17. Let α be a smooth differential one-form and V be a smooth vector field. Assume that

$$L_V\alpha = f\alpha$$

where f is a smooth function. Define the function F as $F := V \rfloor \alpha$, where $\rfloor$ denotes the contraction. Show that $dF = f\alpha - V \rfloor d\alpha$.

Solution 17. Using $L_V\alpha \equiv V \rfloor d\alpha + d(V \rfloor \alpha)$ we have

$$dF = d(V \rfloor \alpha) = L_V\alpha - V \rfloor d\alpha.$$

From $L_V\alpha = f\alpha$ it follows that $dF = f\alpha - V \rfloor d\alpha$.

Problem 18. Let V, W be two smooth vector fields

$$V = \sum_{j=1}^n V_j(\mathbf{x})\frac{\partial}{\partial x_j}, \qquad W = \sum_{j=1}^n W_j(\mathbf{x})\frac{\partial}{\partial x_j}$$

defined on $\mathbb{R}^n$. Assume that $[V, W] = f(\mathbf{x})W$. Let $\Omega = dx_1 \wedge \cdots \wedge dx_n$ be the volume form and $\alpha := W \rfloor \Omega$. Find the Lie derivative $L_V\alpha$. Discuss.

Solution 18. Using the product rule we have

$$\begin{aligned}
L_V\alpha &= L_V(W \rfloor \Omega) = (L_V W) \rfloor \Omega + W \rfloor (L_V\Omega) \\
&= (f(\mathbf{x})W) \rfloor \Omega + W \rfloor (\text{div}(V))\Omega = (f(\mathbf{x}) + \text{div}(V))(W \rfloor \Omega) \\
&= (f(\mathbf{x}) + \text{div}(V))\alpha.
\end{aligned}$$

Thus α is conformal invariant under the vector field V.

Problem 19.   Let $du_1/dt = V_1(\mathbf{u})$, ..., $du_n/dt = V_n(\mathbf{u})$ be an autonomous system of ordinary differential equations, where $V_j(\mathbf{u}) \in C^\infty(\mathbb{R}^n)$ for all $j = 1, \ldots, n$. A function $\phi \in C^\infty(\mathbb{R}^n)$ is called *conformal invariant* with respect to the vector field

$$V = V_1(\mathbf{u})\frac{\partial}{\partial u_1} + \cdots + V_n(\mathbf{u})\frac{\partial}{\partial u_n}$$

if $L_V\phi = \rho\phi$, where $\rho \in C^\infty(\mathbb{R}^n)$. Let $n = 2$ and consider the vector fields

$$V = u_1\frac{\partial}{\partial u_2} - u_2\frac{\partial}{\partial u_1}, \qquad W = u_1\frac{\partial}{\partial u_1} + u_2\frac{\partial}{\partial u_2}.$$

Show that $\phi(\mathbf{u}) = u_1^2 + u_2^2$ is conformal invariant under V and W. Find the commutator $[V, W]$.

Solution 19.   We find $L_V\phi = 0$. Thus $\rho = 0$. We obtain $L_W\phi = 2\phi$. Thus $\rho = 2$. For the commutator we obtain $[V, W] = 0$.

Problem 20.   Let M be differentiable manifold and $\phi : M \to \mathbb{R}$ be a smooth function. Let α be a smooth differential one-form defined on M. Show that if V is a vector field defined on M such that $d\phi = V \rfloor d\alpha$, then

$$L_V\alpha = d(V\rfloor\alpha + \phi).$$

Solution 20.   We start from $L_V\alpha = V\rfloor d\alpha) + d(V\rfloor\alpha)$. Since $V\rfloor d\alpha = d\phi$ and d is linear we obtain $L_V\alpha = d(V\rfloor\alpha + \phi)$.

Problem 21.   Consider the manifold $M = \mathbb{R}^n$ and the volume form

$$\Omega = dx_1 \wedge \cdots \wedge dx_n.$$

Consider the analytic vector field

$$V = \sum_{j=1}^{n} V_j(\mathbf{x})\frac{\partial}{\partial x_j}.$$

Find $\omega = V\rfloor\Omega$ and $L_V\Omega$.

Solution 21.   We obtain

$$\omega = \sum_{j_1,\ldots,j_n=1}^{n} \epsilon_{j_1,\ldots,j_n} V_{j_1} dx_{j_2} \wedge \cdots \wedge dx_{j_n}$$

where $\epsilon_{j_1,\ldots,j_n}$ is the completely antisymmetric *Levi-Civita tensor* on n indices. For the Lie derivative we obtain

$$L_V \Omega = V \rfloor d\Omega + d(V \rfloor \Omega) = d(V \rfloor \Omega) = \left(\sum_{j=1}^n \frac{\partial V_j}{\partial x_j} \right) \Omega = \operatorname{div}(V)\Omega.$$

Problem 22. Consider the autonomous system of first order ordinary differential equations

$$\frac{du_j}{dt} = V_j(\mathbf{u}), \qquad j = 1, \ldots, n$$

where the V_j's are polynomials. The corresponding vector field is

$$V = \sum_{j=1}^n V_j(\mathbf{u}) \frac{\partial}{\partial x_j}.$$

Let f be an analytic function. Then the Lie derivative of f is

$$L_V f = \sum_{j=1}^n V_j \frac{\partial f}{\partial x_j}.$$

A *Darboux polynomial* is a polynomial g such that there is another polynomial p satisfying $L_V g = pg$. The couple is called a *Darboux element*. If m is the greatest of $\deg(V_j)$ $(j = 1, \ldots, n)$, then $\deg(p) \leq m - 1$. All the irreducible factors of a Darboux polynomial are Darboux. If the autonomous system of first order differential equations is homogeneous of degree m, i.e. all V_j are homogeneous of degree m, then p is homogeneous of degree $m - 1$ and all homogeneous components of g are Darboux. The search can be restricted to homogeneous g.

(i) Show that the product of two Darboux polynomials is a Darboux polynomial.

(ii) Consider the *Lotka-Volterra model* for three species

$$\frac{du_1}{dt} = u_1(c_3 u_2 + u_3), \qquad \frac{du_2}{dt} = u_2(c_1 u_3 + u_1), \qquad \frac{du_3}{dt} = u_3(c_2 u_1 + u_2)$$

where c_1, c_2, c_3 are real parameters. Find the determining equation for the Darboux element.

Solution 22. (i) If $L_V g_1 = p_1 g_1$ and $L_V g_2 = p_2 g_2$, then we have using the product rule

$$L_V(g_1 g_2) = (L_V g_1)g_2 + g_1(L_V g_2) = p_1 g_1 g_2 + g_1 p_2 g_2 = (p_1 + p_2)(g_1 g_2).$$

(ii) We obtain

$$L_V g = u_1(c_3 u_2 + u_3)\frac{\partial g}{\partial u_1} + u_2(c_1 u_3 + u_1)\frac{\partial g}{\partial u_2} + u_3(c_2 u_1 + u_2)\frac{\partial g}{\partial u_3}$$

$$= (\epsilon_1 u_1 + \epsilon_2 u_2 + \epsilon_3 u_3)g.$$

Problem 23.   Let $H : \mathbb{R}^{2n} \to \mathbb{R}$ be a smooth Hamilton function with the corresponding vector field

$$V_H = \sum_{j=1}^{n}\left(\frac{\partial H}{\partial p_j}\frac{\partial}{\partial q_j} - \frac{\partial H}{\partial q_j}\frac{\partial}{\partial p_j}\right)$$

and

$$W = \sum_{j=1}^{n}\left(f_j(\mathbf{p},\mathbf{q})\frac{\partial}{\partial q_j} + g_j(\mathbf{p},\mathbf{q})\frac{\partial}{\partial p_j}\right)$$

be another smooth vector field. Assume that

$$[V_H, W] = hW \tag{1}$$

where h is a smooth function of $\mathbf{p}$ and $\mathbf{q}$. Let

$$\Omega = dq_1 \wedge \cdots \wedge dq_n \wedge dp_1 \wedge \cdots \wedge dp_n$$

be the standard volume differential form. Let $\alpha = W \rfloor \Omega$. Show that (1) can be written as

$$L_{V_H}\alpha = h\alpha.$$

Solution 23.   Using the identity

$$L_{V_H}(W\rfloor\Omega) \equiv (L_{V_H}W)\rfloor\Omega + W\rfloor(L_{V_H}\Omega)$$

and $\operatorname{div}(V_H) = 0$ we obtain

$$L_{V_H}\alpha = L_{V_H}(W\rfloor\Omega) = (L_{V_H}W)\rfloor\Omega + W\rfloor(L_{V_H}\Omega)$$
$$= [V_H, W]\rfloor\Omega + W\rfloor(\operatorname{div}(V_H))\Omega = [V_H, W]\rfloor\Omega$$
$$= h(W\rfloor\Omega)$$
$$= h\alpha.$$

Problem 24.   Consider the $2n + 1$ dimensional *anti-de Sitter space* AdS_{2n+1}. This is a hypersurface in the vector space $\mathbb{R}^{2n+2}$ defined by the equation $R(\mathbf{x}) = -1$, where

$$R(\mathbf{x}) = -(x_0)^2 - (x_1)^2 + (x_2)^2 + \cdots + (x_{2n+1})^2.$$

One introduces the even coordinates **p** and odd coordinates **q**. Then we can write

$$R(\mathbf{p}, \mathbf{q}) = -p_1^2 - q_1^2 + p_2^2 + q_2^2 + \cdots + p_{n+1}^2 + q_{n+1}^2.$$

We consider $\mathbb{R}^{2n+2}$ as a symplectic manifold with the canonical symplectic differential form

$$\omega = \sum_{k=1}^{n+1} (dp_k \wedge dq_k).$$

Let

$$\alpha = \frac{1}{2} \sum_{k=1}^{n+1} (p_k dq_k - q_k dp_k).$$

Consider the vector field V in $\mathbb{R}^{2n+2}$ given by

$$V = \frac{1}{2} \sum_{k=1}^{n+1} \left(p_k \frac{\partial}{\partial p_k} + q_k \frac{\partial}{\partial q_k} \right).$$

Find the Lie derivative $L_V R$ and $V \rfloor \omega$.

Solution 24. We obtain $L_V R = R$. Thus R is conformal invariant under V. We find $V \rfloor \omega = \alpha$.

Problem 25. Let V, W be smooth vector fields defined in $\mathbb{R}^n$. Let $f, g : \mathbb{R}^n \to \mathbb{R}$ be smooth functions. Consider now the pairs (V, f), (W, g). One defines a commutator of such pairs as

$$[(V, f), (W, g)] := ([V, W], L_V g - L_W f).$$

Let $n = 2$,

$$V = u_2 \frac{\partial}{\partial u_1} - u_1 \frac{\partial}{\partial u_2}, \qquad W = u_1 \frac{\partial}{\partial u_1} + u_2 \frac{\partial}{\partial u_1}$$

and $f(u_1, u_2) = g(u_1, u_2) = u_1^2 + u_2^2$. Calculate the commutator.

Solution 25. With $[V, W] = 0$, $L_V f = 0$, $L_W g = 2(u_1^2 + u_2^2)$ we obtain the commutator

$$[(V, f), (W, g)] = (0, -2(u_1^2 + u_2^2)).$$

Problem 26. (i) Let $n \geq 1$. The *Heisenberg group* $\mathbb{H}^n$ can be considered as $\mathbb{C} \times \mathbb{R}$ endowed with a polynomial group law $\cdot : \mathbb{H}^n \times \mathbb{H}^n \to \mathbb{H}^n$. Its Lie algebra identifies with the tangent space $T_0 \mathbb{H}^n$ at the identity $0 \in \mathbb{H}^N$. Consider the tangent bundle $T \mathbb{H}^n$, where

$$X_j := \frac{\partial}{\partial x_j} - \frac{y_j}{2}\frac{\partial}{\partial t}, \quad Y_j := \frac{\partial}{\partial y_j} + \frac{x_j}{2}\frac{\partial}{\partial t}, \quad T := \frac{\partial}{\partial t}$$

and $p \in \mathbb{H}^n$. Find the commutators of the vector fields

$$[X_j, Y_k], \quad [X_j, T], \quad [Y_j, T].$$

(ii) Consider the differential one-form

$$\alpha := dt + \frac{1}{2} \sum_{j=1}^{n}(y_j dx_j - x_j dy_j)$$

which is the contact form of $\mathbb{H}^n$. Note that

$$d\alpha = -\sum_{j=1}^{n}(dx_j \wedge dy_j).$$

Find the Lie derivatives $L_{X_j}\alpha$, $L_{Y_j}\alpha$, $L_T\alpha$.

Solution 26.   (i) We have $[X_j, Y_j] = T$, $j = 1, \ldots, n$ and all other commutators vanish.
(ii) We find

$$X_j \rfloor \alpha = 0, \quad Y_j \rfloor \alpha = 0, \quad X_j \rfloor d\alpha = -dy_j, \quad Y_j \rfloor d\alpha = dx_j.$$

Hence we find $L_{X_j}\alpha = -dy_j$, $L_{Y_j}\alpha = dx_j$, $L_T\alpha = 0$.

Problem 27.   For the *Poincaré upper half plane*

$$\mathbb{H}_+^2 := \{ z = x_1 + ix_2 : x_2 > 0 \}$$

the metric tensor field is given by

$$g = \frac{1}{x_2^2}(dx_1 \otimes dx_1 + dx_2 \otimes dx_2).$$

Find the Killing vector fields for g, i.e. $L_V g = 0$, where

$$V = V_1(x_1, x_2)\frac{\partial}{\partial x_1} + V_2(x_1, x_2)\frac{\partial}{\partial x_2}.$$

Solution 27. From $L_V g = 0$ we find the three conditions

$$\frac{\partial}{\partial x_1}V_2 + \frac{\partial}{\partial x_2}V_1 = 0, \qquad \frac{\partial}{\partial x_1}V_1 = \frac{V_2}{x_2}, \qquad \frac{\partial}{\partial x_2}V_2 = \frac{V_2}{x_2}$$

with the solution

$$V_1 = 3c_1(x_2^2 - x_1^2) - 2c_2 x_1 - c_3, \qquad V_2 = -x_2(6c_1 x_1 + 2c_2).$$

Setting (i) $c_1 = 0$, $c_2 = 0$, $c_3 = -1$; (ii) $c_1 = 0$, $c_2 = -1/2$, $c_3 = 0$; (iii) $c_1 = 1/3$, $c_2 = 0$, $c_3 = 0$ we can form the three Killing vector fields

$$K_1 = \frac{\partial}{\partial x_1}$$

$$K_2 = x_1\frac{\partial}{\partial x_1} + x_2\frac{\partial}{\partial x_2}$$

$$K_3 = (x_2^2 - x_1^2)\frac{\partial}{\partial x_1} - 2x_1 x_2\frac{\partial}{\partial x_2}.$$

They form a basis of a Lie algebra. The commutation relations are

$$[K_1, K_2] = K_1, \quad [K_1, K_3] = -2K_2, \quad [K_2, K_3] = K_3.$$

Problem 28. Consider the *Lotka-Volterra equation*

$$\frac{du_1}{dt} = (a - bu_2)u_1, \qquad \frac{du_2}{dt} = (cu_1 - d)u_2$$

where a, b, c, d are constants and $u_1 > 0$ and $u_2 > 0$. The corresponding vector field V is

$$V = (a - bu_2)u_1\frac{\partial}{\partial u_1} + (cu_1 - d)u_2\frac{\partial}{\partial u_2}.$$

Let $\omega = f(u_1, u_2)du_1 \wedge du_2$, where f is a smooth nonzero function. Find a smooth function H (*Hamilton function*) such that $\omega\rfloor V = dH$. Note that from this condition since $d(dH) = 0$ we obtain $d(\omega\rfloor V) = 0$.

Solution 28. We obtain from $d(\omega\rfloor V) = 0$

$$\frac{\partial(A - Bu_2)u_1 f}{\partial u_1} + \frac{\partial(Cu_1 - D)u_2 f}{\partial u^2} = 0.$$

We find $f(u_1, u_2) = 1/(u_1 u_2)$ and

$$H(u_1, u_2) = a\ln(u_2) + d\ln(u_1) - cu_1 - bu_2.$$

Problem 29.   Consider the generalized *Lotka-Volterra model*

$$\frac{dx_j}{dt} = c_j x_j + x_j \sum_{k=1}^{n} c_{jk} x_k, \quad j = 1, \ldots, n$$

and the differential *n*-form

$$\omega = \frac{dx_1 \wedge dx_2 \wedge \cdots \wedge dx_n}{x_1 x_2 \cdots x_n}$$

where $x_j > 0$ for $j = 1, \ldots, n$. Let V be the vector field

$$V = \sum_{j=1}^{n} x_j \left(c_j + \sum_{k=1}^{n} c_{jk} x_k \right) \frac{\partial}{\partial x_j}.$$

Find the condition on the c_j, c_{jk} such that $L_V \omega = 0$.

Solution 29.   From $L_V \omega = 0$ we obtain that $c_{kk} = 0$ for $k = 1, \ldots, n$.

Problem 30.   Let I, f be analytic functions of u_1, u_2. Consider the autonomous system of differential equations

$$\begin{pmatrix} du_1/d\tau \\ du_2/d\tau \end{pmatrix} = \begin{pmatrix} 0 & f(\mathbf{u}) \\ -f(\mathbf{u}) & 0 \end{pmatrix} \begin{pmatrix} \partial I/\partial u_1 \\ \partial I/\partial u_2 \end{pmatrix}.$$

Show that I is a first integral of this autonomous system of differential equations.

Solution 30.   We have

$$\frac{dI}{d\tau} = \frac{\partial I}{\partial u_1}\frac{du_1}{d\tau} + \frac{\partial I}{\partial u_2}\frac{du_2}{d\tau} = \frac{\partial I}{\partial u_1} f \frac{\partial I}{\partial u_2} - \frac{\partial I}{\partial u_2} f \frac{\partial I}{\partial u_1} = 0.$$

Problem 31.   Let $a \in \mathbb{R}$. Consider the autonomous system of first order differential equations

$$\frac{du_1}{d\tau} = u_2(u_1 + u_2) + a u_3$$

$$\frac{du_2}{d\tau} = u_1(u_2 + u_3) - a u_3$$

$$\frac{du_3}{d\tau} = u_3(u_1 + u_2)$$

with the corresponding vector field

$$V = (u_2(u_1 + u_2) + au_3)\frac{\partial}{\partial u_1} + (u_1(u_2 + u_3) - au_3)\frac{\partial}{\partial u_2} + u_3(u_1 + u_2)\frac{\partial}{\partial u_3}.$$

Show that

$$I_1 = \frac{1}{2}(u_1^2 - u_2^2) - au_3$$

is a first integral. Show that

$$I_2 = \frac{u_1 + u_2}{u_3}$$

is a first integral.

Solution 31. We have $L_V I_1 = 0$ or alternatively

$$\frac{d}{d\tau}I_1 = u_1\frac{du_1}{d\tau} - u_2\frac{du_2}{d\tau} - a\frac{du_3}{d\tau} = 0.$$

We have $L_V I_2 = 0$ or alternatively

$$\frac{d}{d\tau}I_2 = \frac{u_3(du_1/d\tau + du_2/d\tau) - (u_1 + u_2)du_3/d\tau}{u_3^2} = 0.$$

Hence we have two constants of motion

$$u_1^2 - u_2^2 - 2au_3 = C_1, \qquad u_1 + u_2 = C_2 u_3$$

which can be used to integrate

$$\frac{du_1}{u_2(u_1 + u_2) + au_3} = \frac{du_2}{u_1(u_1 + u_2) - au_3} = \frac{du_3}{u_3(u_1 + u_2)} = \frac{d\tau}{1}.$$

The constants C_1 and C_2 are fixed by the initial conditions.

Problem 32. The *Kermack-McKendrick model* for epidemics is given by the autonomous system of three first order differential equations

$$\frac{dS}{dt} = -rSI, \qquad \frac{dI}{dt} = rSI - aI, \qquad \frac{dR}{dt} = aI$$

where S is the number of susceptibles, I for those infected and R denotes the removals. The positive constants determine the infection and removal rates of infectives, respectively. The corresponding vector field is

$$V = -rSI\frac{\partial}{\partial S} + (rSI - aI)\frac{\partial}{\partial I} + aI\frac{\partial}{\partial R}.$$

Find the first integrals.

Solution 32. Adding all three equations we find that $F_1 = S + I + R$ is a first integral. Another first integral is

$$F_2 = R + \frac{a}{r} \ln(S)$$

as can be seen as follows

$$\frac{dF_2}{dt} = \frac{dR}{dt} + \frac{a}{r} \frac{dS/dt}{S} = 0.$$

Problem 33. Find the first integrals of the autonomous system of ordinary first order differential equations

$$\frac{dx_1}{d\tau} = x_1 x_2 + x_1 x_3, \qquad \frac{dx_2}{d\tau} = x_2 x_3 - x_1 x_2, \qquad \frac{dx_3}{d\tau} = -x_1 x_3 - x_2 x_3.$$

Solution 33. Addition of all three equations provides the first integral

$$I_1(x_1, x_2, x_3) = x_1 + x_2 + x_3.$$

This means that $L_V I_1 = 0$, where V is the vector field associated with the differential equation. Let U be the open set

$$U := \mathbb{R}^3 \setminus \{ (x_1, 0, x_3) \,|\, x_1, x_3 \in \mathbb{R} \}.$$

Then the function $I_2 \in C^\infty(U, \mathbb{R})$ defined by

$$I_2(x_1, x_2, x_3) = \frac{x_1 x_3}{x_2}$$

is also a first integral.

Problem 34. Consider the nonlinear one-dimensional diffusion equation

$$\frac{\partial u}{\partial t} - \frac{\partial}{\partial x} \left(u^n \frac{\partial u}{\partial x} \right) = 0$$

where $n = 1, 2, \ldots$. An equivalent set of differential forms is given by

$$\alpha = du - u_t dt - u_x dx$$
$$\beta = (u_t - n u^{n-1} u_x^2) dx \wedge dt - u^n du_x \wedge dt$$

with the coordinates t, x, u, u_t, u_x The exterior derivative of α is given

$$d\alpha = -du_t \wedge dt - du_x \wedge dx.$$

Consider the vector field

$$V = V_t \frac{\partial}{\partial t} + V_x \frac{\partial}{\partial x} + V_u \frac{\partial}{\partial u} + V_{u_t} \frac{\partial}{\partial u_t} + V_{u_x} \frac{\partial}{\partial u_x}.$$

Then the symmetry vector fields of the partial differential equation are determined by

$$L_V \alpha = g\alpha, \qquad L_V \beta = h\beta + w\alpha + rd\alpha$$

where g, h, r are smooth functions depending on t, x, u, u_t, u_x and w is a differential one-form also depending on t, x, u, u_t, u_x. Find the symmetry vector fields from these two conditions. Note that we have

$$L_V(d\alpha) = d(L_V\alpha) = d(g\alpha) = (dg) \wedge \alpha + gd\alpha.$$

Solution 34. The first condition $L_V\alpha = g\alpha$ provides 5 equations. The second condition provides 10 equations. Solving these 15 equation yields

$$V_t = c_1 + c_3 t, \quad V_x = c_2 + c_4 x, \quad V_u = \frac{1}{n}(2c_4 - c_3)u$$

$$V_{u_t} = \frac{1}{n}(2c_4 - (n+1)c_3)u_t, \quad V_{u_x} = \frac{1}{n}((2-n)c_4 - c_3)u_x$$

where c_j $(j = 1, 2, 3, 4)$ are constants. Thus we obtain the symmetry vector fields

$$V_1 = \frac{\partial}{\partial t}, \qquad V_2 = \frac{\partial}{\partial x}$$

$$V_3 = -\frac{1}{n}\left(u\frac{\partial}{\partial u} + (n+1)u_t\frac{\partial}{\partial u_t} + u_x\frac{\partial}{\partial u_x}\right) + t\frac{\partial}{\partial t}$$

$$V_4 = \frac{1}{n}\left(2u\frac{\partial}{\partial u} + 2u_t\frac{\partial}{\partial u_t} + (2-n)u_x\frac{\partial}{\partial u_x}\right) + x\frac{\partial}{\partial x}.$$

We have $[V_1, V_3] = V_1$, $[V_2, V_4] = V_2$, $[V_3, V_4] = 0$.

Problem 35. Let M, N be differential manifolds. Let T be a tensor field and $f : M \to N$ and $f : M \to N$ be an orientation preserving diffeomorphism. Then

$$f^*(L_V T) = L_{(f_*V)}(f^*T).$$

Let V be a smooth vector field and γ a smooth differential form. If $f : M \to N$ is an orientation preserving diffeomorphism, then

$$f^*(V \rfloor \gamma) = (f_* V) \rfloor (f^* \gamma).$$

Here $f_* V$ is the *push forward* for vector fields and $f^* \gamma$ is the *pull back* for differential forms. Apply it to

$$V = x_2 \frac{\partial}{\partial x_1}, \qquad T = \gamma = x_1 dx_2$$

and the map $f : \mathbb{R}^2 \to \mathbb{R}^2$, $(x_1, x_2) \mapsto (y_1, y_2)$ given by

$$y_1 = f_1(x_1, x_2) = \sinh(x_2), \quad y_2 = f_2(x_1, x_2) = \sinh(x_1).$$

Hence $x_1 = \text{arcsinh}(y_2)$, $x_2 = \text{arcsinh}(y_1)$.

Solution 35.   We have $d\gamma = dx_1 \wedge dx_2$, $V \rfloor \gamma = 0$, $L_V \gamma = dx_1 \wedge dx_2$. Now

$$f^*(x_2 dx_2) = \text{arcsinh}(y_1) d(\text{arcsinh}(y_1)) \frac{\text{arcsinh}(y_1)}{\sqrt{1 + y_1^2}} dy_1$$

$$f^*(x_1 dx_2) = \text{arcsinh}(y_2) d(\text{arcsinh}(y_1)) \frac{\text{arcsinh}(y_2)}{\sqrt{1 + y_1^2}} dy_1.$$

Note that $\cosh(\text{arcsinh}(\alpha)) = \sqrt{1 + \alpha^2}$. Then we obtain

$$f_*(V) = f_* \left(x_2 \frac{\partial}{\partial x_1} \right) = \text{arcsinh}(y_1) \left(\cosh(x_1) \frac{\partial}{\partial y_2} \right)$$

$$= \text{arcsinh}(y_1) \sqrt{1 + y_1^2} \frac{\partial}{\partial y_2}.$$

Hence for the left-hand side we obtain

$$f^*(L_V \gamma) = f^*(x_2 dx_2) = \frac{\text{arcsinh}(y_1)}{\sqrt{1 + y_1^2}} dy_1.$$

For the right-hand side we have

$$L_{(f_* V)}(f^* \gamma) = \text{arcsinh}(y_1)(\sqrt{1 + y_1^2}) \frac{\partial}{\partial y_2} \rfloor d(\text{arcsinh}(y_2) \frac{1}{\sqrt{1 + y_1^2}} dy_1)$$

$$= \text{arcsinh}(y_1)(\sqrt{1 + y_2^2}) \frac{\partial}{\partial y_2} \rfloor \frac{1}{\sqrt{1 + y_2^2}} \frac{1}{\sqrt{1 + y_1^2}} dy_2 \wedge dy_1)$$

$$= \frac{\text{arcsinh}(y_1)}{\sqrt{1 + y_1^2}} dy_1.$$

Problem 36. Some quantities in physics owing to the transformation laws have to be considered as *currents* instead of differential forms. Let M be an orientable n-dimensional differentiable manifold of class C^∞. We denote by $\Phi_k(M)$ the set of all differential forms of degree k with compact support. Let $\phi \in \Phi_k(M)$ and let α be an exterior differential form of degree $n - k$ with locally integrable coefficients. Then, as an example of a current, we have

$$T_\alpha(\phi) \equiv \alpha(\phi) := \int_M \alpha \wedge \phi.$$

Define the Lie derivative for this current.

Solution 36. The Lie derivative of this current can be defined as

$$L_V T(\phi) = -T(L_V \phi).$$

This definition can be motivated as follows. Assume that $L_V \alpha$ exist. Since

$$(L_V \alpha) \wedge \phi \equiv L_V(\alpha \wedge \phi) - \alpha \wedge (L_V \phi)$$

we obtain

$$T_{L_V \alpha} \phi = \int_M (L_V \alpha) \wedge \phi = \int_M (L_V(\alpha \wedge \phi) - \alpha \wedge (L_V \phi))$$

$$= -\int_M \alpha \wedge (L_V \phi)$$

where we used that

$$\int_M L_V(\alpha \wedge \phi) = \int_M d(V \rfloor (\alpha \wedge \phi)) = 0$$

noting that the $(n-1)$ differential form $V \rfloor (\alpha \wedge \phi)$ has compact support.

Programming Problems

Problem 37. Let $x_0 = ct$ with c the speed of light and a a positive constant with dimension [*meter*]. The *Gödel metric tensor field* is given by

$$g = -dx_0 \otimes dx_0 - e^{x_1/a} dx_0 \otimes dx_2 + dx_1 \otimes dx_1 - e^{x_1/a} dx_2 \otimes dx_0$$
$$- \frac{1}{2} e^{2x_1/a} dx_2 \otimes dx_2 + dx_3 \otimes dx_3$$

with the corresponding 4×4 matrix

$$G = \begin{pmatrix} -1 & 0 & -\exp(x_1/a) & 0 \\ 0 & 1 & 0 & 0 \\ -\exp(x_1/a) & 0 & -\exp(2x_1/a)/2 & 0 \\ 0 & 0 & 0 & 1 \end{pmatrix}.$$

Counting starts at 0 for the matrix. Show that

$$V = -2a\exp(-x_1/a)\frac{\partial}{\partial x_0} + x_2\frac{\partial}{\partial x_1} + \left(a\exp(-2x_1/a) - \frac{x_2^2}{2a} \right)\frac{\partial}{\partial x_2}$$

is a *Killing vector field*. Apply computer algebra. The other Killing vector fields for the Gödel metric tensor field are

$$\frac{\partial}{\partial x_0}, \quad \frac{\partial}{\partial x_2}, \quad \frac{\partial}{\partial x_3}, \quad a\frac{\partial}{\partial x_1} - x_2\frac{\partial}{\partial x_2}.$$

They form a basis of a Lie algebra under the commutator.

Solution 37.   We also find the inverse of the matrix G. In SymbolicC++ df(.) is the differentiation operator.

```
// killing1.cpp
#include <iostream>
#include "symbolicc++.h"
using namespace std;

int main(void)
{
  int N = 4;
  Symbolic a("a");
  Symbolic g("g",N,N), Lg("Lg",N,N), V("V",N), x("x",N);
  g(0,0) = -1; g(0,1) = 0; g(0,2) = -exp(x(1)/a); g(0,3) = 0;
  g(1,0) = 0; g(1,1) = 1; g(1,2) = 0; g(1,3) = 0;
  g(2,0) = -exp(x(1)/a); g(2,1) = 0;
  g(2,2) = -exp(2*x(1)/a)/2; g(2,3) = 0;
  g(3,0) = 0; g(3,1) = 0; g(3,2) = 0; g(3,3) = 1;
  V(0) = -2*a*exp(-x(1)/a); V(1) = x(2);
  V(2) = a*exp(-2*x(1)/a)-x(2)*x(2)/(2*a); V(3) = 0;
  for(int j=0;j<N;j++)
   for(int k=0;k<N;k++)
   { Lg(j,k) = 0;
      for(int l=0;l<N;l++)
      Lg(j,k) += V(l)*df(g(j,k),x(l)) + g(l,k)*df(V(l),x(j))
           + g(j,l)*df(V(l),x(k)); }
```

```
cout << g << endl; cout << V << endl; cout << Lg << endl;
Symbolic gi("gi",N,N);
gi = g.inverse();
cout << gi << endl;
return 0;
}
```

The output is

```
[          -1              0        -e^(x1*a^(-1))          0]
[           0              1              0                 0]
[  -e^(x1*a^(-1))          0    -1/2*e^(2*x1*a^(-1))        0]
[           0              0              0                 1]
```

```
[              -2*a*e^(-x1*a^(-1))           ]
[                      x2                    ]
[a*e^(-2*x1*a^(-1))-1/2*x2^(2)*a^(-1)]
[                       0                    ]
```

```
[0 0 0 0]
[0 0 0 0]
[0 0 0 0]
[0 0 0 0]
```

```
[          1             0      -2*e^(-x1*a^(-1))          0]
[          0             1            0                    0]
[ -2*e^(-x1*a^(-1))      0     2*e^(-2*x1*a^(-1))          0]
[          0             0            0                    1]
```

Problem 38. Consider the *Lorenz model*

$$\frac{du_1}{dt} = -\sigma u_1 + \sigma u_2, \qquad \frac{du_2}{dt} = -u_1 u_3 + r u_1 - u_2, \qquad \frac{du_3}{dt} = u_1 u_2 - b u_3.$$

For certain values of the bifurcation parameters σ, b and r the system admits explicitly the time-dependent first integrals. We insert the ansatz

$$I(\mathbf{u}(t), t) = (u_2^2 + u_3^2) \exp(2t)$$

into the system in order to find the conditions on the coefficients σ, r, b such that $I(\mathbf{u}(t), t)$ is a first integral.

Solution 38.

```
// firstintegral.cpp
#include <iostream>
```

```
#include "symbolicc++.h"
using namespace std;

int main(void)
{
 Symbolic u("u",3), v("v",4);
 Symbolic term, sum, I, R1, R2;
 Symbolic t("t"), s("s"), b("b"), r("r");
 // Lorenz model
 v(0) = s*u(1)-s*u(0); v(1) = -u(1)-u(0)*u(2)+r*u(0);
 v(2) = u(0)*u(1)-b*u(2); v(3) = 1;
 I = (u(1)*u(1)+u(2)*u(2))*exp(2*t); // ansatz
 sum = 0;
 for(int i=0;i<3;i++) sum += v(i)*df(I,u(i));
 sum += v(3)*df(I,t);
 cout << "sum = " << sum << endl; cout << endl;
 R1 = sum.coeff(u(2),2); R1 = R1/(exp(2*t));
 cout << "R1 = " << R1 << endl;
 R2 = sum.coeff(u(0),1); R2 = R2.coeff(u(1),1);
 R2 = R2/(exp(2*t));
 cout << "R2 = " << R2 << endl;
 return 0;
}
```

It is obvious that for R1, R2 to be equal to zero, we need $b = 1$ and $r = 0$.

5.3   Supplementary Problems

Problem 1.   Consider the function $f : \mathbb{R}^{n+1} \to \mathbb{R}$

$$f(x_1, \ldots, x_{n+1}) = x_1^2 + \cdots + x_n^2 - x_{n+1}^2.$$

Find linear vector fields such that $L_V f = 0$.

Problem 2.   We consider the manifold $\mathbb{R}^2$.
(i) Given the vector field

$$V = x_1 \frac{\partial}{\partial x_2} - x_2 \frac{\partial}{\partial x_1}.$$

Find all metric tensor fields

$$g = \sum_{j,k=1}^{2} g_{jk}(\mathbf{x}) dx_j \otimes dx_k$$

such that $L_V g = 0$.

(ii) Consider the metric tensor fields

$$g_1 = dx_1 \otimes dx_2 + dx_2 \otimes dx_1$$
$$g_2 = dx_1 \otimes dx_2 - dx_2 \otimes dx_1$$
$$g_3 = dx_1 \otimes dx_1 - dx_2 \otimes dx_2.$$

Find the vector fields

$$V_j = f_{j1}\frac{\partial}{\partial x_1} + f_{j2}\frac{\partial}{\partial x_2}, \quad j = 1, 2, 3$$

such that $L_{V_j}g_j = 0$, $j = 1, 2, 3$.

(iii) Consider the smooth vector field

$$V = V_1(\mathbf{u})\frac{\partial}{\partial u_1} + V_2(\mathbf{u})\frac{\partial}{\partial u_2}.$$

Let $f_1(u_1)$, $f_2(u_2)$ be smooth functions. Calculate the Lie derivative

$$L_V\left(f_1(u_1)du_1 \otimes \frac{\partial}{\partial u_1} + f_2(u_2)du_2 \otimes \frac{\partial}{\partial u_2}\right).$$

Find the condition arising from setting the Lie derivative equal to 0. Calculate the Lie derivative

$$L_V\left(f_1(u_1)du_1 \otimes du_1 + f_2(u_2)du_2 \otimes du_2\right).$$

Find the conditions arising from setting the Lie derivative equal to 0. Compare the conditions to the conditions from (i).

Problem 3. Let V be a smooth vector field in $\mathbb{R}^2$. Assume that

$$L_V(dx_1 \otimes dx_1 + dx_2 \otimes dx_2) = 0, \quad L_V\left(\frac{\partial}{\partial x_1} \otimes \frac{\partial}{\partial x_1} + \frac{\partial}{\partial x_2} \otimes \frac{\partial}{\partial x_2}\right) = 0.$$

Can we conclude that

$$L_V\left(dx_1 \otimes \frac{\partial}{\partial x_1} + dx_2 \otimes \frac{\partial}{\partial x_2}\right) = 0 ?$$

Problem 4. Let $M = \mathbb{R}^2$ and let x_1, x_2 denote the Euclidean coordinates on $\mathbb{R}^2$. Consider the differential one-form

$$\alpha = \frac{1}{2}(x_1 dx_2 - x_2 dx_1)$$

and the vector field defined on $\mathbb{R}^2 \setminus \{\mathbf{0}\}$

$$V = \frac{1}{x_1^2 + x_2^2} \left(x_1 \frac{\partial}{\partial x_1} - x_2 \frac{\partial}{\partial x_2} \right).$$

Find $V \rfloor d\alpha$ and the Lie derivative $L_V \alpha$. Note that $d\alpha = dx_1 \wedge dx_2$, i.e. the volume form in $\mathbb{R}^2$ and

$$V \rfloor d\alpha = \frac{1}{x_1^2 + x_2^2} (x_1 dx_2 + x_2 dx_1), \quad V \rfloor \alpha = -\frac{x_1 x_2}{x_1^2 + x_2^2}.$$

Problem 5.   Consider the tensor fields in $\mathbb{R}^2$

$$T_1 = \sum_{j,k=1}^{2} t_{jk}(\mathbf{x}) dx_j \otimes dx_k$$

$$T_2 = \sum_{j,k=1}^{2} t_{jk}(\mathbf{x}) dx_j \otimes \frac{\partial}{\partial x_k}$$

$$T_3 = \sum_{j,k=1}^{2} t_{jk}(\mathbf{x}) \frac{\partial}{\partial x_j} \otimes \frac{\partial}{\partial x_k}.$$

Find the condition on the vector field

$$V = \sum_{\ell=1}^{2} V_\ell(\mathbf{x}) \frac{\partial}{\partial x_\ell}$$

such that $L_V T_1 = 0$, $L_V T_2 = 0$, $L_V T_3 = 0$. Simplify for the case $t_{jk}(\mathbf{x}) = 1$ for all $j, k = 1, 2$.

Problem 6.   Consider the vector space $\mathbb{R}^3$ and the smooth vector field

$$V = V_1(\mathbf{x}) \frac{\partial}{\partial x_1} + V_2(\mathbf{x}) \frac{\partial}{\partial x_2} + V_3(\mathbf{x}) \frac{\partial}{\partial x_3}.$$

Given the differential two-forms

$$\beta_1 = x_1 dx_2 \wedge dx_3, \quad \beta_2 = x_2 dx_3 \wedge dx_1, \quad \beta_3 = x_3 dx_1 \wedge dx_2.$$

Find the conditions on V_1, V_2, V_3 such that the following three conditions are satisfied

$$L_V \beta_j \equiv V \rfloor d\beta_j + d(V \rfloor \beta_j) = 0, \quad j = 1, 2, 3.$$

Then solve the initial value problem of the autonomous system of first order differential equations corresponding to the vector field V.

Problem 7. Given a smooth vector field in $\mathbb{R}^3$

$$V = V_1(\mathbf{x})\frac{\partial}{\partial x_1} + V_2(\mathbf{x})\frac{\partial}{\partial x_2} + V_3(\mathbf{x})\frac{\partial}{\partial x_3}.$$

(i) Consider the differential two-form

$$\beta = dx_1 \wedge dx_2 + dx_2 \wedge dx_3 + dx_3 \wedge dx_1.$$

Find $V \rfloor \beta$ and $d(V \rfloor \beta)$. Thus find $L_V\beta$. Find solutions of the partial differential equations given by $L_V\beta = 0$.
(ii) Consider the differential two-form in $\mathbb{R}^3$

$$\beta = x_3 dx_1 \wedge dx_2 + x_1 dx_2 \wedge dx_3 + x_2 dx_3 \wedge dx_1.$$

Find $d\beta$. Find the Lie derivative $L_V\beta$. Find the condition on the vector field V such that $L_V\beta = 0$.

Problem 8. Let V, W be smooth vector fields in $\mathbb{R}^3$. Then

$$L_V(dx_1 \wedge dx_2 \wedge dx_3) = (\text{div}(V))dx_1 \wedge dx_2 \wedge dx_3$$
$$L_W(dx_1 \wedge dx_2 \wedge dx_3) = (\text{div}(W))dx_1 \wedge dx_2 \wedge dx_3.$$

Calculate $L_{[V,W]}(dx_1 \wedge dx_2 \wedge dx_3)$.

Problem 9. Consider the vector field

$$V = \sigma(-u_1 + u_2)\frac{\partial}{\partial u_1} + (-u_1 u_3 + ru_1 - u_2)\frac{\partial}{\partial u_2} + (u_1 u_2 - bu_3)\frac{\partial}{\partial u_3}$$

associated with the *Lorenz model*

$$\frac{du_1}{dt} = \sigma(u_2 - u_1)$$
$$\frac{du_2}{dt} = -u_1 u_3 + ru_1 - u_2$$
$$\frac{du_3}{dt} = u_1 u_2 - bu_3.$$

Let $\alpha = u_1 du_2 + u_2 du_3 + u_3 du_1$. Calculate the Lie derivative $L_V\alpha$. Discuss. Find the Lie derivatives $L_V(du_1 \wedge du_2)$, $L_V(du_2 \wedge du_3)$, $L_V(du_3 \wedge du_1)$. Discuss.

Problem 10.   The *Kustaanheimo-Stiefel transformation* is defined by the map from $\mathbb{R}^4$ (coordinates u_1, u_2, u_3, u_4) to $\mathbb{R}^3$ (coordinates x_1, x_2, x_3)

$$x_1(u_1, u_2, u_3, u_4) = 2(u_1 u_3 - u_2 u_4)$$
$$x_2(u_1, u_2, u_3, u_4) = 2(u_1 u_4 + u_2 u_3)$$
$$x_3(u_1, u_2, u_3, u_4) = u_1^2 + u_2^2 - u_3^2 - u_4^2$$

together with the constraint

$$u_2 du_1 - u_1 du_2 - u_4 du_3 + u_3 du_4 = 0.$$

(i) Show that $r^2 = x_1^2 + x_2^2 + x_3^2 = u_1^2 + u_2^2 + u_3^2 + u_4^2$.
(ii) Show that

$$\Delta_3 = \frac{1}{4r}\Delta_4 - \frac{1}{4r^2}V^2$$

where

$$\Delta_3 = \frac{\partial^2}{\partial x_1^2} + \frac{\partial^2}{\partial x_2^2} + \frac{\partial^2}{\partial x_3^2}, \qquad \Delta_4 = \frac{\partial^2}{\partial u_1^2} + \frac{\partial^2}{\partial u_2^2} + \frac{\partial^2}{\partial u_3^2} + \frac{\partial^2}{\partial u_4^2}$$

and V is the vector field

$$V = u_2 \frac{\partial}{\partial u_1} - u_1 \frac{\partial}{\partial u_2} - u_4 \frac{\partial}{\partial u_3} + u_3 \frac{\partial}{\partial u_4}.$$

(iii) Consider the differential one-form

$$\alpha = u_2 du_1 - u_1 du_2 - u_4 du_3 + u_3 du_4.$$

Find $d\alpha$ and $L_V \alpha$.
(iv) Let $g(x_1(u_1, u_2, u_3, u_4), x_2(u_1, u_2, u_3, u_4), x_3(u_1, u_2, u_3, u_4))$ be a smooth function. Show that $L_V g = 0$.

Problem 11.   (i) Consider the metric tensor field

$$g = d\theta \otimes d\theta + \sin^2(\theta) d\phi \otimes d\phi.$$

Show that g admits the Killing vector fields

$$V_1 = \sin(\phi)\frac{\partial}{\partial \theta} + \cos(\phi)\cot(\theta)\frac{\partial}{\partial \phi}, \qquad V_2 = \cos(\phi)\frac{\partial}{\partial \theta} - \sin(\phi)\cot(\theta)\frac{\partial}{\partial \phi}$$

$$V_3 = \frac{\partial}{\partial \phi}.$$

Is the Lie algebra given by the vector fields semi-simple?

(ii) Consider the metric tensor field

$$g = \frac{1}{4}(d\theta \otimes d\theta + d\psi \otimes d\psi + d\phi \otimes d\phi + \cos(\theta)(d\psi \otimes d\phi + d\phi \otimes d\psi)).$$

Show that the Killing vector fields are given by

$$V_1 = 2\left(\sin(\psi)\frac{\partial}{\partial\theta} - \frac{\cos(\psi)}{\sin(\theta)}\frac{\partial}{\partial\phi} + \frac{\cos(\psi)}{\tan(\theta)}\frac{\partial}{\partial\psi}\right)$$

$$V_2 = -2\left(\cos(\psi)\frac{\partial}{\partial\theta} + \frac{\sin(\psi)}{\sin(\theta)}\frac{\partial}{\partial\phi} - \frac{\sin(\psi)}{\tan(\theta)}\frac{\partial}{\partial\psi}\right)$$

$$V_3 = 2\frac{\partial}{\partial\psi}$$

with

$$[V_1, V_2] = -2V_3, \quad [V_2, V_3] = -2V_1, \quad [V_3, V_1] = -2V_2.$$

Problem 12. Can one construct a metric tensor field in $\mathbb{R}^3$ which admits the Killing vector fields

$$V_1 = \frac{\partial}{\partial x_2}, \quad V_2 = \frac{\partial}{\partial x_3}$$

$$V_3 = -a\frac{\partial}{\partial x_1} + x_2\frac{\partial}{\partial x_2} + x_3\frac{\partial}{\partial x_3}, \quad V_4 = -x_3\frac{\partial}{\partial x_2} + x_2\frac{\partial}{\partial x_3}.$$

Problem 13. Let $x_0 = ct$. Consider the metric tensor field

$$g = dx_0 \otimes dx_0 - e^{P_1(x_0)}dx \otimes dx - e^{P_2(x_0)}dx_1 \otimes dx_1 - e^{P_3(x_0)}dx_3 \otimes dx_3$$

where P_j $(j = 1, 2, 3)$ are smooth functions of x_0. Find the Killing vector fields.

Problem 14. Let $x_0 = ct$. Consider the metric tensor field

$$g = -dx_0 \otimes dx_0 + dr \otimes dr + r^2 d\theta \otimes d\theta + r^2 \sin^2(\theta)d\phi \otimes d\phi$$

and the vector field

$$V = \frac{1}{\sqrt{1 - \omega^2 r^2 \sin^2(\theta)/c^2}}\left(\frac{\partial}{\partial t} + \omega\frac{\partial}{\partial\phi}\right)$$

where ω a fixed frequency. Find the Lie derivative $L_V g$.

Problem 15. Let $x_0 = ct$. Consider the metric tensor field

$$g = dx_0 \otimes dx_0 - dv \otimes dv - kx_1 dx_0 \otimes dx_2 - kx_1 dx_2 \otimes dx_0$$
$$+ (k^2 x_1^2 - e^{kv}) dx_2 \otimes dx_2 - e^{-kv} dx_1 \otimes dx_1$$

the differential two-form

$$\beta = \frac{1}{\sqrt{2}} k e^{ikv} (dv \wedge dx_0 + kx_1 dx_2 \wedge dv + i dx_1 \wedge dx_2)$$

and the vector fields

$$V_1 = \frac{\partial}{\partial x_0}, \quad V_2 = \frac{\partial}{\partial x_2}, \quad V_3 = ky \frac{\partial}{\partial x_0} + \frac{\partial}{\partial x_1}, \quad V_4 = \frac{\partial}{\partial v} + \frac{1}{2} kx_1 \frac{\partial}{\partial x} - \frac{1}{2} kx_2 \frac{\partial}{\partial x_2}.$$

Show that

$$L_{V_1} g = L_{V_2} g = L_{V_3} g = L_{V_4} g = 0, \quad L_{V_1} \beta = L_{V_2} \beta = L_{V_3} \beta = L_{V_4} \beta = 0.$$

Problem 16. Let (M, g) be a smooth Riemannian manifold. We denote by d the exterior derivative and by δ the codifferential on differential forms. We know that

$$d(L_V \gamma) = L_V (d\gamma).$$

Find the commutator $[L_V, \delta]$. Show that in general $[L_V, \delta] \neq 0$.

Problem 17. A *de Sitter universe* may be represented by the hypersurface

$$x_1^2 + x_2^2 + x_3^2 + x_4^2 - x_0^2 = R^2$$

where R is a real constant. This hypersurface is embedded in a five-dimensional flat space whose metric tensor field is

$$g = dx_0 \otimes dx_0 - dx_1 \otimes dx_1 - dx_2 \otimes dx_2 - dx_3 \otimes dx_3 - dx_4 \otimes dx_4.$$

Find the Killing vector fields V of g, i.e. the solutions of $L_V g = 0$.

Problem 18. Find the Killing vector fields for the metric tensor field

$$g = d\alpha \otimes d\alpha - R^2 \sin(\alpha)(d\theta \otimes d\theta + \sin^2(\theta) d\phi \otimes d\phi)$$

where $0 \leq \alpha \leq \pi$, $0 \leq \theta \leq \pi$, $0 \leq \phi \leq 2\pi$. Note that for $R \to \infty$, $\alpha \to 0$ with $R\alpha \to r$ we obtain the metric tensor field

$$\widetilde{g} = dr \otimes dr - r^2 (d\theta \otimes d\theta + \sin^2(\theta) d\phi \otimes d\phi).$$

Problem 19. Consider the *wave equation*

$$\left(-\frac{\partial^2}{\partial x_0^2} + \sum_{j=1}^{n} \frac{\partial^2}{\partial x_j^2} \right) u = 0$$

with underlying metric tensor field

$$g = -dx_0 \otimes dx_0 + \sum_{j=1}^{n} (dx_j \otimes dx_j).$$

The wave equation could also be written as $d(\star du) = 0$, where $\star$ is the Hodge duality operator. The conformal vector fields V for the metric tensor field form a Lie algebra. These vector fields are defined by

$$L_V g = \rho_V g.$$

Show that the vector fields are given by

$$T = \frac{\partial}{\partial x_0}, \quad \rho_T = 0$$

$$P_j = \frac{\partial}{\partial x_j}, \quad \rho_{P_j} = 0, \quad j = 1, \dots, n$$

$$R_{jk} = x_j \frac{\partial}{\partial x_k} - x_k \frac{\partial}{\partial x_j}, \quad \rho_{R_{jk}} = 0, \quad j, k = 1, \dots, n$$

$$L_j = x_j \frac{\partial}{\partial x_0} + x_0 \frac{\partial}{\partial x_j}, \quad \rho_{L_j} = 0, \quad j = 1, \dots, n$$

$$S = \sum_{j=0}^{n} x_j \frac{\partial}{\partial x_j}$$

$$I_j = 2x_j \sum_{j=0}^{n} x_j \frac{\partial}{\partial x_j} - \left(\sum_{k=1}^{n} x_k^2 - x_0^2 \right) \frac{\partial}{\partial x_j}, \quad \rho_{I_j} = -4x_j, \quad j = 1, \dots, n$$

$$I_0 = 2x_0 \sum_{j=1}^{n} x_j \frac{\partial}{\partial x_j} + \sum_{j=1}^{n} x_j^2 \frac{\partial}{\partial x_0}, \quad \rho_{I_0} = 4x_0.$$

Problem 20. Consider the differential p-form in $\mathbb{R}^n$

$$\gamma = \sum_{1 \leq i_1 < \cdots < i_p \leq n} \gamma_{i_1 \dots i_p}(\mathbf{x}) dx_{i_1} \wedge \cdots \wedge dx_{i_p}$$

and $V = \partial/\partial x_1$. Show that

$$L_V\gamma = \sum_{1 \leq i_1 < \cdots < i_p \leq n} \frac{\partial \gamma_{i_1\ldots i_p}(\mathbf{x})}{\partial x_1} dx_{i_1} \wedge \cdots \wedge dx_{i_p}.$$

Problem 21.   Consider the smooth vector field in $\mathbb{R}^3$

$$V = f_1(\mathbf{x})\frac{\partial}{\partial x_1} + f_2(\mathbf{x})\frac{\partial}{\partial x_2} + f_3(\mathbf{x})\frac{\partial}{\partial x_3}.$$

Then $L_V(dx_1 \wedge dx_2 \wedge dx_3) = \mathrm{div}(V)dx_1 \wedge dx_2 \wedge dx_3$. Find

$$L_V(dx_1 \wedge dx_2), \quad L_V(dx_2 \wedge dx_3), \quad L_V(dx_3 \wedge dx_1)$$

$$L_V(dx_1 \wedge dx_2 + dx_2 \wedge dx_3 + dx_3 \wedge dx_1)$$

$$L_V(dx_1 \otimes dx_1), \quad L_V(dx_2 \otimes dx_2), \quad L_V(dx_3 \otimes dx_3)$$

$$L_V(dx_1 \otimes dx_1 + dx_2 \otimes dx_2 + dx_3 \otimes dx_3)$$

$$L_V(dx_1 \otimes dx_2 + dx_2 \otimes dx_3 + dx_3 \otimes dx_1)$$

$$L_V\left(\frac{\partial}{\partial x_1} \otimes \frac{\partial}{\partial x_2} \otimes \frac{\partial}{\partial x_3}\right)$$

$$L_V\left(dx_1 \otimes \frac{\partial}{\partial x_1} + dx_2 \otimes \frac{\partial}{\partial x_2} + dx_3 \otimes \frac{\partial}{\partial x_3}\right)$$

$$L_V\left(dx_1 \otimes \frac{\partial}{\partial x_2} + dx_2 \otimes \frac{\partial}{\partial x_3} + dx_3 \otimes \frac{\partial}{\partial x_1}\right).$$

Apply these results to the vector fields given by the *Lorenz model*

$$\frac{dx_1}{dt} = -\sigma x_1 + \sigma x_2, \quad \frac{dx_2}{dt} = -x_1x_3 + rx_1 - x_2, \quad \frac{dx_3}{dt} = x_1x_2 - bx_3$$

the Rikitake two-disc dynamo

$$\frac{dx_1}{dt} = -\mu x_1 + x_2x_3, \quad \frac{dx_2}{dt} = -\mu x_2 + (x_3 - a)x_1, \quad \frac{dx_3}{dt} = 1 - x_1x_2$$

and the integrable system

$$\frac{dx_1}{dt} = (c_3 - c_2)x_2x_3, \quad \frac{dx_2}{dt} = (c_1 - c_3)x_1x_3, \quad \frac{dx_3}{dt} = (c_2 - c_1)x_1x_2.$$

Both the Lorenz model and the Rikitake two-disc dynamo can show chaotic behaviour.

Problem 22. Consider the manifold $\mathbb{R}^n$ and the smooth vector field and metric tensor field

$$V = \sum_{j=1}^{n} V_j(\mathbf{x})\frac{\partial}{\partial x_j}, \quad g = \sum_{j,k=1}^{n} g_{jk}(\mathbf{x})dx_j \otimes dx_k.$$

Show that

$$L_V g =$$

$$\sum_{j,k=1}^{n}\left((L_V g_{jk})\, dx_j \otimes dx_k + g_{jk}\sum_{\ell=1}^{n}\frac{\partial V_j}{\partial x_\ell}dx_\ell \otimes dx_k + g_{jk}\sum_{\ell=1}^{n}\frac{\partial V_k}{\partial x_\ell}dx_j \otimes dx_\ell\right).$$

Problem 23. Consider the system of first order partial differential equations

$$\frac{\partial u_1}{\partial t} = c_1\frac{\partial u_1}{\partial x} + f_1(u_1, u_2), \quad \frac{\partial u_2}{\partial t} = c_2\frac{\partial u_2}{\partial x} + f_2(u_1, u_2)$$

where $f_1, f_2 : \mathbb{R}^2 \to$ are smooth functions. Find the condition such that

$$V = a(x,t,u_1,u_2)\frac{\partial}{\partial x} + b(x,t,u_1,u_2)\frac{\partial}{\partial t} + c(x,t,u_1,u_2)\frac{\partial}{\partial u_1} + d(x,t,u_1,u_2)\frac{\partial}{\partial u_2}$$

is a *Lie symmetry vector field*. Note that the *vertical vector field* V_v is given by

$$V_v = (-au_{1,x} - bu_{1,t} + c)\frac{\partial}{\partial u_1} + (-au_{2,x} - bu_{2,t} + d)\frac{\partial}{\partial u_2}.$$

The *contact forms* are

$$\Theta_1 = du_1 - u_{1,x}dx - u_{1,t}dt, \quad \Theta_2 = du_2 - u_{2,x}dx - u_{2,t}dt$$

with

$$V_v\rfloor\Theta_1 = c - au_{1,x} - bu_{1,t}, \quad V_v\rfloor\Theta_2 = d - au_{2,x} - bu_{2,t}$$

and $V_v\rfloor\Theta_1 = V\rfloor\Theta_1$, $V_v\rfloor\Theta_2 = V\rfloor\Theta_2$. The prolonged vertical vector field is given by

$$\tilde{V}_v = V_v + (D_t g_1)\frac{\partial}{\partial u_{1,t}} + (D_x g_1)\frac{\partial}{\partial u_{1,x}} + (D_t g_2)\frac{\partial}{\partial u_{2,t}} + (D_x g_2)\frac{\partial}{\partial u_{2,x}}$$

where

$$D_t := u_{1,t}\frac{\partial}{\partial u_1} + u_{2,t}\frac{\partial}{\partial u_2} + u_{1,tx}\frac{\partial}{\partial u_{1,x}} + u_{2,tx}\frac{\partial}{\partial u_{2,x}} + u_{1,tt}\frac{\partial}{u_{1,t}} + u_{2,tt}\frac{\partial}{\partial u_{2,t}}$$

$$D_x := u_{1,x}\frac{\partial}{\partial u_1} + u_{2,x}\frac{\partial}{\partial u_2} + u_{1,xx}\frac{\partial}{\partial u_{1,x}} + u_{2,xx}\frac{\partial}{\partial u_{2,x}} + u_{1,xt}\frac{\partial}{\partial u_{1,t}} + u_{2,xt}\frac{\partial}{\partial u_{2,t}}$$

and

$$g_1 = -au_{1,x} - bu_{1,t} + c, \quad g_2 = -au_{2,x} - bu_{2,t} + d.$$

The invariance condition is

$$L_{\widetilde{V}_v} F_1 \hat{=} 0, \quad L_{\widetilde{V}_v} F_2 \hat{=} 0$$

where

$$F_1 = -u_{1t} + c_1 u_{1x} + f_1(u_1, u_2), \qquad F_2 = -u_{2t} + c_2 u_{2x} + f_2(u_1, u_2).$$

Bibliography

[1] Anderson L. and Ibragimov N.H. (1979) *Lie-Bäcklund Transformations in Applications*, SIAM, Philadelphia

[2] Baumslag B. and Chandler B. (1968) *Group Theory*, Schaum's Outline Series, McGraw-Hill, New York

[3] Bluman G.W. and Kumei S. (1989) *Symmetries and Differential Equations*, Applied Mathematical Science **81**, Springer Verlag, New York

[4] Bott R. and Tu L. W. (1982) *Differential Forms and Algebraic Topology*, Graduuate Text in Mathematics 82, Springer Verlag, New York

[5] Cartan E. (1967) *Formes différentielles*, Hermann, Paris

[6] Chern S. S., Chen W. H. and Lam K. S. (1999) *Lectures on Differential Geometry*, World Scientific, Singapore

[7] Choquet-Bruhat Y., DeWitt-Morette C. and Dillard-Bleick M. (1978) *Analysis, Manifolds and Physics* (revised edition) North-Holland, Amsterdam

[8] Clarkson P. A. (1993) *Applications of Analytic and Geometric Methods to Nonlinear Differential Equations*, Kluwer, Dordrecht

[9] Crampin M. and Pirani F. A. E. (1986) *Applicable Differential Geometry*, Cambridge University Press, Cambridge

[10] Davis Harold T. (1962) *Introduction to Nonlinear Differential and Integral Equations*, Dover Publication, New York

[11] Do Carmo M. (1976) *Differential Geometry of Curves and Surfaces*, Prentice-Hall

[12] Edelen D.G.B. (1981) *Isovector Methods for Equations of Balance* Sijthoff & Nordhoff, Alphen an de Rijn

[13] Edelen D. G. B. (2005) *Applied Exterior Calculus*, Dover Publication, New York

[14] Eisenhart E. P. (1961) *Continuous Groups of Transformations*, Dover, New York

[15] Flanders H. (1963) *Differential Forms with Applications to the Physical Sciences*, Academic Press, New York

[16] Gilmore R. (1974) *Lie Groups, Lie Algebras, and Some of Their Applications*, Wiley-Interscience, New York

[17] Göckeler, M. and Schücker T. (1987) *Differential geometry, gauge theories, and gravity*, Cambridge University Press, Cambridge

[18] Gomes A., Voiculescu I., Jorge J., Wyvill B. and Galbraith C. (2009) *Implicit Curves and Surfaces: Mathematics, Data Structures and Algorithms*, Springer Verlag, London

[19] Goriely A. (2001) *Integrability and Nonintegrability of Dynamical Systems*, World Scientific, Singapore

[20] Greub W., Halperin S. and Vanstone R. (1973) *Connections, Curvature and Cohomology*, Academic Press, New York, vols. 1 and 2

[21] Guggenheimer H. W. (1977) *Differential Geometry*, Dover, New York

[22] Guillemin V. and Pollack A. (1974) *Differential Geometry*, AMS Chelsea Publishing

[23] Guillemin V. and Sternberg S. (1984) *Symplectic Techniques in Physics*, Cambridge University Press, Cambridge

[24] Hardy A. and Steeb W.-H. (2008) *Mathematical Tools in Computer Graphics with* C# *Implementations*, World Scientific, Singapore

[25] Hawking S. W. and Ellis G. F. S. (1973) *The Large Scale Structure of Space-Time*, Cambridge University Press, Cambridge

[26] Helgason S. (1979) *Differential Geometry, Lie Groups, and Symmetric Space*, Volume 80, Academic Press

[27] Helgason S. (1984) *Groups and Geometric Analysis, Integral Geometry, Invariant Differential Operators and Spherical Functions*, Academic Press

[28] Ibragimov N. H. (1993) *Handbook of Lie group analysis of differential equations*, Volume I, CRC Press, Boca Raton

[29] Ince E. L. (1956) *Ordinary Differential Equations*, Dover, New York

[30] Isham C. J. (1989) *Modern Diffferential Geometry*, World Scientific, Singapore

[31] Jacobson N. (1962) *Lie Algebras*, Interscience Publisher, New York

[32] Klingenberg W. (1978) *A Course in Differential Geometry*, Springer Verlag, Berlin

[33] Kobayashi S. (1972) *Transformation Groups in Differential Geometry* Springer Verlag, New York

[34] Kobayashi S. and Nomizu K. (1963) *Foundations of Differential Geometry*, Wiley, New York

[35] Kowalski K. and Steeb W.-H. (1991) *Nonlinear Dynamical Systems and Carleman Linearization*, World Scientific, Singapore

[36] Kramer D., Stephani H., Hertl E. and MacCallum M. A. (1980) *Exact Solutions of Einstein's Field Equations*, Cambridge University Press, Cambridge

[37] Kreyszig E. (1991) *Differential Geometry*, Dover Publications, New York

[38] Lie S. (1927) *Gesammelte Abhandlungen*, B. G. Teubner, Leipzig

[39] Lipschutz M. M. (1969) *Schaum's Outline of Differential Geometry*, McGraw-Hill

[40] Matsushima Y. (1972) *Differential Manifolds*, Translated by Kobayashi E.T., Marcel Dekker Inc., New York

[41] Michor P. W. (2008) *Topics in Differential Geometry*, American Mathematical Society

[42] Miller W. Jr. (1972) *Symmetry Groups and their Applications*, Academic Press, New York

[43] Millman R. S. and Parker G. D. (1977) *Elements of Differential Geometry* Prentice-Hall, New Jersey

[44] Misner C. W., Thorne K. S. and Wheeler J. A. (1973) *Gravitation*, Freeman

[45] Munkres J. R. (1991) *Analysis on Manifolds*, Westview Press

[46] Nakahara M. (2003) *Geometry, Topology and Physics*, second ed., Taylor and Francis, New York

[47] Ohtsuki T. (2002) *Quantum Invariants*, World Scientific, Singapore

[48] Olver P. J. (1986) *Applications of Lie Groups to Differential Equations*, Springer Verlag, New York

[49] Robbin J. W. and Salamon D. A. (2013) *Introduction to Differential Geometry*, http://www.math.ethz.ch/ salamon/PREPRINTS/ diffgeo.pdf

[50] Rogers C. and Ames W. F. (1989) *Nonlinear Boundary Value Poblems in Science and Engineering*, Academic Press, New York

[51] Rogers C. and Shadwick W. F. (1982) *Bäcklund Transformations and their Applications*, Academic Press, New York

[52] Rosen G. (1969) *Formulations of Classical and Quantum Dynamical Theory*, Academic Press, New York

[53] Sattinger D.H. and Weaver O.L. (1986) *Lie Groups and Algebras with Applications to Physics, Geometry, and Mechanics*, Applied Mathematical Science, **61**, Springer Verlag, New York

[54] Sharpe R. W. (1996) *Differential Geometry*, Springer Verlag, Berlin

[55] Sneddon I. (1957) *Elements of Partial Differential Equations*, McGraw-Hill, Singapore

[56] Spiegel M. R. (1968) *Mathematical Handbook of Formulas and Tables*, Schaum Outline Series, McGraw-Hill

[57] Spivak M. (1999) *A Comprehensive Introduction to Differential Geometry*, Publish or Perish, Berkeley, vols. 1-5

[58] Steeb W.-H. (1978) "The Lie derivative, Invariance Conditions, and Physical Laws", *Z. Naturforsch.* **33a**, 724–748

[59] Steeb W.-H. (1978) "A note on the Lorentz transformation", *J. Math. Phys.* **20**, 1684–1686

[60] Steeb W.-H. (1980) "Symmetries and vacuum Maxwell's equations" *J. Math. Phys.* **21**, 1656–1658

[61] Steeb W.-H., Erig J. and Oevel W. (1981) "A Note on the Nonrelativistic and Relativistic Classical Mechanics and the Constants of the Motion", *Lettere Al Nuovo Cimento*, **30**, 421–426

[62] Steeb W.-H., Erig W. and Strampp W. (1982) "Similarity Solutions of Nonlinear Dirac Equations and Conserved Currents", *J. Math. Phys.* **23**, 145–153

[63] Steeb W.-H. (1983) "Constants of Motions in Relativistic and Non-Relativistic Classical Mechanics" *Hadronic J.* **6**, 68–70

[64] Steeb W.-H. (1993) *Invertible Point Transformation*, World Scientific, Singapore

[65] Steeb W.-H. and Euler N. (1988) *Nonlinear Field Equations and Painlevé Test*, World Scientific, Singapore

[66] Steeb W.-H. and Strampp W. (1982) "Diffusion Equations and Lie and Lie-Bäcklund Transformation Groups", *Physica* **114 A**, 95–99

[67] Steeb W.-H. (2009) *Problems and Solutions in Theoretical and Mathematical Physics*, Volume II, Third edition, World Scientific, Singapore

[68] Steeb W.-H., Hardy Y. and Tanski I. (2011) "Vector Product and an Integrable Dynamical System", Commun. Theor. Phys. **56**, 992–994

[69] Steeb W.-H. and Hardy Y. (2015) "Eigenvalue Problem, Spin Systems, Lie Groups, and Parameter Dependence", *Z. Natur. A* **70**, 605–609

[70] Sternberg S. (1983) *Lectures on Differential Geometry*, New York: Chelsea

[71] Stoker J. J. (1969) *Differential Geometry*, Wiley-Interscience, New York

[72] Thorpe J. A. (1979) *Elementary Topics in Differential Geometry*, Springer Verlag, New York

[73] Vaisman I. (1973) *Cohomology and Differential Forms*, Marcal Dekker, New York

[74] Von Westenholz C. (1981) *Differential Forms in Mathematical Physics*, (Revised edition) North-Holland, Amsterdam

[75] Wald R. M. (1984) *General Relativity*, University of Chicago Press

[76] Whittaker E. T. (1937) *A treatise on the analytical dynamics of particles and rigid bodies*, (4th edition), New York: Dover Publication

[77] Yano K. (1957) *The Theory of Lie derivatives and its Applications*, North Holland

[78] Zwillinger D. (1990) *Handbook of Differential Equations*, Academic Press, Inc. Boston

Index

Printed in the United States
By Bookmasters